ALSO BY BENSON BOBRICK

Testament:
A Soldier's Story of the Civil War

Wide as the Waters:
The Story of the English Bible and the Revolution It Inspired

Angel in the Whirlwind:
The Triumph of the American Revolution

Knotted Tongues:
Stuttering in History and the Quest for a Cure

East of the Sun:
The Epic Conquest and Tragic History of Siberia

Fearful Majesty:
The Life and Reign of Ivan the Terrible

Labyrinths of Iron:
Subways in History, Myth, Art, Technology, and War

Parsons Brinckerhoff:
The First Hundred Years

Astrology in History

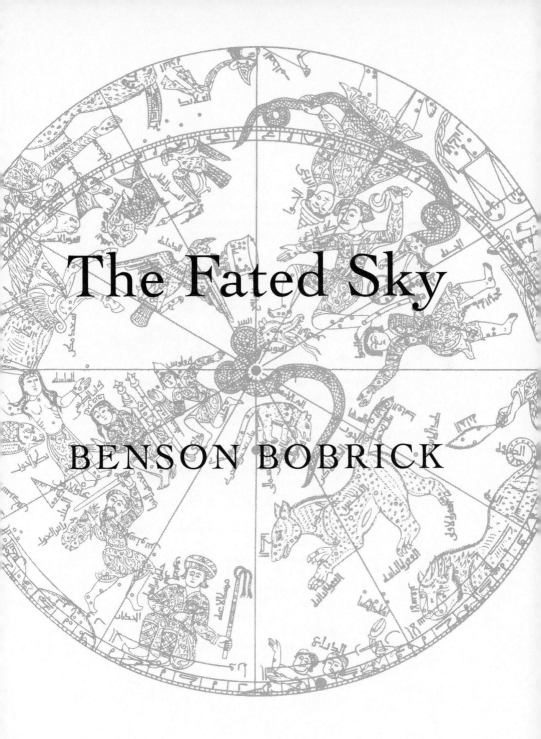

The Fated Sky

BENSON BOBRICK

Simon & Schuster

New York   London   Toronto   Sydney

SIMON & SCHUSTER
Rockefeller Center
1230 Avenue of the Americas
New York, NY 10020

SIMON & SCHUSTER and colophon are registered trademarks
of Simon & Schuster, Inc.

For information about special discounts for bulk purchases,
please contact Simon & Schuster Special Sales at
1-800-456-6798 or business@simonandschuster.com

Book design by Ellen R. Sasahara

Manufactured in the United States of America

1   3   5   7   9   10   8   6   4   2

Library of Congress Cataloging-in-Publication Data
Bobrick, Benson, date.
The fated sky : astrology in history / Benson Bobrick.
p.   cm.
Includes bibliographical references and index.
1. Astrology—History.   2. History—Miscellanea.   I. Title.
BF1729.H57B63 2005
133.5'09—dc22   2005051674
ISBN-13: 978-0-7432-2482-6
ISBN-10: 0-7432-2482-5

ACKNOWLEDGMENTS

My grandfather, who was a devoutly religious scholar and a bishop of the Methodist Church, acknowledged in his old age that he had once been persuaded, reluctantly, to consult an astrologer about the where-abouts of a wallet he had lost. The wallet had all sorts of valuables in it and he had looked everywhere for it in vain. The astrologer cast a chart for the question—"Where is my wallet?"—and after examining the planets, told him, correctly, where it could be found. That both pleased the bishop and annoyed him. Many years later, he was still clearly abashed. He had too honest a mind to dismiss it, but he couldn't explain it either. And so in my own mind, too, this story curled a question mark over the entire subject which, some forty years later, I at last began to explore. As a student of history, I took the long view, which in the end proved indispensable for getting my bearings right. This book is the result.

Astrology may be a suspect subject, at least to some; the history of astrology can scarcely be to anyone who cares about the history of ideas. In recent years, that history in all its aspects has benefited from the work of a number of fine scholars—John D. North, Michael Molnar, Robert Zoller, James Herschel Holden, Patrick Curry, David Plant, Nicholas Campion, J. Lee Lehman, Tamsyn Barton, Anthony Grafton, William R. Newman, Hilary Carey, Annabella Kitson, Demetra George, Laura Ackerman Smoller, Robert Hand, and John Frawley, among the more prominent—who have made it possible to glimpse the true sweep and compass of the art. Invaluable work is also being done daily by Project Hindsight, dedicated to the recovery and translation of the classic Greek, Latin, Arabic, and Hebrew works; and by ARHAT, the Archive for the Retrieval of Historical Astrological Texts. Both deserve to be better known. All my cordial contacts with the American Federation of Astrologers and the Astrological Association of Great Britain were also fruitful of results.

I have indeed relied on the labors of many, but those who helped in any direct or coordinate way with the making of this book are absolved of any errors it contains. Among family, friends, colleagues, associates, and helpful correspondents, I am happy to name: Eleanor Bach, James and Peter Bobrick, Robin Brownstein, Herschel Farbman, John Frawley, Svetlana Gorokhova, Nancy Griffin, Peter Guttmacher, Hagop Merjian, Gloria Mulcahy, Peter Murkett, Pamela Robertson, George and Gene Rochberg, David Roell, Lora Sharnoff, Edward W. Tayler, P. L. Travers, Richard and Bea Wernick, and Danielle Woerner, among others, who over the years contributed something of value, however indirectly, to the text. My dedicated agent, Russell Galen, stood foursquare behind the book from the start; my editor, Bob Bender, was exemplary as always in allowing me to work in my own way. His assistant, Johanna Li, helpfully attended to details. Three hometown haunts—Mocha Joe's, Amy's Bakery Arts Cafe, and The Cafe Beyond—often provided sustenance and a home away from home. Throughout, Hilary and the twins, Zuzu & Jasper, did much to keep my spirits up. I am grateful to them all.

CONTENTS

To Hilary & the Twins
and
In Memory of my learned friend
George Rochberg,
whose ear was tuned to the Music of the Spheres

"There are more things in heaven and earth, Horatio,
Than are dreamt of in your philosophy. But come. . . ."

Hamlet, Act 1, Scene 5, ll.187–88

PART ONE

---

*The universe, eternity, the infinite are typified by the sphere . . .
On a sphere every point is a center, and every point is the highest
point, and this explains the puzzle of time and space. There never
was a beginning of time, and there never will be an end. Time
always is. Any number of trillions of years hence, and any number . . . past, and you are just as near the end, or the beginning, of
Time as now, and no nearer. This moment is the center of Time;
this instant is the highest point in the revolving sphere. The same
with that other form of Time, Space. There is no end to Space,
and no beginning. This point where you now stand, this chair,
this tree, is the center of Space; it all balances from this point. Go
to the farthest fixed star and . . . you have only arrived at Here.
Your own doorstep is just as near the limit, and no nearer. This is
the puzzle of puzzles, but it is so.*

—JOHN BURROUGHS, *Journals,* January 13, 1882

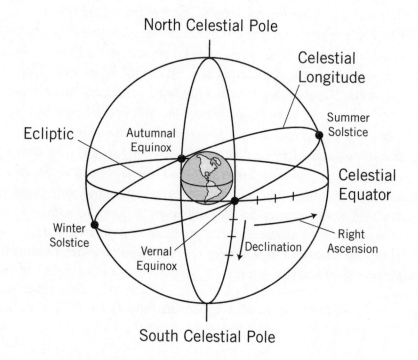

North Celestial Pole

Celestial
Longitude

Ecliptic

Autumnal
Equinox

Summer
Solstice

Celestial
Equator

Winter
Solstice

Vernal
Equinox

Declination

Right
Ascension

South Celestial Pole

Chapter 1

AMERICA WOULD NEVER have been discovered by Christopher Columbus in 1492 had it not been for the thought of Arab astrologers in Baghdad in the 9th century A.D. When Columbus set sail on the great western voyage that carried him to America's shores, he had biblical prophecy to inspire him, Arab astrology to guide him, and various practical aids that three continental astrologers, who were also mathematicians, had supplied: the planetary tables of Regiomontanus; a map drawn up by Paolo Toscanelli; and an ephemeris prepared by Samuel Zacuto, who later made the splendid astrolabe of iron used by Vasco da Gama in his voyage around the Cape of Good Hope. These were all of use to Columbus in his celestial calculations and his navigation of the open sea. He also used an astrolabe and quadrant to determine the altitude of stars, set his hourglass by the transits of the Sun, depended on the North Star to fix magnetic north, and judged the time of night by the constellation of the Great Bear. He overawed the natives of one island by his ability to predict a lunar eclipse, and drew with some success on astrological lore to predict the weather—taking his ships to shelter, for example, in the port of Santo Domingo because an aspect between Jupiter and Mercury seemed to portend a tropical storm. Yet Columbus could not proceed solely by the sky. Knowledge of celestial navigation in Europe was wanting, and so, for the most part, he relied on a magnetic compass to measure his course or direction, and on his own method of "dead" or deduced reckoning to estimate his position on the main.

But it was the stars that led him on. Columbus understood that the world was a globe and believed that by sailing directly west he would eventually reach the shores of Asia (or the "Indies"). He could not know, of course, that America intervened. But it was not the fabled wealth of the Indies that held him most in thrall. For the voyage itself was spurred on by an astrological idea. That idea was the "great conjunction" theory of history, as first set forth in the writings of the Persians, elaborated by the Arabs, and adopted by the Latin West. Columbus had encountered it in the work of the French cardinal, theologian, and astrologer Pierre d'Ailly.

According to this theory, important historical events such as the rise and fall of empires, the birth of religions, and cultural transformations were marked by the "great planetary conjunctions" of Jupiter and Saturn as they revolved through their cycles in the sky. Such great conjunctions occurred once every 960 years—a principal source of our idea of the millennium—as the planets completed a circuit of the zodiac, combining and recombining in the signs. In the course of that round, the two conjoined—that is, occupied the same degree of celestial longitude—forty-eight times. For d'Ailly, human history was explained by the unfolding impact of these conjunctions, according to their scale. Shifts between triplicities or elements (earth, air, fire, and water, by which the signs of the zodiac were grouped) were associated with dynastic change; the greater or near-millennial conjunctions were linked to epochal change as well as natural disasters such as earthquakes and overwhelming floods. In d'Ailly's view, such great conjunctions had heralded or coincided with the Great Flood, the fall of Troy, the death of Moses, the foundation of Rome, and the advent of Christ. "All astronomers are agreed in this," he declared, "that there never was one of those conjunctions without some great and notable change in this world."

D'Ailly's work had convinced Columbus that the end of the world was near, and that it would be accompanied by the conversion of all heathenkind to Christ. For that reason, he called himself Christophorus (or "Christo-ferens," as he came to sign his name), "the Christbearer," and conceived himself the agent of God's work as the world approached its final days. All this he explained in a letter to his royal patrons, Ferdinand and Isabella of Spain. He wrote of the Indies: "These vast realms are peopled with immortal souls, for whose redemption Christ, the Son of God, has made an atoning sacrifice. It is the mission which God has assigned to me to search them out, and to carry to them the Gospel of Salvation." He took as his text Isaiah 11:10–12—"The Lord shall . . . recover the remnant of his people . . . and gather together the dispersed . . . from the four corners of the earth"—and his historic first voyage itself seemed emblematic of that charge.

On the morning of August 3, 1492, Columbus set sail with three small ships—the *Niña,* the *Pinta,* and the *Santa María*—from Palos, Spain, and steered for the Canary Islands, where he reprovisioned before striking due west. After a difficult voyage of two months with a near-mutinous crew, on October 12 he at length sighted land. At two o'clock in the morning, a gun was fired to give the signal. All three vessels then took in their sails and laid to, "waiting impatiently for the dawn." Upon making landfall, "the voice of prayer and the melody of praise rose from his ships," and his own first action was to prostrate himself upon the ground. To Columbus, his journey's end was heaven-sent. For their part, the natives on the small Bahamian island were not wholly mistaken, perhaps, when they cried out at dawn to their brethren, "Come see the people from the sky."

Columbus would later say that he owed all he had achieved to the grace of God and "God-given" arts of astrology, geometry, navigation, and arithmetic.

His own heavily annotated copy of d'Ailly's work, *Treatise on the Image of the World,* may still be seen in the Columbine Library at Seville.

✳   ✳   ✳

ACCORDING TO AN ANCIENT TRADITION, common to both Gnostic and Syriac Christians as well as to the Persians and Jews, Adam received the doctrines and mysteries of astrology directly from the Creator, and by knowledgeably scanning the constellations in the skies foretold that the world would one day be destroyed by water, then by fire. As a memorial to those who came after him, he (or his descendants, Seth and Enoch) had this knowledge engraved upon two pillars, one of brick, the other of stone. According to Flavius Josephus, the Jewish historian and near contemporary of Christ, the second pillar could still be seen in Syria in A.D. 63.

Astrology is the oldest of the occult sciences. It is also the origin of science itself. From astrology are derived astronomy, calculation of time, mathematics, medicine, botany, mineralogy, and (by way of alchemy) modern chemistry, among other disciplines. Logarithms were originally devised to simplify the calculations necessary in casting horoscopes; the ray theory of vision—the foundation of modern optics—developed from astrological theories of the effect of stellar rays on the soul. For five thousand years, from ancient Sumeria and Babylonia to the present day, the stars have been viewed as shaping, by divine power, the course and destiny of human affairs. Indeed, according to the *New Catholic Encyclopedia,* the earliest symbol of deity known to us—the cuneiform sign for "god"—was a star (∗).

Astrological terms permeate our language: *conjunction, opposition, forecast, aspect, lunatic, venereal, disaster, influence*—as in *influenza,* since all epidemics were once ascribed to celestial effects; we speak of "mercurial," "saturnine," or "jovial" temperaments; and people thank their "lucky stars," or consider a person "ill-starred" if his luck is bad. The Hebrew word *mazzal* means "sign" or constellation; so "Mazzal tov" (the colloquial "Congratulations!") really means, "May you have good stars!" The term *fall* is astrological, for the fall or autumn equinox marks the descendant of the zodiac year; and *revolution* is taken from an astrological calculation called a "solar return." The star-shaped halo that once encompassed the Roman emperor's posthumous image—according to the belief that he ascended to heaven as a star—was later transformed into the halo of the Christian saint. The pharmaceutical symbol R_x—commonly said to be an abbreviation for the Latin verb *recipere* (from which we get *recipe* or compound)—is derived from the ancient symbol for the Roman god Jupiter, based on the "Eye" of Horus, an Egyptian god with magical healing powers.

∗   ∗   ∗

ASTRONOMY STUDIES THE heavenly bodies in order to formulate the natural laws that govern them and to understand how the physical structure of the universe evolved; astrology describes the influence of those bodies upon human character and life. Or, as Ralph Waldo Emerson reputedly remarked, "Astrology is astronomy brought down to earth and applied to the affairs of men." It is an applied science, insofar as it is based on astronomy; an exact science, insofar as its judgments are based on mathematical calculations; and an empirical science, insofar as its deductions are based on data gathered over the course of time.

Its method is a horoscope, which is a map or diagram of the heavens cast for a particular moment of time, and read according to well-established rules. Those rules, if properly applied, are free from the elements of chance or divination; moreover, they are substantially based on a written tradition that derives its authority not just from dogma and belief, but from thousands of years of observation. The idea at the heart of astrology is that the pattern of a person's life—or character, or nature—corresponds to the planetary pattern at the moment of his birth. Such an idea is as old as the world is old—that all things bear the imprint of the moment they are born.

Whether this is true or not may be subject to debate. But the belief that it is has proved to have enduring power.

Astrology in modern times has undergone a remarkable resurgence, and is now (as Carl Jung predicted it would) knocking again at the doors of academe. Astrologers are attempting to verify traditional doctrine by scientific methods and in general to meet the demand of Johannes Kepler (one of its true believers) that they "separate the gems from the slag." In a number of countries, including England, France, Russia, Germany, and the United States, astrology is once again being taught at the university level, for the first time since the Renaissance. In England, courses in the subject are now offered at Brasenose College, Oxford; Bath Spa University College; the University of Southampton; and the University of Kent. It can also be studied at Cardiff University in Wales, the Bibliotheca Astrologica in France, the University of Zaragoza in Spain, Dogus University in Turkey, Benares Hindu University in northern India, and at Kepler College in the United States, among other schools. Scholarly journals such as *Culture and Cosmos* (A Journal of the History of Astrology and Cultural Astronomy), the *Dublin Astrologer* (The Journal of the Dublin Astrological Centre), and *Apollon* (The Journal of Psychological Astrology), have begun to establish themselves, while the prestigious Warburg Institute in London recently created a "Sophia Fellowship" for astrological research.

For the past thirty years or so, polls have shown that from 30 to 40 percent of Americans (or about 100 million people) "believe in astrology and think their lives are governed by the stars." An estimated ten million people have paid an astrologer to cast their horoscope, while almost everybody seems to know their own "sign." Astrology columns are carried by most of the nation's daily newspapers and hundreds of magazines, and can be found on numerous Internet sites. Yahoo alone lists about 1,700 of the latter, while Amazon.com counts 3,155 books on the subject in print. Most large bookstores today devote an entire section to the field. According to one recent estimate, there are some 15,000 full-time and 225,000 part-time astrologers today in the United States.

There can be no doubt that the subject maintains an unshakeable hold on the human mind.

★   ★   ★

THE BIBLE IS RICH with astrological allusion. It opens with the pronouncement that the "lights in the firmament of the heavens" were established in part "for signs," and in Psalm 19, for example, we read: "The heavens declare the glory of God; and the firmament showeth his handiwork. Day unto day uttereth speech, and night unto night showeth knowledge. There is no speech nor language where their voice is not heard." According to rabbinical tradition each of the twelve tribes of Israel represented a zodiac sign, and the astrological symbols for the four fixed signs—a lion, a man, a bull, and an eagle—were carried as totems in the Egyptian desert by the Hebrew host. These same symbols made up the composite creature we call the Egyptian Sphinx, and in accordance with Ezekiel's vision came to stand for the four great Christian evangelists—Matthew, Mark, Luke, and John. Of the twelve precious stones that adorned the breastplate of Aaron as high priest, Josephus wrote, "whether we understand by them the months, or the like number of the signs of that circle which the Greeks call the Zodiac, we shall not be mistaken." The seven-branched candlestick, he tells us, also symbolized the seven planets, and the twelve loaves of shewbread in the temple the twelve signs. It is said that each of the twelve disciples of Christ likewise stood for (or embodied) a sign—an idea that was carried over into medieval romance, where the twelve knights of King Arthur's Round Table (a symbol of the zodiac) also stood for the twelve astrological types. The idea that those types together constitute a complete circle of humanity is also carried over into our jury system, which is supposed to ensure that a man is properly tried by a representative assessment (or complete cross-section) of his peers. That means, in theory, that they will combine their experience to perfect the judgment of a case. The

Hindus also say twelve is the number of completeness, which is why the Bible tells us that at the age of twelve, Jesus was able to confute the doctors in the temple, because his knowledge was already complete.

Throughout antiquity, the constellations and planets were honored by shrines and temples of learning. There were twelve great Mystery religions, "each one paying homage to or deriving its authority from a zodiac sign." The rites of Aries, or the Celestial Ram, so Manly Hall tells us, "were celebrated in the temple of Jupiter Ammon in the Libyan desert; the rites of Taurus in the Egyptian Mysteries of Serapis, or the tomb of the Heavenly Bull; the rites of Gemini in Samothrace, where Castor and Pollux—the Dioscuri—were worshipped; the rites of Cancer in Ephesus, where Diana (goddess of the Moon) was revered; the rites of Leo in the Bacchic and Dionysiac orgies of the Greeks," and so on.

The ecclesiastical calendars of all known religions are also linked astrologically with the major phases of the Sun and Moon. Passover, for example, begins on the first full Moon after the vernal equinox; Easter Sunday, which marks the end of Lent, is usually the first Sunday after that;✳ the Christian Sabbath is the day of the Sun; and the Jewish New Year, Rosh Hashanah, starts at sunset on the day of the new moon closest to the autumn equinox. The first day of Ramadan is set by the new Moon in Libra, which begins the most holy period for Moslems of fasting and prayer. In Vietnam, the New Year begins at the time of the first full Moon after the Sun enters Aquarius and is termed Tet. Hanukkah is set by the new Moon in Capricorn, and Purim by the full Moon in Pisces. Christmas was "coopted by the Church from pagan celebrations at the winter solstice, which was also the festival of the Persian Sun god Mithras. The rebirth of the Sun god was thus replaced in Christianity by the birth of God the Son."

The names of the days of the Western week, of course, are those of the star-gods, as derived from Roman and Norse mythology. Sunday is the Sun's day; Monday the Moon's; Tuesday the day of Tiw, the pagan god of war, akin to Mars; Wednesday belongs to Woden, akin to Mercury (in French, Mercredi); Thursday to Thor, or Jupiter; and Friday

---

✳ Except when the ecclesiastical full moon (determined from tables) and the astronomical full moon do not correspond.

to the goddess Freya, or Venus (in French, Vendredi). Saturday is Saturn's day and rounds out the cycle.

Our seven-day week itself derives from a convergence around the 2nd century B.C. of the Sabbath cycle of the Jews, in which the seventh day was held to be holy, and an astrological week based upon the planets (which included the Sun and Moon) according to which each day was ruled by one of the seven planetary gods. Each hour of each day was also so ruled, hence the cycle of planetary hours. Following Egyptian practice, there were twenty-four hours in a day, but before clock time they were not all of equal length: the twelve daytime hours were equally divided from sunrise to sunset, the twelve nighttime hours from sunset to dawn. In sequence, the hours belonged to Saturn, Jupiter, Mars, the Sun, Venus, Mercury, and the Moon, in an endless circle, with each one in turn serving as regent or ruler for that day. This was the Ptolemaic order of the planets, according to their perceived speed and distance from the earth.

The planets also gave us the seven liberal arts, and, by number and type, the seven deadly sins: sloth (Saturn), pride (Jupiter), anger (Mars), gluttony (the Sun), lust (Venus), avarice (Mercury), and envy (the Moon). Like the signs, the planets inspired worship and adulation, and each of the Seven Wonders of the ancient world, according to one scholar, arose in homage to one of the planets then known. The Colossus of Rhodes was an altar to the Sun; the temple of Diana at Ephesus to the Moon; the Great Pyramid at Giza to Mercury; the hanging gardens of Semiramis to Venus; the mausoleum of Halicarnassus to Mars; the temple of Olympian Zeus to Jupiter; and the Pharos of Alexandria to Saturn.

Many of the myths of the ancients, moreover, can be unlocked only by an astral key. One key is the vernal equinox, the Sun's annual "crossing" (or "passover") from the southern to the northern hemisphere—which enters a new constellation every 2,160 years. Some 4,700 years ago, for example, when it entered Taurus, the Egyptian god Osiris assumed the form of the Celestial Bull, while in the Egyptian desert the children of Israel made offerings to a golden calf. When the equinox later entered Aries, "the solar deity was commonly represented by a golden-haired youth cradling a lamb in his arms and holding a shepherd's crook." When the equinox entered Pisces, the Savior of the World "appeared as the Fisher of Men."

In the symbolism of this great story, the Ram (Aries, from the Greek *eras,* meaning lamb) and the lamb are one. After their Exodus from Egypt, the Jews sacrificed a lamb at the Passover festival; this paschal sacrifice later became the Easter Lamb of sacrifice and crucifixion in the Christian faith. The idea of Passover itself is linked to the Sun's equinoctial passing or crossing over, which also underlies the symbol of the Cross. We may go deeper. After Christ was crucified, dead, and buried, he ascended into heaven after three days. Just so, the Sun remains for three days in transit at the equinoctial point before it begins its ascent into the northern hemisphere.

This is not to say that the story of the Passover, or the life of Christ, is a mere allegory of a celestial event. God forbid! Astrologically speaking, in the divine scheme of things, it is rather the other way around.

★ ★ ★

ADORATION OF THE HEAVENS as the face of the divine was perhaps the beginning of true worship, and the mythology of the Egyptians and Greeks often involved parables or stories of the stars. The Trojan War, Homer tells us in the *Iliad,* was provoked by "Jove and Latona's son," that is, Jupiter (Zeus in Homer), and Apollo, the Sun. Arrayed on the side of the Trojans are Apollo, Venus, and Mars; on that of the Greeks, Neptune, the Moon, Vulcan, Athena, and Jupiter. Mercury is not mentioned by Homer, but Iris, the rainbow goddess, is his female form. As a messenger, she acts with strict neutrality, but "every scene on earth is a reflex, outcome, or willed event of some previous celestial scene." "What do you think Homer and Vergil had in mind," wrote one Renaissance astrologer, Girolamo Cardano, "when they continually made the gods quarrel or fight, the Homeric ones for the Greeks or Trojans, the Vergilian for Turnus or Aeneas? Clearly that some of the stars favored one party, others the other. That is the explanation of those numerous meetings and counsels of the gods . . . Therefore when they said that Venus favored Aeneas because he was very handsome, or that Juno, that is, fortune, and the Moon favored Turnus, or that Apollo, or the Sun, favored Hector because he was strong and just, they had in mind, concealed under the veil of fable, the genius or star that ruled each one at birth." The Twelve Labors of Hercules are also a figurative description of the Sun's passage through the twelve signs and are akin to the stories the Babylonians told about their own solar hero, Gilgamesh, "whose life unfolded in twelve epic songs."

If the pagan myths are astrological allegories of a sort, so too may be some of the biblical tales—for example, that of Samson, whose name in Hebrew means "belonging to the Sun." His long hair, like the lion's mane, was his pride and symbolized his strength; it is his encounter with a lion as a young man that first proves his might. Delilah is his opposite in every sense. If he is Leo-like, the root of *her* name in Hebrew is the word for Aquarius—the opposite sign. This is a story in which astrological opposites meet, mate, and clash.

✷　✷　✷

WE MAY EVENTUALLY KNOW what everything is, but we will never know what everything means. Ludwig Wittgenstein once touchingly glanced at this idea in one of his mournful moods when he wrote, "We feel that when all possible scientific questions have been answered, the problems of life remain completely untouched." Religion occupies the sacred heart of all those questions to which the problems of life give rise, and astrology is the most venerable branch of that inner knowledge from which religion springs.

This is not a book for or against astrology, but a book about its impact on history and on the history of ideas. That impact has been large, and without a competent knowledge of the subject it is almost impossible to accurately trace or construe much of history itself or the ideas that have governed its course. For it runs, and has run, like an underground river through human affairs. Indeed, until the middle of the 17th century at least, astrology "entered into the councils of princes, guided the policy of nations, and ruled the daily actions of individuals," great and small. Astrological predictions often affected the course of events, while those in power based their actions on astrological advice. It is said that the Incas submitted to the Spanish almost without a fight because the arrival of the conquistadors happened to coincide with an astrological prophecy that their civilization was coming to an end. Depending on how one cares to interpret this, the prophecy fulfilled itself or, by acquiescence, was fulfilled. Either way, astrology had power.

The very idea of a "period" of history (to which the Incas belonged) is astrological, and based on the conjunction theory Columbus embraced. That theory brought the otherwise indistinguishable flow of time into an ordered sequence, and made history intelligible by identifying its hectic course with celestial events. It also helped to explain why history often seemed to repeat itself, as imaged in the repetitions in the sky. Modern science, like modern history, tends to disregard it, but this is a senseless bias or neglect. The history of science itself is so beholden to astrology that it owes it a debt of respectful attention if not abundant gratitude. "Do you believe then that the sciences would ever have arisen and become great," Friedrich Nietzsche once remarked, "if there had not beforehand been magicians, alchemists, astrologers, and wizards, who thirsted and hungered after abscondite and forbidden powers?" Astrology, of course, possesses its own kind of knowledge, which has nothing to do with what modern science reveres. But in some sense, it is also true that magic and science originally advanced side by side. The desire to understand the secret workings of nature created an intellectual environment favorable to experiment and induction; alchemy gave birth to chemistry; Neoplatonic and hermetic ways of thinking led to the heliocentric hypothesis of Copernicus, Kepler's laws of planetary motion, and Harvey's discovery of the circulation of the blood. The mystical conviction that number contained the key to all mysteries fostered the development of mathematics—and subsequently revived it in the wake of the Dark Ages when knowledge of the subject had waned.

The irreverent scorn in which astrology is sometimes held is ultimately based on a superstition, one "all the more dangerous," as Theodore Roosevelt once remarked (in an essay entitled, "The Search for Truth in a Reverent Spirit"), "because those suffering from it are profoundly convinced that they are freeing themselves from superstition itself. No medieval superstition . . . could be more intolerant . . . than that . . . which not merely calls itself scientific but arrogates to itself the sole right to use the term." Surely a degree of humility is not unbecoming in any attempt to assess the value of a doctrine—or "teaching"—that has survived for thousands of years.

✶　✶　✶

LIKE THE BONES OF Columbus himself, those of astrology have been stirred so often as almost to acquire a life of their own. Exhumed and reinterred at least half a dozen times over the course of three centuries, from Vallodolid to Santo Domingo, from Havana to Genoa to Seville, the explorer's remains have seemed to multiply like the relics of a saint, and today can be found in at least three sites in both the Old World and the New. If astrology is dead and buried, as some would have it, its grave is as unquiet as that of Columbus, and as indeterminate as his tomb.

Sir Elias Ashmole (for whom the Ashmolean Museum at Oxford is named) once remarked: "There are in Astrologie (I confess) shallow Brooks, through which young Tyroes may wade; but withal there are deep Fords, over which the Giants themselves must swim." There is far more to the subject than tends to meet the modern eye. Its story, at the very least, is enlarged with remarkable lives, including some of the most illustrious (and infamous) in human history, and draws its line through the whole chronology and range of human culture, from the back alleys of imperial Rome, where fortune-tellers plied their trade, to the inner circles of secular and religious power.

Chapter 2

THE APPARENT BIRTHPLACE of astrology was Mesopotamia, the land between the Euphrates and the Tigris rivers, now occupied by Iraq. From the "Chaldean East," as it came to be called, which encompassed the realms of Chaldea, Babylonia, and Assyria, astrology spread to Egypt, and thence to the ancient world of Greece. The Greeks believed the Egyptians and Babylonians had invented it. Plato, in the *Epinomis,* specifically credited Assyria and Egypt. Marcus Manilius, the Roman author of *Astronomica,* an astrological poem written in the age of Augustus, held that the origins of astronomy could be found in the lands of the Euphrates and the Nile. "Very deep is the well of the past," wrote Thomas Mann. "Should we not call it bottomless?" We may never know for certain whether the Egyptians owed the fundamentals of their astrological knowledge to the Babylonians, or whether, as some continue to insist, the Babylonians imbibed it at least in part from some Egyptian source. But just as religious ideas, and their symbolic language, seem to belong to a kind of received understanding worldwide, so those of astrology, which are religious in nature, belong to the immemorial past. Even so, there are markers in time that allow some of its progress to be traced.

Four thousand years before Christ, the Babylonians and Assyrians scanned the heavens for omens of their fate, and from atop their ziggurats, or multitiered towers, mapped the course of the planets and from their observations began to make predictions about the weather, the harvest, drought, famine, war, peace, and the fates of kings. Some of the earliest known towers were at Uruk and Ur. The biblical prophet Abraham, father of the Jewish people, was born in the city of Ur of the Kasdim (a phrase meaning "light of the astrologers") when the rulers of Mesopotamia were said to be astrologer-kings. Astrology flourished in the reign of Assurbanipal (called Sardanapalus by the Greeks), who reigned at Nineveh in the middle of the 7th century B.C. Assurbanipal was the son of Esarhaddon, who had succeeded Sennacherib, the ruler of Assyria mentioned in Isaiah and 2 Kings. In the time of the biblical prophet Daniel, it was still customary under the Assyrian monarchs for the general in the field to be accompanied by his *asipu,* or "prophet," on whose interpretation of the signs of heaven the movements of the army relied. Whether or not these Assyrians worshipped the planets themselves as gods—or regarded their patterned flight as the agents of some higher power—they began to trust in their import and recorded their observations on calcite and greenstone cylinder seals. The Sun was depicted as a rayed disk, the Moon as a crescent, and Venus as an eight-pointed star.

A large collection of cuneiform tablets, known as *Enūma Anu Enlil,* survive from the ancient archives of Nineveh and include many observations made before Assurbanipal's reign. Most took the form of celestial omens, which accurately noted the rising and setting of Venus with predictions based on its appearance and location in the sky. One typical omen read: "When Venus appears in Dilgan (Virgo), rains in heaven, floods on [earth], the crops of Aharru will prosper; and men will reinhabit ruined homes." Or: "If Venus appears in the east in the month of Airu and the Great and Small Twins surround her, all four of them, and she is dark, then the King of Elam will sicken and die." Again: "When the fiery light of Venus illuminates the breast of Scorpio, then rain and floods will ravage the land." Other planets were also assessed. For example, "If a halo encircle the Moon, and Jupiter is found within it, animals will perish and the king of Akkad will be besieged." Or: "When Jupiter stands in front of Mars, there will be corn in the fields and men will be slain . . . When Mars approaches Jupiter, there will be great devastation . . . In that year the king of Akkad will die."

Before the ancient twelve-month calendar emerged, the different seasons were identified with particular stars that rose as the seasons turned. At the time the constellations were established (about 3000 B.C.), the Sun at the spring equinox was near Aldebaran, the brightest star of Taurus; at the summer solstice, near Regulus, the brightest star of Leo; at the autumnal equinox, near Antares, the brightest star of Scorpio; and at the winter solstice, near Fomalhaut, the brightest star of Aquarius. These four stars were called "royal," and the signs (or constellations) in which they were placed were said to be "fixed" because they were close to the four fixed points in the Sun's seeming path among the stars. By 700 B.C., the Assyrians at Nineveh had more or less traced the ecliptic; divided it into four parts according to the seasons; drawn up the list of constellations whose heliacal rising corresponded to the various months; distinguished the planets from the fixed stars; followed their course; and "approximately determined the duration of their synodic revolutions." That enabled them to predict eclipses of the Sun and Moon and to "accurately fix the duration of the lunar month at a little more than twenty-nine and one half days." The earliest star maps, in fact, were lists of stars charted in their relation to the Moon. Early tablets therefore referred to the "stars in the path of the Moon," and the earliest calendars were lunar, a month lasting either from first crescent to first crescent or from full Moon to full Moon. This subsequently evolved into a lunar zodiac of twenty-eight "mansions" or divisions, which roughly traced the distance traveled by the Moon each day. These mansions were sometimes thought of as the temporary resting places of the Sun, Moon, and planets in their journey across the sky. Over time, the annual lunar calendar or cycle was linked to the now standard zodiac of twelve signs or constellations, as the great astronomer Johannes Kepler once explained: "[In ancient times] the farmers had to seek their calendar in the sky . . . When the Moon was full, they could easily see, for example, that the first full Moon appeared in the Ram's horns, the second near the Pleiades, the third near the Twins, etc. and finally that the thirteenth again appeared in the first constellation, the Ram's horns. Thus the full Moons divided the whole circle into twelve parts." The twelve constellations were eventually mapped and formed into a zodiac round (about the 6th century B.C.) and the signs in turn (as distinct from the constellations) were established as twelve 30-degree arcs over the course of the next two hundred years.

Such are the ascertainable beginnings of Chaldean lore. The word *Chaldean* was originally a geographical term from the Assyrian *Kaldu* and referred to Chaldea, or lower Mesopotamia, near the Persian

Gulf. It eventually came to refer to members of the Babylonian priesthood, then to Greek astrologers "directly or indirectly affiliated with Babylonian schools," and ultimately to "all those who professed to foretell the future according to the stars." The Roman historian Diodorus Siculus tells us that the Chaldeans called the planets "the Interpreters" because their course and relative positions revealed to men the will of the gods. He adds: "The star which the Greeks name Cronos [Saturn] they call the 'star of the Sun' [or the 'Sun of the night'], because it gives the most numerous and important predictions."

Babylonian star lore migrated to Egypt with the Persian conquests in the 6th century B.C., though the positions of many of the stars had already been mapped out by ancient Egyptian astronomers as early as the 13th century B.C. In the tomb of Ramses II, who lived about 1292–25 B.C., the 19th-century French Egyptologist Jean Champollion found massive circles of wrought gold divided into 365 degrees, each of which marked the rising and setting of the stars for the day. In the tomb of Ramses V, moreover, he found papyri giving tables of constellations and their influences on human beings for every hour of every month of the year. The different hours of different constellations in the ascendant, for example, were believed to rule different parts of the body—the ears, heart, arms, and so on—according to an astrological tradition that persists to this day. This type of astrology was "not only common among the Chaldeans," writes Ellen McCaffery, "but entered the oral tradition of the Hebrews, which seemed to give support to the statement of the *Sephir Yetzirah* (composed in the early Christian centuries) that astrological knowledge had been handed down by Abraham, born among the Chaldees."

★ ★ ★

THE EARLIEST SURVIVING birth chart was drawn for a child born in the region of modern Iraq, just south of Baghdad on April 29, 410 B.C.—"when Socrates was about sixty and Plato seventeen years old." All in all, some 205 individual horoscopes, most of them Greek, have been preserved from ancient times. It is sometimes said that natal astrology was invented by the Greeks, but that can hardly be, for as early as 1300 years before Christ we have a Hittite translation of a Babylonian omen text offering personal predictions according to the month in which a child is born. Herodotus, in his *Histories,* tells us that long before his own time (the 5th century B.C.), "the Egyptians had learned how to foretell by the date of a man's birth his character, his fortunes, and the day of his death."

But the astrological knowledge of the Greeks also seems to go far back. Philostratus, writing in the early Christian era, tells us that astrology was known in Greece as early as 1184 B.C.; Plutarch claims that Hesiod, the Greek poet who lived eight hundred years before Christ, was an adept; and various constellations, such as Orion, the Pleiades, and the Great Bear were certainly familiar to the Greeks when the *Odyssey* of Homer (in which they are named) was composed. There is also an astrological allusion in which Euripides (480–406 B.C.) refers to a prediction based on the rising of the stars. Pliny, Plutarch's contemporary (in the first century A.D.), tells us that many early Greek astronomers were also astrologers, including Thales (born ca. 640 B.C.), who is said to have been familiar with Egyptian lore. Thales, not incidentally, described the Earth as a sphere, as did most Greek astronomers, and determined the ratio of the Sun's diameter to its apparent orbit. According to Herodotus, Thales also predicted that the total eclipse of the Sun in 609 B.C. would herald the end of the war between the Medes and Lydians. The eclipse occurred when the battle between these two nations was at its height, and "darkness falling on both armies, the war ceased." Anaximander, a disciple and friend of Thales, developed various instruments, including the sundial, for making celestial calculations; Plato alludes to astrology in his *Timaeus* and was tutored by a Chaldean astrologer in his old age; Theophrastus, who succeeded Aristotle as head of the Lyceum, held that astrology could not only predict world events but also the course of a person's life. And so on. It is perhaps as futile to attempt a fixed timeline for astrological knowledge as it is for the history of faith.

One thing seems certain: that astrological knowledge came to include Pythagorean concepts of equation and proportion ("common to all branches of mathematics"); the mystical meaning of number; and the idea that man is a microcosm or miniature version of the universe, bound by a system of affinities to the stars. Such ideas are perhaps immemorial, and probably did not originate with Pythagoras himself, a Greek of the 6th century B.C. who settled in Crotona in southern Italy. But in his teaching, he gave them a local habitation and a name.

With the conquests of Alexander the Great, such occult learning spread, as both Mesopotamia and Egypt came under Greek rule. It was in Hellenistic Egypt, indeed, that Western astrology really developed out of an amalgam of Babylonian, Greek, and Egyptian elements. There it cohered into a system with its own doctrines and traditions, and was joined to Greek mathematical astronomy, which gave a scientific dimension to the Pythagorean teachings it embraced.

That system had signs; sign rulers; fixed stars; a doctrine of comets and eclipses; decans (a division of each sign into thirds); dodecatemoria (a division of each sign into twelfths); the four elements; the planetary week; the five essential dignities of sign, exaltation, triplicity, term, and face; celestial houses; planets in signs, houses, and aspect; and the parts or lots—for example, the Part of Fortune, an arithmetically derived point on the ecliptic obtained by adding the celestial longitudes of the Moon and the ascendant and subtracting the longitude of the Sun. The efflorescence and development of the art helps to explain the profusion of planetary and other zodiac symbols that began to find expression in architecture and painting, and to appear on contemporary coins. Egyptian star maps adorned temples such as those at Esna or Denderah, and a relief found in Samosata depicts a conjunction of planets in Leo that marked the coronation of Egypt's Antiochus I. The earliest known astrological handbook was also a Greco-Egyptian work known as the *Nechepso-Petosiris,* drawn up in Alexandria around 150 B.C. Indeed, throughout the whole period there was a vibrant convergence (or conversation) of ideas east and west—from India to Egypt—across the Greek world.

★   ★   ★

ON THE MORNING of July 14, A.D. 479, a worried client sought out an Egyptian astrologer in Smyrna by the name of Palchus and, at 8:30 A.M., asked whether a ship that he was expecting from Alexandria, now way overdue, would eventually arrive safely, and, if so, when. The astrologer drew up a figure for the time the question was asked (according to a venerable branch of astrology less well-known today called horary, meaning, "of the hour"), and gave his expert judgment of the case: "Finding that the lords of the day and hour," he began, "being Saturn and Mars, are both in the Ascendant, and observing that the Moon is applying to [moving toward] an aspect of Saturn, it seems the ship has encountered a violent storm, but has escaped, inasmuch as Venus and the Moon are beheld by [in aspect to] Jupiter . . . And observing that the Ascendant is in a bi-corporeal sign . . . the ship's company has survived and had passed from one vessel to another. And as Venus, who has dominion over birds, is in opposition to Sagittarius, they will bring some birds with them. And because the Moon is in the house of Mars and in the terms of Mercury [a sign division ruled by that planet], they will probably bring some books and papers [papyri], and some brass vessels on account of the Moon also being in Scorpio . . . And observing that Aesculapius [the constellation Ophiuchus, the Serpent Bearer] will rise with the Moon, their cargo will include medicines, too." It is said that all that Palchus carefully described proved out. Moreover, he apparently predicted that the ship would arrive when the Moon entered Aquarius. And so it did.

Some astrologers were better than others, of course, and the half-baked knowledge of a pretender could produce disastrous results. For example, in A.D. 484, two Byzantine court astrologers persuaded an aspiring noble by the name of Leontius to set himself up as emperor in opposition to Zeno, at that time the legitimate head of state. They "elected" a time for his coup (according to another venerable branch of astrology, known as "elections") that had the Sun, Jupiter, and Mars all rising, with Mercury succedent to them, and with the Moon in friendly aspect both to Saturn and Jupiter (that is, by sextile or trine). One would think, according to the doctrine of elections, that Leontius had it made. Not so. For according to a contemporary source, after he was crowned at the time selected he "at once lost both his kingdom and his luck." The two astrologers, we learn, had failed to note that Mercury, "ruler of the day and of the next ensuing hour, was in an evil state; for he was at his greatest elongation from the Sun, which signified a violent death, and his only aspect was to Saturn, the great malefic of the planetary system in the sky." Also Venus, "being isolated," could not bestow its beneficent rays upon the occasion, "being intercepted by the Sun." Nor had they taken proper note of the fact that the Moon, being the dispositor (in effect, supplanter, according to certain rules) of the Sun, the ascendant, Jupiter, Mars, and the preceding new Moon, was in her "fall" and afflicted. As a result, the fact that the Sun, Jupiter, and Mars were all rising was not enough to outweigh the malevolent force of the rest. Indeed, the fact that the Moon was in her fall (opposite the sign in which she is exalted) and applying to a conjunction of Saturn was a "sure sign" Leontius would fail. Tricky business, this. In point of fact, after an illusory moment of success Leontius and his followers were vanquished, and so fled, first to Antioch in Syria and then to Papirius in Isauria (in modern Turkey), where they were besieged in a fortress for four years. Before the siege ended, they butchered the astrologers who had given them so much hope—though that could scarcely help them in their straits. Leontius himself was eventually caught and met the violent death that Mercury had shown. In 488 he was beheaded at Seleucia on the Calycadnus, and his head taken to Constantinople, where it was impaled on the city walls.

★   ★   ★

TRADITIONAL ASTROLOGY had four main branches: mundane, natal, horary, and elections. The first, mundane (meaning "world"), pertained to society as a whole and made predictions about the weather, harvests, epidemics, politics, and war; natal concerned the character and destiny of an individual; horary (or interrogations) dealt with a question by means of a chart cast for when it was asked; and elections sought to determine the most propitious moment for an action or enterprise. The three great circles fundamental to the art were, and are, the ecliptic (the Sun's apparent path around the earth), the celestial equator, and the horizon. The celestial equator is simply the Earth's equator projected onto the celestial sphere. Similarly, the celestial horizon is simply a circle parallel to the visible or sensible horizon, with its center the center of the Earth. That sensible horizon changes from place to place. In a horoscope it is determined by the place of birth, or the location of some other event for which a chart was drawn. Unlike the ecliptic and celestial equator, the horizon is unique to the chart in question. The point in space directly overhead and at right angles to the horizon is the Midheaven (or Medium Coeli), and that directly below it the Imum Coeli. The ascendant is the point where the ecliptic intersects the horizon in the east (horoscopes are drawn so that the east is on the left); opposite is the descendant, where the ecliptic meets the horizon in the west.

The natural beginning for the zodiac (a Greek word meaning "circle of animals") is the vernal equinox or First Point of Aries, where the planes of the ecliptic and the equator intersect. Opposite is the First

Point of Libra. These two mark the beginnings of spring and fall. At the same time, the two points where the ecliptic and the equator are farthest apart mark midsummer and midwinter (the solstices), and the latitudes on which they lie are called the tropics (from a Greek word meaning "turning point"). It is the angle of the ecliptic to the equator that therefore gives us seasonal change.

But the Earth is not a perfect sphere. There is a slight bulge at the equator, so the gravity of the Moon, and, to a lesser extent, the Sun, causes the Earth's axis to wobble and describe a cone in space. Because of precession, as it is called, the equinoctial point appears to shift slowly backward through the zodiac by about 1 degree of arc every seventy-two years. For that reason, the signs and constellations do not correspond. Precession, however, has no bearing on the signs, which are based on the seasons. The signs never change, though once every twenty-six thousand years or so the signs and constellations roughly coincide.

As above, so below. That is the sacred heart of the astrological idea. "Overhead," as one astronomer put it, "the glorious procession, so regular and unfaltering, of the silent, unapproachable stars: below, in unfailing answer, the succession of spring and summer, autumn and winter, seedtime and harvest, cold and heat, rain and drought." These are "the ordinances of heaven" that the Book of Job speaks of, and the evidence of "the dominion [of the stars] thereof" on Earth. God's fiercely pointed questions to Job were meant to remind him that such is the order of creation, beyond Job's power to affect. "Canst thou bring forth Mazzaroth [the twelve zodiac signs]," He asks, in their seasons? Or bind or loosen the influence of the stars on winter or spring? Just so, astrology tied the character and procession of events in life to nature, as governed by the heavens, whose order and motion in turn were ordained by God.

The Hebrews knew and revered the zodiac long before the Torah or Pentateuch, the First Five Books of Moses, were inscribed. At the beginning of the Christian Era, Ptolemy referred to it as "of unquestioned authority, unknown origin, and unsearchable antiquity." However carefully traced, its origins (like the true age and origin of man) remain obscure enough to warrant our wonder, and to merit the belief, as some insist, that the knowledge itself was "revealed." The names of the signs and constellations are in any case "prehistoric," in the sense that there is no record of their coining, though it is popularly supposed that the various signs were arbitrarily christened after some imagined image in the sky. On the face of it that seems unlikely, since few constellations—or star clusters within them—correspond to their names. One would be hard put, for example, to find the likeness of a ram in the stars of Aries, or a goat in Capricorn. Nevertheless, one typical writer on the matter offers this quaint conjecture as to how the names came about:

> A Sumerian priest was likely talking to two farmers one night. "I am designing a calendar," he said. "I want to name the star cluster through which the gods pass each season so that we can have reference points. I need names for points of reference." He paused and picked up the reed stylus. "Let's start with midsummer." "Name them scorpion," said Sin-Samuh. "One stung my foot last year at this time." "It was later," said Zakir. "He bit you in the fall." Sin-Samuh thought. "Maybe it was later." "We may use that name, then, for later," said the priest. "What about a lion?" asked Zakir. "One big guy chased Ben's wife the other day, but I think he was looking for water. The creek is dry." "Lion, huh?" The priest thought for a moment. "Not bad." They looked at the quiet, starry night sky. "I see a paw." "I see the body, over there." "No, that's the mane." Sin-Samuh moved his hand in the air. "See how it curves down." "I think you're right, Sam," said the priest. He pressed the stylus into the wet clay. "Then lion it will be."

"This is the reason," the writer concludes, "you are a Leo, I a Scorpio. Nothing more."

More likely, the opposite is true: that the constellations were given names that symbolized ideas. Just as Anubis, the Dog Star of Egypt, was so named because its annual appearance gave warning of the flooding of the Nile, so among all ancient peoples Libra, which marks the fall equinox, when the days and nights are of equal length, is invariably represented by a balance or scale. If a plausible image could then afterwards be found by selectively connecting certain stars, that ultimately served to aid the memory in recalling the meaning it expressed.

A traditional rhyme makes a charming wreath of the signs:

The Ram, the Bull, the Heavenly Twins,
And next the Crab the Lion shines,
　The Virgin and the Scales,
The Scorpion, Archer, and Sea-Goat,
The Man that bears the Watering-Pot,
　The Fish with glittering tails.

The seven planets ruled these twelve, according to a time-honored order and arrangement: Aries by Mars, Taurus by Venus, Gemini by Mercury, Cancer by the Moon, Leo by the Sun, Virgo by Mercury, Libra by Venus, Scorpio by Mars, Sagittarius by Jupiter, Capricorn by Saturn, Aquarius by Saturn, and Pisces by Jupiter. Depending on their angular position, there were five ways for these planets to be in aspect, or regard each other and mingle their rays: by conjunction, sextile, square, trine, and opposition—that is, at 0, 30, 90, 120, and 180 degrees. In general, trine and sextile aspects were said to be "easy" because they linked signs of the same element or temperature, the square and opposition "hard" because they joined incompatible or opposing signs.

Whereas the signs were divisions of the heavens relative to the vernal equinox, as connected to the revolution of the Earth about the Sun (or, indifferently, to the Sun about the Earth), the twelve houses of the horoscope were divisions of the heavens relative to birthplace, as connected to the rotation of the Earth on its axis as it revolved. In chart interpretation, they encompassed every aspect of human life. In the simplest description, they governed (1) the self; (2) movable assets and finance; (3) communication, siblings, and short journeys; (4) the home and immovable property, such as land; (5) pleasure and children; (6) ill health and servants; (7) open enemies, partnership, and marriage; (8) death; (9) learning, religion, and long journeys; (10) career and social standing; (11) friends, hopes, and wishes; (12) secret enemies, prisons, hospitals, and self-undoing. In any birth chart, the nature of the activities and experiences represented by the twelve were indicated by the condition of the planets that occupied them. The strength or weakness of a planet depended upon its sign position, placement, aspect, and motion. A planet without essential dignity was rudderless, without power, or "peregrine." Signs were distinguished by quality (cardinal, fixed, mutable) as well as element; the north and south nodes of the Moon—the two points where the lunar orbit and the plane of the ecliptic intersect (also known as the Dragon's Head and the Dragon's Tail)—were also of importance to any chart.

In the briefest possible synopsis, this was but the elementary grammar of astrology as practiced in the ancient world. That astrology emphasized the dignity or strength of the planets in the signs, receptions between them, and, in prediction, aspects as showing the occasion of events—a very different thing from the Sun-sign astrology of modern times. It also affirmed (once and for all) that the structure of Nature was evidenced by the visible sky, that the quality of time differed not only from time to time but place to place, that everything bore the celestial imprint of the moment of its birth, and that from that imprint or pattern its fate could be construed. Indeed, as baffling as the calculations and pronouncements of Palchus and other adepts might seem, their procedures were orthodox and standard; their judgments, if competent, governed by the doctrines and traditions (leavened perhaps by experience) they had learned.

Chapter 3

ASTROLOGY SWEPT ACROSS the Greco-Roman world, reaching all races, nations, and types and classes of men: rulers, scholars, the poor and wealthy; invaded the sciences of medicine, botany, chemistry (via alchemy), and mineralogy; occupied a central place in the "mystery religions," especially that of Mithras; and convinced all but the Epicureans and Skeptics.

Some part of its hegemony may be traced to the Hellenizing influences that followed the second Punic War (201 B.C.), though it had already begun to seep into the body politic a century before, when Roman legions first came into contact with various Eastern cults. At the outset it had attracted those primarily at the low end of the social scale, but its influence increased mightily until, by the end of the 1st century B.C., it had taken firm hold. Backstreet astrologers and other pretenders to the art flooded Rome, prompting a Praetorian Edict in 139 B.C. to expel them—the first of several promulgated over the next 150 years. Some set up their fortune-telling stalls under the shadows of the Circus Maximus, Rome's vast ancient sporting arena in the valley between the Palatine and Aventine hills; others were new places of assignation, such as the temples of Isis, or in the seedy alleyways and slums of the Suburra, and on the outskirts of Rome. By then astrology had also begun to touch the seats of power. In 86 B.C., for example, one Roman consul was misled by an astrological forecast and slain in a popular rebellion. On his dead body was found the diagram that had lured him to his death.

Meanwhile, astrology gained steadily in stature thanks to the popularity of Stoic thought and the fatalistic doctrine, astrological and otherwise, that it seemed to condone. It was taught at some of the foremost schools, (such as the one founded by Posidonius on the island of Rhodes); embraced by rationalists convinced that nature was governed by immutable laws; and cultivated by the learned elite. Many political leaders in the Roman Republic had confidence in its teachings, while throughout the imperial period up through the reign of Domitian, astrologers were said to be the power behind the throne. Political and social turmoil, marked by sudden, unexpected reversals of fortune, always helped to foster it; along with any predicted outcome—as epitomized by the fate of Julius Caesar on the Ides of March in 44 B.C.

An immensely popular figure in the Republic's waning days, Caesar had shaped a career that exalted its signal virtues and ensured its grim demise. As a member of the democratic party, he had exposed corruption and promoted reform; as a member of the First Triumvirate (with Pompey and Crassus), he had conquered the whole of Gaul in a series of brilliant military campaigns. But when Pompey, supported by the Senate, turned against him, Caesar refused to disband his legions, in defiance of republican law, crossed the Rubicon, and civil war began. After two years of hard fighting, in battles ranging from Egypt to Spain, he emerged triumphant and began to assume the trappings of autocratic power. Scornful of the Senate, he issued coins with his own likeness, allowed his statues to be adorned like those of the gods, wore the imperial purple garb in public, and in February of 44 B.C. was named *dictator perpetuus,* or dictator for life. One month later, at a meeting of the Senate, he met his predicted end.

The fearful dreams of his wife, Calpurnia, had concerned him, and the warnings of the astrologer Vestritius Spurinna (the soothsayer of Shakespeare's play) had given him some pause—though Spurinna had not always been right. In 46 B.C., for example, he had cautioned Caesar against crossing into Africa in advance of the winter solstice, but nothing in that campaign had gone awry. Even so, according to historian Cassius Dio, Caesar's disregard for all forecasts was really due to his fatalistic view of life: "Despite the warnings he received, he felt that if it was fated to die at some appointed time, he saw no point trying to defy the hour." A few minutes after taking his seat in the Senate chamber, on the Ides of March, a flurry of daggers struck him down.

The great orator and statesman Marcus Tullius Cicero had not been party to the plot, but in its aftermath he became, with Marc Antony, one of the two leading men of Rome—Cicero as spokesman for the Senate, Antony as consul and executor of Caesar's will. The two soon had a falling out, however, and Cicero attacked Antony in a series of dramatic speeches, or philippics, from the Senate floor. Meanwhile, Antony had allied himself with Marcus Lepidus and Octavian (the future Augustus) in a new triumvirate and denounced Cicero as an enemy of the state.

Starting out as a lowly but brilliant lawyer with unmatched rhetorical skills, Cicero had risen rung by rung in public life from quaestor to aedile to praetor to consul, the highest office in the land. As consul in 63 B.C., he had exposed the conspiracy by Catiline (an ex-praetor and governor of Africa) to seize power by force, but in the civil war between Pompey and Caesar he had cast his lot with the losing side. Caesar afterward forgave him, but there was no such charity in Antony's heart. As the walls closed in, yet another prediction played itself out.

Cicero's own horoscope indicated that he would one day be murdered, and for that reason alone, understandably enough, he had been loathe to credit the art. He questioned how twins could have different destinies although born under the same constellation, and why all those born under the same constellation would not have the same fate. "Were all those who perished at the battle of Cannae," he demanded, referring to Hannibal's defeat of the Romans in 216 B.C., "born under the same star?" Yet something in him seems to have acquiesced. He counted two leading astrologers—Publius Nigidius Figulus and Lucius Tarrutius of Firmum—among his closest friends; commended Jupiter as "a star that brings prosperity and health"; feared Mars as a "planet that bodes ill to men"; accepted the astrological conceit that the world had begun with all the planets in Aries; believed that Babylonian astrological records were "based on empirical observations which went back over 470,000 years"; and gave credit to Vestritius Spurinna for warning Caesar about the Ides. Now, in the face of Antony's blistering wrath, his heart must have sunk as he fled for his life toward the coast. Yet nothing could save him. A few days later he was caught—"his hair long and disheveled, his face pinched with fear"—and beheaded on December 7, 43 B.C. As a gruesome warning to others, his hands were afterward severed and displayed on the rostrum of the Forum; and in a final indignity to his corpse, Marc Antony's wife pulled out his tongue and pierced it with her hatpin in revenge upon the eloquence that had blackened her husband's name.

Antony's own exaltation was short-lived. In a reprise of the conflict between Pompey and Caesar, Antony and Octavian soon parted ways. Once again civil war erupted and after a decade of fighting, Octavian at last routed the followers of Antony at the naval battle of Actium and emerged as master of the Roman world.

Not long after Octavian, now Augustus, was crowned in A.D. 31, a comet appeared and shone for seven consecutive days during a festival of public games. The games had been held to commemorate Caesar's life, and the new emperor skillfully took advantage of its sighting to christen it "the Star of Julius" and promote the idea that upon his death Caesar had been transformed into a star. Such a metamorphosis had been a conspicuous feature of Greek religion and mythology, in which figures like Hercules, Perseus, Andromeda, and Castor and Pollux had assumed the form of constellations. But Caesar was the first Roman to be so exalted, inaugurating a tradition that subsequent rulers were only too eager to embrace. Be that as it may, in matters of divination Augustus was certainly a believer. His own horoscope had been cast at birth (by the illustrious astrologer and Roman senator Nigidius Figulus, who had predicted his rise to power) and years later, during his student days at Apollonia, he had gone with his devoted friend Marcus Agrippa to consult one Theogenes, a local astrologer of note. Agrippa was the first to try his fortune, and when "a great and almost incredible career was predicted for him" (accurately, for he later became the emperor's right-hand man), Augustus, with some diffidence, sought to learn his own fate. He feared in truth that it might be less. But after making his calculations, Theogenes reportedly gasped and threw himself at his feet.

What did Theogenes see? Born at sunrise, on September 22, 63 B.C., Augustus had a truly royal and majestic chart. His Sun was conjunct his ascendant in Libra; his fifth-house Capricorn Moon was waxing and moving toward Jupiter, which was exalted in the tenth. A modern astrologer, finding Mars and Saturn in the eighth, would be apt (mistakenly in this instance) to ramble on about a violent death. But what counted here was something else. As one classical textbook tells us, "If Jupiter in his own house or exaltation or in signs in which he rejoices comes into aspect with the waxing Moon, or when the Moon is moving toward Jupiter, the result is unconquerable generals who govern the whole world." The judgment therefore was "He will wield power over life and death. He will lay his yoke upon the countries and will proclaim the laws. Before him cities and kingdoms will bow and be ruled by the rod of a single man."

From that time on, Augustus became an open and devout believer, with such faith in his destiny that he never looked back. A remarkable confidence marked all his actions as emperor, most of which were crowned with success, and early in his reign he erected an Egyptian obelisk as part of a sundial on the Campus Martius, or Field of Mars, to commemorate the stars at his birth. Though from time to time he introduced curbs on astrological and other divinatory practices out of concern that reckless predictions might stir social unrest, when it was rumored in 27 B.C. that he might meet some untoward end he made his horoscope public and issued a silver coin stamped with the sign of Capricorn.

The birth of the empire under Augustus "ushered in an era of almost universal acceptance of astrology by the Roman elite." Evidence would have been hard to escape, in peace or war. Some of the Roman legions took Taurus as their emblem; each of the twelve chariot stables of the imperial race track stood for a zodiac sign, and the seven lanes were dedicated to the seven known planets (including the Sun and Moon). Astrological notions permeated the literature of the day. The greatest ancient treatise on architecture, written by Vitruvius, an engineer in Augustus's employ, included a statement of complete faith in astrology, and specifically in the Chaldean traditions of "astronomical calculations . . . of great skill and subtlety" from which "the past and the future can be explained." Horace, Persius, Propertius, Ovid, Juvenal, and other poets all wrote of the planets as exerting their own peculiar influence for good or ill; Pliny the Elder in his *Natural History* accepted the rule of the planetary gods; and Vergil in his *Georgics* rehearsed the astrological rules for predicting the weather and for planting by the Sun and Moon. Based on horoscope analysis, Horace accurately predicted his own death and that of his great patron, Maecenas; Propertius pored over the chart of his unfaithful mistress with the lovelorn yearnings of his jilted heart; Ovid could hardly contain his joy when he discovered that one of his enemies was doomed by the sky to a wretched life. "You were born to be unfortunate," he wrote to him with gusto. "Not a single star was propitious and gentle at your birth. Venus did not shine, nor Jupiter [in favorable aspect]. Neither did the Sun or Moon. Mars is so placed as to bring you constant sorrow; Saturn blackens your chart."

As a satirist, Juvenal viewed the stars with a sardonic eye:

> If fortune will, she may a rhetorician make
> Into a consul; and she may the same man take
> And bring him down unto his former state.
> What was Ventidius, or what Tully [Cicero]? Fate
> And the stars alone may tell thee this,
> They doom to misery, or give thee bliss.

At the same time, he poked fun at high-born women who were always consulting their charts—loathe even to apply ointment to a sore without first checking to see if the aspects were right—as well as "old whores in their off-shoulder dresses" who regularly ran to consult some quack "by the dolphin-columns and the public stands."

A more decorous Vergil implored the Muses for help in understanding the profounder reaches of the art:

> Give me the ways of wandering stars to know
> The depths of heaven above, and earth below,
> Teach me the various labors of the Moon,
> And whence proceed the eclipses of the Sun . . .

The "wandering stars," of course, are the planets that crisscross the sky; the "labors of the Moon" those tasks appropriate to each lunar mansion, as measured in equal segments along the ecliptic, of which there were twenty-eight. "To everything there is a season, and a time to every purpose under heaven." What are those seasons, and those times? Here, as often, Scripture and classical astrology unite. For the Hebrew prophet of Ecclesiastes names them, and the times of the naming are twenty-eight—one for every mansion of the Moon.

★   ★   ★

AUGUSTUS DIED PEACEFULLY at Nola in A.D. 14. His successor, Tiberius, was a practicing astrologer, and most of the later emperors, including Caligula, Claudius, Nero, Galba, Otho, Vitellius, Vespasian, and Vespesian's sons, Titus and Domitian, were convinced believers. Vespasian may have tempered his faith in the stars with some circumspection, but even those like Vitellius, who persecuted astrologers with unrelenting fervor, did so in part out of fear of their divinatory powers. From one reign to the next, the life of many a Roman noble hung upon the words of some imperial adept who might or might not depict him as a threat to the throne.

Under Tiberius the office of court astrologer was held by Thrasyllus, an Alexandrian Greek scholar who had made a name for himself in learned circles on the island of Rhodes. That name was richly deserved, for it is to this very man that we owe the definitive edition of the complete works of Plato, the greatest philosopher of ancient times. Tiberius had met Thrasyllus during an extremely painful time when, estranged from Augustus, he had decamped to Rhodes in a self-imposed exile that seemed likely to cost him any chance for the throne. The circumstances of that estrangement are involved, but in brief, Augustus had forced Tiberius into a loveless marriage with Julia, Augustus's wayward daughter, whose own two sons by a previous marriage Augustus had adopted as his own. To the proud Tiberius, a former consul, governor of Transalpine Gaul, and commander on the Rhine, it appeared that his intended role was merely to act as the official guardian of the true heirs.

Tiberius and Thrasyllus formed a strong bond, but it might have been otherwise. For if the story told about their first meeting is true, nothing could have been more precariously conceived. Tiberius, it seems, had recently undertaken an earnest search for an astrologer to assist his career. At his hilltop villa overlooking the sea, he would interview prospective candidates on a little verandah set on the edge of the cliff. His questioning was tough, and it is said that if he suspected some fellow or other was merely flattering his prospects, he had him thrown into the sea. Eventually, it was Thrasyllus's turn to prove his skill. Without hesitation, he predicted that Tiberius would one day rule the Empire in spite of the rivals who then stood in the way. Tiberius then asked him what he could possibly foresee about his own impending fate. Thrasyllus begged leave to consider his own horoscope, then paled and announced: "It seems my own life is right now hanging by a thread." By that Tiberius knew he had an accurate (if crafty) prognosticator, and stayed his hand. In time the rivals of Tiberius fell away and Augustus proclaimed him his heir.

Tiberius consolidated the imperial power Augustus had established but gave it a more autocratic stamp. A stern and secretive ruler, he disparaged the Senate and strengthened the palace guard. Even so, he lived in constant fear of his life. Often he struck out blindly at those he supposed might challenge his power. Thrasyllus did merciful work on behalf of some, but Tiberius, who pored over the horoscopes of potential rivals, was a hard man to appease. One such possible rival was Servius Sulpicius Galba, a distinguished political and military commander who had served as praetor, governor of Aquitania, and a general in Gaul. Tiberius thought to make him consul but first scrutinized his chart. It was said to show eminence, even touching the throne, which in fact had been noticed long before by Augustus when Galba was a child. According to Suetonius, when Galba and other patrician youths called to pay their respects, as was the custom, Augustus had pinched his cheek and said: "You too, sonny, will one day have a nibble at this power of mine." However, as Thrasyllus now correctly pointed out, that would not happen until Galba was an old man. Accordingly, Tiberius raised him in rank.

Another figure of concern to Tiberius was his nephew Caius Caesar, known to history by his childhood nickname of Caligula, or Little Boots. Tiberius recognized him as a fledgling monster and in pondering the succession asked Thrasyllus if he should exclude him from the line. In perhaps the greatest blunder of his career, Thrasyllus assured him that Caligula "had as much chance of becoming emperor as he had of driving a chariot across the bay of Baiae." Upon the death of Tiberius in A.D. 37 (one year after Thrasyllus himself had died), Caligula took power and in his perversity and madness exceeded everyone's worst fears. He promptly squandered the large budget, ransacked private estates to support his lavish court, reveled in sadistic cruelties, mocked civil government itself, it is said, by making his horse a consul, and indulged in incestuous affairs. His only military venture of note was to plunder Gaul. Even so, he demanded to be worshipped as a god. Meanwhile, to annul the prediction Thrasyllus had made about him, he spanned the Bay of Baiae with a pontoon bridge and drove his chariot across. He was so impressed with this accomplishment that in his harangue to an assembled multitude he compared himself to Darius, who had bridged the Danube, and Xerxes, who had crossed the Hellespont.

On January 24, A.D. 41, Caligula's six-year reign came to an end with his assassination, which brought his uncle Claudius to the throne. Born Tiberius Claudius Drusus in Lugdunum (modern-day Lyons) on August 1, 10 B.C., Claudius's beginnings were both blessed and cursed. On the one hand, his father was the Roman governor of Gaul and the stepson of Augustus, his mother, the daughter of Marc Antony. With such a pedigree he was the automatic recipient of wealth and power. However, he was also a sickly child, with multiple ills that have been variously ascribed to hydrocephalus, epilepsy, or a form of infantile paralysis associated with premature birth. His own mother complained that he was "unfinished by nature," and his stutter, nervous tremors, tics, and other infirmities persuaded his family that he was mentally deficient and, as he grew to manhood, "incapable of acting with that dignity expected of a prince of the house." Suetonius says that he staggered when he walked, nervously tossed his head from side to side, laughed uncontrollably when merry, or when enraged "slobbered horribly at the mouth." Augustus, fearing that Claudius would make himself ridiculous at social functions, had made sure he always had an escort; and at one point subjected him to the supervision of a muleteer in the hope that brute discipline might help him shape up.

Claudius remained unreconstructed, yet gradually his capacities also made themselves known. He demonstrated intellectual curiosity by his diligent study of history, encouraged by the historian Livy, and was the last person we know of who could speak Etruscan. In time he also displayed an aptitude for affairs of state. In public, he learned to speak without always stuttering (even though his conversation remained "unclear"), and in A.D. 41, when he was unexpectedly enthroned, he proved worthy of his elevation and the role destiny had assigned. Though sometimes portrayed as a dithering fool whose inept administration contributed to the decline of the empire, Cassius Dio tells us that Claudius was an industrious, public-spirited ruler. Pliny in his *Natural History* praised him as a master builder and patron of the arts. Most of the lesser-known Roman historians credit him with character and judgment above his faults.

As emperor, he extended Roman rule in North Africa, made Britain a province of the empire, and in his domestic policies seemed to appreciate that the welfare of the people reposed in his care. His public works included a new aqueduct for the capital and a new port at Portus, near Ostia. Though he expelled the Jews from Rome during one troubled juncture of his reign, "elsewhere he confirmed Jewish rights and privileges."

The apparent discrepancy between his early lack of promise and the accomplishments of his rule lends credence to a tradition, inadvertently begun by Suetonius and developed most recently in a pair of historical novels by Robert Graves, that Claudius had deliberately exaggerated his own infirmities in order to appear negligible to those ruthlessly striving for power. "His health," Suetonius wryly reminds us, "was wretched until he succeeded to the throne, when it suddenly became excellent."

But all his shrewdness could not save him, and in October A.D. 54, he was poisoned by his second wife, Agrippina, in a plot to advance her own son, Nero, to the throne.

If Claudius had a court astrologer, it was Balbillus, Thrasyllus's son. Balbillus was a man of many parts, and under Claudius also served as head of the engineer corps for a time in Britain. He later took a top post as prefect in the administration of Egypt, where he was in charge of all imperial buildings and public works. But his mind was perverse, and under Nero he would conspicuously thrive. The two were well-matched. Nero, born at sunrise on December 15, A.D. 37, had an unaspected Mercury in Sagittarius (that is, a mind without anchor or direction in an expansive sign opposite one it ruled) with Mars conjunct his Sun on the ascendant squaring an elevated Saturn in Virgo in the tenth. The Moon was also in Leo (an exhibitionist and self-centered placement) in square to Jupiter in Scorpio—all of which marked him out as a person of inherent and random violence prepared to do anything for power. Indeed, Balbillus had confided to Nero's mother, Agrippina, that her son would one day kill her. But she was so besotted with ambition she exclaimed, "Oh, let him kill me so long as he gets to rule!"

In the hope that his nature might be tamed, Nero was tutored by the illustrious Stoic Seneca, who tried to make the most of his sorry pupil, though he seems to have known deep down that all his efforts were in vain. Writing to a friend who had recently lost her son, Seneca revealed his belief in the immutability of fate:

Imagine I were coming to advise you at the moment of your birth: You are about to enter a vast community . . . that embraces the universe, that is bound by fixed and eternal laws that holds the celestial bodies as they whirl through their unwearied rounds. You will see there the gleaming of countless stars, you will see one star flooding everything with its light—the Sun . . . You will see the Moon taking its place by night . . . borrowing from the Sun a pale reflected light . . . You will see the five planets pursuing their different courses . . . On even the slightest motions of these hangs the fortunes of nations, and the greatest and smallest happenings are shaped to accord with the progress of a kindly or unkindly star.

For Seneca, the stars and planets, comets, the patterned flight of birds, even the arrangement of the entrails of an animal probed for divination were all signs rather than causes of impending things. "Whatever happens," he once wrote, "is a sign of something that will happen." For example, "A comet is not a sign of storm in the same way that there is a sign of coming rain when the oil in the lamp sputters, or the way a rough sea is forecast when the sea-coots play on dry land. Rather, it is a sign in the way that the equinox is a sign of the year turning to hot or cold, or as the things the Chaldeans predict, the sorrow or joy that is established for a person at birth by a star." For "how do things indicate future events unless they are sent [by God]? The roll of fate is unfolded on that principle, sending ahead everywhere indications of what is to come." For what something is, in a natural sense, or how it comes to be, by a material cause, is not the same thing as what it means. Pliny similarly held that meteors and comets were heralds of misfortune— "not that the misfortune occurred because the meteor appeared, but the meteor appeared because the misfortune was about to," as he explained.

✳   ✳   ✳

WHEN NERO ASSUMED POWER at the age of seventeen he was still under Seneca's guidance. And as long as that guidance lasted, Nero acquitted himself with a certain enlightened aplomb. The first five years of his reign were actually auspicious, marked by a series of regulations that eased the general burden of taxation and brought a new and noticeable degree of fairness to the Empire's provincial rule. But nature will out, and when he turned twenty-three he poisoned a potential rival and had his mother beaten to death. More atrocities followed, as Seneca, in his faltering role as regent, became the captive of a perverted youth whose nature he had failed to tame. Over the next few years, Nero dispatched various members of the royal Augustan line, killed a number of others whom his mistress chanced to dislike, and at night "let it all hang out" when he cruised the back streets of Rome with thuggish companions, raping girls who caught his fancy and cutting the throat of any poor fellow that drew his evil eye. On other occasions, he loved to lounge till dawn in the festering stews and low dens of the slums, "where slave minions, sailors, and the obscene priests of the Great Mother roistered about in some orgiastic debauch." Meanwhile, he staffed the government with hacks and cronies, maintained an extravagant court, and rolled up huge deficits even more appalling than those of Caligula, who had set a pretty high mark.

Fear gripped the capital as a period of secret tribunals and police-state terror also took hold.

Under Claudius, Balbillus had been a sober and moderate voice in state councils, perhaps because Claudius himself had been a man of some restraint. But under Nero he emerged as a man of unquenchable malice, quite unlike his father, with a ruthless heart. On one occasion, for example, when a comet appeared, Nero grew anxious and asked Balbillus what to do. The latter advised him that "although the comet portended grave events, Nero could deflect them from his own person by executing a number of great men." According to his argument, the sum total of lesser victims, offered as scapegoats, would prove "a substitute acceptable to the Fates for the life of a mighty prince." Nero followed this grim advice and proceeded to destroy the ruling class. In selecting his victims, any pretext would do. Not surprisingly, perhaps, among the nobles slain was Seneca himself. This measure was dolefully repeated whenever a comet appeared, and so helped to confirm the tradition that a comet brought catastrophe upon the body politic.

In A.D. 64, Nero, in his iniquity, surpassed even himself and burned Rome to the ground. For six days the fire ravaged the closely built quarters and narrow, hut-lined streets; many thousands perished in the blaze. Not surprisingly, he blamed the disaster on the Christians—whose doctrine of earthly self-denial he took as a personal affront. Three years later, in the ensuing persecution, St. Peter and St. Paul were both martyred in Rome. Not long afterward Nero killed his pregnant mistress, Poppaea, in a rage. He then set out on a pleasure tour through lower Italy and Greece, where at various local stops he showed off his hoped-for gifts—as an actor, singer, and musician—in grotesque theatrical skits.

At this weird juncture, Galba's appointed hour came round at last. Popular rebellions had recently broken out at numerous places in the Empire, and scattered legions had begun to look for a new leader to support. That had a ripple effect. During Nero's prolonged absence, his own praetorian guard defected, which emboldened the Senate to declare him an outlaw and dethroned. Galba had grown old in state service but was considered reliable and was popular with the army. Scarcely had his banner been raised in a march toward Rome when Nero, in hiding on June 9, A.D. 68, cut his own throat.

No one could say that Galba hadn't earned his moment in the sun. Though of noble stock, he was unconnected by either birth or adoption with any of the former emperors and had advanced by his own merit through the ranks. In the course of his career, he had served ably as praetor and then consul, and subsequently as a firm but impartial administrator in the military governments of Germany, Africa, Spain, and Gaul. He had also shown himself selflessly loyal when power prematurely beckoned—most notably upon Caligula's death, when he had rebuffed his partisans and sworn allegiance at once to Claudius, whom he also served with energy and skill.

As emperor, however, he proved inept at handling the reins of power. The prediction had always been that his tenure would be brief, and before long he managed to alienate not only his former allies in the government, but even the populace itself. In a misbegotten attempt to show resolve, he executed a number of senators who had wavered in their support, and failed to gratify the people with the usual handouts and festivals and games. Perhaps advanced age had bereft him of the vitality and judgment he needed. After just six months in office he was challenged by Marcus Salvius Otho, a former provincial governor of Lusitania, who engineered his murder by the palace guard.

Otho had little to commend him to the title he claimed. He had been one of Nero's decadent companions, and had even allowed his wife to become Nero's mistress in return for various imperial perks. But he understood better than Galba how court politics worked. In short order Otho reinstated some former officials (whom Galba had demoted or estranged), pampered the praetorian guard to ensure their allegiance, and promised the public entertainment on a grand scale. But it was the favor of the army he needed, and that he didn't get. Not a single legion was beholden to him. Their turbulent ranks opted to promote Lucius Vitellius instead. Vitellius was a general of considerable standing and head of the army of the lower Rhine. Otho tried to avoid civil war by a power-sharing agreement, but was curtly rebuffed. Once the fighting began he was outmaneuvered, and forced to make a stand at Cremona, where on April 14, A.D. 69, his forces were smashed. Two days later—after a reign of just three months—he went the way of Nero and took his own life.

Vitellius had been told as a child that his horoscope was royal but also, as was the case for Galba, that his reign would be brief. He professed a strong personal dislike for astrologers, in spite of his general faith in divination, but that did not save him from their predictions coming true. In the rapidly revolving door of imperial power, he was killed by the adherents of Titus Flavius Vespasian before the year was out.

Vespasian had served as quaestor in the Levant, with the army in Germany (as almost all aspirants at some point did), and as the head of an expeditionary force sent to Britain in A.D. 64. As proconsul in Africa, with jurisdiction over Palestine, he had also led the war against the Jews. In A.D. 66, after rioting broke out in Caesarea and Jerusalem against the brutality of Roman rule, he invaded Judea and three years later laid siege to Jerusalem. In the midst of this carnage, the army offered him the crown. As he set out from Palestine for Rome, he turned to the astrologer Ptolemy Seleucus to direct his course. Seleucus predicted that Vespasian would have no trouble taking power, and that, once enthroned, would rule for ten years. That sounded good to Vespasian and as a result he "was so tranquil and assured about his own fate and that of his family, as written in the stars, that, in spite of the perpetually fomented plots against him, he had the audacity of firmly proclaiming to the senate that no one but his sons would succeed him in Rome." Even when a comet appeared in A.D. 79, and predictions of his demise were rampant, he brushed them all aside.

Given his past history, Vespasian's reign was surprisingly benign. He restored dignity to the Roman Senate; abrogated the law of treason, a legal cloak for tyranny; reformed the courts; and once again placed the imperial administration on a sound financial footing. He also consolidated Roman authority in Gaul and Britain, and began construction of the Colosseum as part of a program to restore the splendor of by-gone days. As predicted, he survived all intrigues against him and held power for a decade, until his death from undulant fever on June 24, A.D. 79. He was then succeeded by his eldest son, Titus, who had served with his father in both Britain and Palestine. It was Titus who had completed the terrible siege and capture of Jerusalem, with its destruction and desecration of the Temple and wholesale slaughter and enslavement of the Jews. It is said that a million people perished during the six months of fighting, which eliminated the Jewish state for almost two thousand years.

As emperor, Titus advanced the public works program his father had begun, committed vast sums to completing the Colosseum, built new imperial baths to the south, began work on the celebrated Arch of Titus, strengthened Roman forts in the East, and built new roads through the Empire's expanding domain. A popular ruler, he also dealt with the huge disaster relief problems created by the eruption of Mount Vesuvius in a prompt and generous fashion, and when he died on September 1, A.D. 81, after a reign of two years, "the common people went into mourning as though they had suffered a personal loss." Yet if the people mourned his passing, Titus also mourned his life. His last days were cheerless and embittered, and he is said to have repented of some of his earlier, sacrilegious deeds, including his destruction of the Temple and the carnage he had wrought.

Titus was succeeded by his younger brother, Domitian, whom he had always justly feared. An utter cynic, Domitian at once proclaimed himself the champion of "traditional family values" and launched a crusade against Christians, atheists, and Jews. Under his aegis only the worship of Roman gods was sanctioned, and generally in its more superstitious forms. While greatly embellishing the capital's architectural grandeur, he exhibited a weird taste in public entertainments, as exemplified in the so-called Capitoline Games. Begun in A.D. 86, these were held every four years in the early summer and consisted of chariot races, athletic and gymnastic competitions, and contests in the arts. To these he added such novel events as gladiatorial combat in the half-light of evening, when combatants could barely see each other, and fights to the death between women and dwarves. Although Domitian lacked military prowess, he hoped to acquire a reputation for himself by engaging in what proved to be a series of costly but ineffective campaigns along the Danube and the Rhine. The Senate occasionally tried to curb his will, but in response to their complaints he executed a number of their leaders to show his contempt for their role in state affairs. In fact the whole theme of his tenure was fear. And no one was more fearful than Domitian himself.

From his youth he had known from his own horoscope (commissioned by his father, Vespasian, at his birth) "the very year, day, hour, and even the manner of his death." Indeed, Suetonius writes, at dinner one evening when Domitian was still a boy his father had made fun of his reluctance to eat some mushrooms, saying, "Oh, come on, lad, it isn't mushrooms you have to worry about but some day someone's knife. But that won't happen for a long time."

But once he knew, he knew. Haunted by the certainty of his own bloody end, Domitian lived a doleful life of dread. That dread increased with age, as the time appointed, once but "a dark if distant threat, came ever more to oppress him as the years tapered into the familiar funnel of middle age." In the end, foreboding cooperated with fate. Out of a fear of weaponry of all kinds, he refused to surround himself with an armed guard, forsaking the retinue other rulers had enjoyed. In anxious seclusion, he lined the walls of his private quarters with plaques of highly polished moonstone, which, mirrorlike, reflected everything going on behind his back. In vain he also executed a number of relatives and nobles whom he suspected of seeking his life. He even tried to discredit astrology itself—though his morbid knowledge of it conditioned all he said and did. On one occasion (reminiscent of the encounter Tiberius had had with Thrasyllus years before) he challenged an astrologer named Ascletarion to predict what form his own death would take. Ascletarion replied that he would ultimately be torn apart by dogs. To spite the prediction, and perhaps to demonstrate that fate could be outmaneuvered in some way, the emperor had him killed at once and prepared to cremate him on a funeral pyre. The pyre was lit, but as the smoke began to rise, a stiff wind overturned it and some wild dogs lurking nearby devoured the half-burned corpse.

At the time of his birth, Domitian's murder had been predicted for five in the morning on September 18, A.D. 96. Inexorably the fateful day arrived. The day before, he said to one of his aides, "There will be blood on the Moon in Aquarius tomorrow and a deed done that will be talked about throughout the whole world." Even so, that evening he declined to eat some apples, a food he relished as a treat. "I'll have them tomorrow," he exclaimed, "if I am spared." He tried to sleep, but at midnight leapt out of bed in a sweat. Breathing was like lifting weights. As the first light came, he clawed with nervous tension at a wart on his brow, drawing blood. "Oh, if only that were all!" he cried. Time passed. He could stand it no longer and asked one of his attendants for the hour. The man (a conspirator) lied and said: "The sixth."

In a frenzy of relief, Domitian devoured a gourmet truffle and, throwing off his robes, readied for his bath. Just at that moment there was a knock at the door. A trusted aide appeared with an urgent message. As Domitian read it, the man stabbed him in the groin.

Everything that happened to Domitian that morning was objectively confirmed in the sky. At the time of his assassination, Mars was setting, and the Moon, in Aquarius, squared to Mars (the "red planet," hence "bloody") and conjunct Saturn at the nadir of his natal chart. His body was carried away on a common litter by the public undertakers, as though he were a pauper, and cremated in a garden on the Latin Way.

Domitian was succeeded by Marcus Cocceius Nerva, a former consul, whose royal horoscope helped persuade the deciding Senate of the rightness of his claim. Domitian, of course, had long known of Nerva's stellar prospects, since he "had not failed to take careful note of the days and the hours when the foremost men had been born." But no one ever succeeds in killing his successor, and though Domitian had meant to dispatch him, he failed to sign the decree. Nerva did his best to restore dignity to his high office, and upon his own death of a stroke sixteen months later (on January 28, A.D. 98), was succeeded by Trajan, whose base of support lay with the army at Cologne.

Under Trajan, the good government and prosperity that Nerva had begun to establish was wonderfully sustained. Upon taking office, he renamed the Palace of Domitian the "House of the People," pledged concord with the Senate, released a number of political prisoners, and called many exiles home. In his cultivation of a modest and dignified court, he refused to allow gold or silver statues of himself to be created; balanced the budget even while cutting taxes on both farmers and the poor; he staffed his administration with dedicated civil servants;

and increased veterans benefits, all while maintaining strength both at home and abroad. In war and peace, he was almost a second Augustus. As a man of war, he humbled the Parthians in a drive to the Persian Gulf; extended Roman rule into Arabia, which became a new province of the Empire; and fulfilled the long-standing imperial dream of pacifying the German frontier. In one spectacular Danube campaign, Trajan spanned the "Iron Gates" (a gorge on the Danube dividing the Carpathian from the Balkan mountains) by cantilevering a road from the sheer face of the rock so that the army appeared to be walking on water as it crossed. In a subsequent expedition, he built a great bridge over the Danube supported on sixty stone piers. As a man of peace, he was scarcely less. He made one-man rule compatible with individual freedom; built the Forum, with a central Basilica (the largest wood-roofed building in the Roman world); great libraries for Greek and Latin scrolls; huge imperial baths; and new and improved harbors, including the hexagonal port at Ostia, which replaced Puteoli in importance with its direct link by the Tiber to Rome.

Upon his death on August 8, A.D. 117, he was succeeded by Hadrian, his adopted son and heir. Hadrian surpassed even Trajan in the majesty and benediction of his reign. Born in Italica (in what is now Spain) on January 24, A.D. 76, he had joined the army as a youth, rising through its ranks to become the military tribune of three legions. He was made suffect consul at the age of thirty-two, and then legate of Syria in the wake of the Parthian War. When Nerva adopted Trajan as his heir apparent in A.D. 97, Hadrian had carried the congratulations of the legions to him on the Rhine. Three years later he had married Trajan's grand-niece. For the next seventeen years he was the emperor's confidant and favorite, and almost his companion on the throne.

As emperor, Hadrian's principal aim was to consolidate the administration of an overstrained empire and to bring its perimeters to within manageable bounds. Some Eastern conquests were sensibly abandoned, since they had proved beyond the means of the legions to hold, but in taking due measure of his delicate task, he assessed the situation for himself. The Roman historian Cassius Dio tells us that he "traveled through one province after another, visiting the various regions and cities and inspecting all the garrisons and forts . . . He personally viewed and investigated absolutely everything," including the living conditions, quarters, and habits of the officers and men in camp. In the process, he instituted new methods of discipline and training, and won the soldiers' respect by living as they did. For, says Dio, "he everywhere led a rigorous life and either walked or rode on horseback on all occasions, never once at this period setting foot in either a chariot or a four-wheeled vehicle. He covered his head neither in hot weather nor in cold, but alike amid German snows and under scorching Egyptian suns he went about with his head bare. In fine, both by his example and by his precepts he so trained and disciplined the whole military force throughout the entire Empire that even today [Dio was writing a century later] the methods then introduced by him are adhered to on campaign."

In his domestic policies, Hadrian also eliminated the public debt, took steps to secure Rome's grain supply, lent money to local communities to boost their economies, helped maintain orphans, and in general pursued a frugal yet humane program of caring for those he ruled. Quite apart from his fitness for the throne, his personal gifts were abundant in a number of fields. He was a good poet, an excellent architect, an accomplished speaker, a fine prose stylist, a sage in philosophy and science, and a great patron of the arts. His villa in Tibur at the foot of the Sabine Hills is representative of a number of his achievements, with its building of a hundred structures, some among "the most daring attempted in ancient times."

Hadrian was also devoted to astrology and had no need of experts to cast or interpret a chart. He did that himself on a regular basis and every year wrote out predictions for himself in advance. In so doing, he forecast each of the major events in store for him, and accurately foretold the hour of his own death in A.D. 138. His own horoscope, like that of Augustus, was unmistakably majestic, with two planets—the Moon and Jupiter—conjunct the ascendant, Venus exalted in Pisces, Mars and Saturn both in signs they ruled. Mercury was in Capricorn. The Moon was also speeding toward its own conjunction with a bright fixed star conferring majesty and command. Altogether these indicated, for one born into the highest circles of worldly power, a destiny to rule. Hephaestio of Thebes, an ancient astrologer who examined the horoscope, tells us: "In this chart Saturn is the Lord of the house of the Moon. Being in his own house he gives death at the age of 56 years. Inasmuch as Venus favors him, she adds another 8 years to his life, making a total of 64. After 61 years and 10 months, however, the Horoscopal Point [ascendant] and the Moon move into a square with Saturn, although that does not kill him, because Venus aids him. So his life will endure to about 62 years and 6 months." In fact, he died when he was sixty-two, after a reign of twenty-one years.

Astrology also played a role in the selection of Hadrian's successor, and apparently determined the fate of one of his disinherited relatives, Pedanius, who sought the scepter for himself. According to Antigonus of Nicaea, a contemporary astrologer and physician, the ill-fated young man "was born to become, at the age of about twenty-five, the cause of his own destruction and that of his parents . . . He was misguided in his judgment because Mercury and Saturn were both in a male sign. And he was killed because the Moon was in its fall in Scorpio, squared by Mars."

The man whom Hadrian wished to succeed him did. This was Antoninus Pius, so-called because, once enthroned, he refused to act on a list of those slated for execution that was immediately placed on his desk. That set the tone for his regime. A former praetor and consul in Rome, proconsul in Asia, and member of Hadrian's cabinet, he ruled in his mentor's spirit, and his twenty-three-year tenure was "unquestionably the most peaceful and the most prosperous in the history of Rome." No wars were undertaken, except those necessary to suppress insurrections or protect frontiers, and his only colonial venture was a new wall, the Wall of Antonine, at the foot of the Caledonian mountains in Scotland between the Firth and the Clyde. Trade and commerce flourished; pioneering routes were opened; new cities sprang up across the Roman world. Legislation was enacted to protect the downtrodden; and numerous charitable institutions received government funding to assist the poor.

In one brief campaign against some Germanic tribes on the Danube, an astrologer played a part. The event was later commemorated on the carved base of an imperial column that depicts Jupiter Pluvius—Jupiter, the ruler of Pisces and the Roman god of rain. The story goes that after a revolt against Roman rule broke out, Antoninus hastily collected an army to quell it, but his troops were green. He lost every engagement, until he found himself completely surrounded on a hill. His retreat was cut off; destruction appeared certain. The enemy began to advance. An astrologer, Julianus the Chaldean, approached and asked leave to speak to him and told him not to fear. At noon, he said, help would come in the form of a terrific storm, and this would happen just before the enemy struck their lines. Antoninus looked helplessly upward. At the time, there wasn't a cloud in sight. But as noon drew near, the sky began to darken and thunder crashed overhead. Then hail fell in clattering cascades upon the mountains, followed by rain in such torrents that it stopped the charge of the Germans and flushed them down the hill.

Antoninus died in peace, after a short illness, at Lorium on March 7, 161, and was succeeded by Marcus Aurelius, a man of personal sanctity and wisdom devoted to the Platonic ideal of the philosopher-king. As consul, tribune, magistrate, and priest, he had managed successfully to combine the active life of public service with the contemplative pursuits of the scholar, thinker, and devotee of the arts. Upon his ascent to power, the statue of Fortuna, according to custom, was transferred to his hands.

Fortune, however, contrived for him a long and troubled reign. The Empire had begun to crumble at the margins and after the king of Parthia placed his own puppet on the Armenian throne, Aurelius was immediately drawn into a five-year war. At first the fighting went badly for the Romans, but by 166 the Parthian strongholds of Seleucia on the Tigris and Ctesiphon had been reduced, and the legions swept eastward along the paths once trodden by Alexander the Great. The exuberance of victory was spoiled, however, by an outbreak of smallpox, which ravaged some of the provinces just secured. Meanwhile, the perennial rising of Germanic tribes took a more frightening turn, as an organized onslaught burst through Roman defenses at Aquileia, in the first invasion of Italy since the 2nd century B.C. Aurelius hastened north to take charge of the army, and was obliged to spend most of the rest of his tenure on the northern frontier.

A series of inconsequential rulers followed Aurelius as Rome's decline began, and in a sense astrology declined with them as Western learning waned. One of the last to fully embrace it in all its sophistication was Septimius Severus, who reigned from 193 to 211. At one time or another, he had served as military legate on the Euphrates, a tribune in Rome, commander of a legion in Syria, and governor of a number of provinces from the Danube to the Seine. When the emperor Pertinax was killed by mutinous soldiers, Severus, then governor of

Pannonia Superior, was proclaimed emperor by a faction of the army in his place. He prevailed in the ensuing civil war but his real power base lay in foreign provincial troops. He died on an expedition to Britain with his two sons, Geta and Caracalla, on February 4, 211.

In Severus's youth, an astrologer had given him a glimpse of his future greatness, and when he came to power he adorned the ceilings of his imperial apartments with frescoes depicting the stars under which he was born. But he did not clearly depict his ascendant, so that no one could know when he would die. He also turned the roof of his palace into an observatory, with an assortment of astronomical instruments; became adept at casting charts; married his second wife, Julia Domna, a Syrian, because she had a royal horoscope; and knew by his own that when he set out to suppress a rebellion in the Roman colony of Britain, he would never return.

With his learning, Severus forms a useful contrast to Commodus, Aurelius's degenerate son, who was far more typical of those who presided over the Empire's decline. As Cassius Dio put it, Commodus turned a once-golden kingdom to one of iron and rust. Raised as a soldier, he had accompanied his father on his northern campaigns, but was quite unlike him, with an unwarranted self-importance joined to a craven heart. In his own honor, he renamed the twelve months of Roman years after his own imperial titles, and rechristened the Senate, army legions, and even Rome herself after weird variations of his own name. Not content with the usual deification that followed an emperor's death, he fancied himself a kind of reborn Hercules in life, and even took to dressing up in lion skins and carrying a club. In such garb, he took part in thousands of gladiatorial bouts, usually against poorly armed opponents, disabled veterans, or helpless beasts. In other respects, he proved almost a second Nero, and even (it is suspected) set Rome ablaze. Not surprisingly, several attempts were made on his life, until in 192 one orchestrated by his mistress and others at last succeeded, when an athlete named Narcissus strangled him in his bath.

Commodus had no use for astrology, which might suggest, some modern historians notwithstanding, that in explaining the tribulations of the Empire astrology itself was not particularly at fault. Divination of one sort or another, in any case, had always been a part of ancient politics and life. Even Cicero had belonged to the Society of Augurs, which specialized in reading the entrails of sacrificed beasts. Yet scholars are prone to blame all the horrors of the tyrants on their predilections for the occult. The estimable Samuel Dill, for example, tells us:

> The emperor's fears and suspicions were immensely aggravated by the adepts in the dark arts of the East. The astrologers were a great and baneful power. They inspired illicit ambitions, or they stimulated them, and they often suggested to a timorous prince the danger of conspiracy. These venal imposters were always being banished, but they always returned. For the men who drove them into temporary exile had the firmest faith in their skill. The prince would have liked to keep a monopoly on it, while he withdrew from his nobles the temptation which might be offered to their ambition by the mercenary adept . . . Holding such a faith as this, it is little wonder that the emperors should dread its effect on rivals who were equally credulous, or that superstition, working on ambitious hopes, should have been the nurse of treason. Thus the emperor's uncertain position made him ready to suspect and anticipate a treachery which may often have had no existence. The objects of his fears in their turn were driven into conspiracy, sometimes in self-defense, sometimes from the wish to seize a prize which seemed not beyond their grasp.

All this is only partly true, and fails to account for Augustus, Seneca, Vespasian, Hadrian, Severus, and others for whom astrology was a religious doctrine, not a science of power. In any case, no serious thinker of the time questioned the influence of the heavens on sublunary affairs. The indisputable power of the Sun over all living things, the Moon's effect on the tides, the repetition of weather patterns such as the stormy periods around the equinoxes, or the coincidence between the rising of Sirius, the Dog Star, and the summer's hottest weeks— helped to make plausible the impact of the planets and stars.

Then as now, in every art and science the level of learning and practice varied a great deal. Some of the horoscopes from Roman times are almost as primitive as Babylonian omens. For example, "If [Sirius] rises when the moon is in Sagittarius . . . grain in the field . . . If [Sirius] rises when Saturn is in Sagittarius: The king [will] fight," which comes from Roman Egypt; or this from Rome herself in 4 B.C.: "Year 27 of Caesar [Augustus], about the third hour of the day. Sun in Libra, Moon in Pisces, Saturn in Taurus, Jupiter in Cancer, Mars in Virgo,

[Venus in Scorpio], [Mercury in Virgo], [Scorpio rising], [Leo at Midheaven], [Taurus] setting. There are dangers. Take care for forty days because of Mars." That was the kind of reading one might get from a fortune-teller in the back streets of Ostia or Alexandria, whose clients belonged to the jaded, superstitious crowds glimpsed in the columbarium reliefs of the Isola Sacra or the pages of the *Satyricon.*

Sophisticates expected more. Although most of the charts drawn up by imperial astrologers such as Thrasyllus, Balbillus, and Ptolemy Seleucus have not survived, one can get a fair idea of their methods from the work of other professionals that have. In combination they provide "a comprehensive idea of the sort of astrology which would have been practiced by the scholars of Alexandria, the astrologers of the Roman court, and the Hellenized Jews of Palestine at the time of Christ."

★ ★ ★

THE FUNDAMENTAL WORK on astrology in classical antiquity was written in the 2nd century A.D. by Claudius Ptolemy, born at Ptolemaïs Hermii, a Greek city in Upper Egypt, and a government-salaried research professor at the great library at Alexandria, the most famous institute of learning in classical times. At one time or another it contained up to 700,000 papyrus rolls, all carefully catalogued by a series of illustrious librarians (such as Callimachus, the foremost elegiac poet of Hellenistic Greece), who made their own lasting mark on science and the arts. It was there, for example, that Euclid, Aristarchus, Hipparchus, Eratosthenes, Hippocrates, and others as well as Ptolemy produced their seminal and enduring works. Aristarchus (in the 3rd century B.C.) had hypothesized that the Earth revolved around the Sun, and that the light of the Moon was a reflection of the Sun's beams; Hipparchus (about 140 B.C.) assembled the first great catalogue of stars, established the modern form of the tropical zodiac, and discovered and accurately measured the annual precession of the equinoxes to within four seconds of arc; Eratosthenes determined the circumference of the Earth; Euclid wrote his *Elements,* the single most important work on mathematics ever composed. Scarcely to be surpassed, Ptolemy discoursed in Greek on a variety of topics, including astronomy, astrology, geography, and optics (in particular, the refraction of light); drew up maps with coordinates based on Roman army encampments; and devised predictive mathematical models for the observed motion of the planets other than the Sun and Moon. In his astronomical work, he expanded upon the industry of Hipparchus, and based in part on the latter's data, identified forty-eight constellations and mapped the longitude of over a thousand stars. In his geocentric model of the universe (which held sway until the heliocentric model of Copernicus displaced it), he gave a plausible account of the actual phenomena of the heavens as they appear to a spectator on Earth. This was set forth in a compendious book on mathematical astronomy called the *Mathematike Syntaxis* (the *Mathematical Composition*), afterward better known by its Arabic name, the *Almagest.*

As a companion piece, he wrote the *Tetrabiblos* (meaning a book in four parts), which codified some of the principles of applied astronomy, or astrology, and admirably summarized much of what Hellenistic astrology had achieved. In composing it, he drew upon the library's vast collection of related materials—tablets and papyri on Chaldean, Persian, Egyptian, and Greek divination—and claimed to have access to eclipse records dating back to the time of the Greek philosopher Pythagoras, or a thousand years. Still in certain respects *the* classic textbook on the subject—the stem and branch of astrological teaching in the West—the *Tetrabiblos* established the tropical zodiac as canonical, laid down the rules for drawing up a chart, identified the influence of various fixed stars, described the astrological rulership of nations, and gave a method for determining the length of a person's life.

At the Library, Ptolemy achieved all this under rather congenial conditions, "surrounded by exotic gardens, in rooms furnished with washstands of marble and gold chalices filled with oil." The perfume of a thousand flowers "drifted through the rooms." From time to time, we are told, he dutifully trudged up the library steps and, "after settling into the great domed reading room, with parchments stacked in front of him . . . released one of those soulful sighs that only scholars know."

Here is Ptolemy on the worldly fortune of "Parents":

In conformity with nature, the Sun and Saturn are allotted to the person of the father; and the Moon and Venus to that of the mother: and the mode in which these luminaries and planets may be found posited, with reference to each other, as well as other planets and stars, will intimate the situation of affairs affecting the parents.

Thus, for example, the degree of their fortune and wealth will be indicated by the satellites or attendants of the luminaries. If the luminaries are accompanied (either in the same signs in which they themselves are placed, or in the signs next following) by the benefics, and by such stars and planets as are of the same tendency as themselves, as conspicuous and brilliant fortune is presaged: especially, should the Sun be attended by matutine [morning] stars, and the Moon by vespertine [evening], and these stars are also well established in the prerogatives before mentioned. Likewise, if Saturn or Venus be matutine, and in proper face [when the aspect each holds to the Sun or Moon corresponds to that which its own house bears to their houses, that is, quintile for Saturn, sextile for Venus], or in an angle, it foreshadows the prosperity of either parent respectively according to the scheme [that is, Saturn being applicable to the father, Venus to the mother]. If, however, the luminaries hold no connection with the planets, and are unattended by any satellite, adverse fortune, humble state and obscurity, are then denoted.

This is how Ptolemy begins his discourse (apt enough for Roman times) on "[Morally] Monstrous or [Physically] Defective Births":

It will be found that, at a birth of this description, the luminaries [Sun and Moon] are either cadent [in the third house] from the Ascendant, or else not in any manner configurated with it; while, at the same time, the angles [the 1st, 4th, 7th, or 10th house] are occupied by the malefics [Saturn and Mars]. It therefore becomes necessary . . . to observe forthwith the preceding New or Full Moon and its ruler [the planet ruling the sign it is in]; as well as the rulers of the luminaries . . . [and if they are not in aspect with the Moon and the ruler of its sign] the birth will be monstrous. And if . . . the luminaries are also posited in quadrupetal or bestial signs, and the two malefics in angles, the birth will in that case not be human. And should the luminaries be not at all supported by any benefic planet, but only by malefics, the creature born will be wholly indocile, wild, and of evil nature . . .

When the luminaries may be in signs of human shape, while other circumstances in the scheme of the nativity may exist as before described, the creature born will then be human, or will partake of human nature, although it will still be defective in some peculiar quality. And, in order to ascertain the nature of that defect, the shape and form of the signs found on the angles occupied by the malefics, as well as of those wherein the luminaries are situated, must be taken into consideration: and, if in this instance also, no benefic planet should lend support to any one of the prescribed places, the offspring produced will be utterly void of reason, and indeed indefinable. If, however, it should happen that Venus or Jupiter give support, the defect will be veiled by a specious outward appearance, similar to that of hermaphrodites, and of those persons called Harpocratiaci [mute], or others of like imperfections. And should Mercury also give support, in addition to that of Jupiter or Venus, the offspring will then become an interpreter of oracles and divinations; but, if Mercury support alone, it will be deaf and dumb, although clever and ingenious in its intellect.

For Ptolemy astronomy and astrology were two sides of the same coin. In astronomy, he writes, "we apprehend the movements of the sun, moon and stars in relation to each other and to the earth"; in astrology, "the changes that [the heavens] bring about in that which they surround." His love of both was ecstatic: "Mortal as I am, I know that I was born for only one day. But when I see the stars circling in their orbits, my feet no longer touch the ground."

Though he was emphatic about the power of astrology to predict, he cautioned against trying to be too precise: "It is not possible that particular forms of events," he wrote, "should be declared by any person, however scientific; since the understanding conceives only a certain general idea of some sensible event, and not its particular form." (For example, an astrologer might foresee an accident, like a fall, but not necessarily whether the fall would take place from the roof of a house or from a cliff.) He added that to be truly specific in such matters one would have to be divinely inspired. Later astrologers, however, argued that greater precision was, in fact, possible, if the technical aspects and principles of the art were mastered completely and diligently practiced over time.

Ptolemy also claimed for general catastrophes an unequivocal precedence over individual lives. If a catastrophe befalls a city, with prodigious loss of life, it is the city's chart that is apt to point the way, though the horoscope of every person to perish should in some way bear the imprint of his fate. But in so far as they belong to a larger entity (with its own horoscope) their lives may also be gathered up, and strewn, through some collective fate. So when Cicero, for example, scornfully wondered aloud whether everyone who fell to Hannibal at Cannae had been born under the same star, he made a rhetorical point, but not a technical one: for no knowledgeable astrologer would ever have held that the whole destiny of a person was sealed by a single planet or constellation, considered on its own. With respect to death itself, in any case, the sign had little to do with it, since the time and type of death was conditioned (among other things) by the terminal houses—the fourth, eighth, and twelfth, and their rulers—while all house positions in turn were determined by the birth moment, which fixed the "horoscope" or ascendant sign and its degree.

Though the best known of all ancient astrological texts, Ptolemy's was one of many, and in some ways incomplete. For example, he did not write about the influence of the planets in the signs or houses, the use of the parts (other than the Part of Fortune), or methods of reading a chart by means of derived houses or house rulers. It would be left for others to fill in those gaps. He also seems to have borrowed informally from work that went before. His *Tetrabiblos,* for example, had included sections on the astrological rulership of nations, which were said to be governed by certain signs. Toward the end of the Augustan Age, Marcus Manilius, in his great astrological poem *Astronomica,* which was based on Greek and Latin translations of Babylonian texts, had also voiced this doctrine and ascribed the laws and customs of different nations to characteristics of the signs under which they happened to lie. Similarly, he explained the affection or enmity between peoples and states by the aspects that were formed by their ruling signs. Certain alliances were therefore naturally strong, others strained. Nations and peoples were destined to clash repeatedly or enjoy bonds of friendship according to the force of their stars. For Manilius, destiny was all. Nothing and no one could evade it: "Fate rules the world, and everything stands firm by law." To understand this was to understand the implacable course of life: "No man by prayer may seize fortune if it demur, / Or escape if it draw nigh: each one must bear his appointed lot."

Some, like Lucian of Samosata, another astrologer of note, viewed such fatalism in a positive light: "Astrology, it is true, cannot make good that which is evil. It can effect no change in the course of events, but it produces joy by anticipation and allows a person to fortify himself against misfortune in advance." At the same time, by elections and by answers sought to horary questions, there was an implied attempt to take advantage of the knowledge one could have. In the *Carmen Astrologicum,* a textbook on horary astrology by Dorotheus of Sidon, a 1st-century adept, we have this guide to assessing the marriage prospects of some hopeful groom:

Look at Venus where it is and which are the first, second, and third lords of its triplicity [element]. If they are with Venus or in an angle or in trine to her, that is a good sign. So, too, if the triplicity lords are in the houses that follow the angles [the succedents, that is, the second, fifth, eighth, or eleventh], rejoicing in their light and direct in motion. But if you find them in a cadent house [third, sixth, ninth, or twelfth], or corrupted, or under the rays of [conjunct] the Sun or near the West, then predict differently, because the native will be one of those who never marry or whose marriage is with slave girls, disgraced old women, or whores.

Another astrologer, Rhetorius the Egyptian, also put together a useful handbook in which, among other things, he provided a thorough synopsis of each house. This is part of his discussion of the eighth:

It is the sign that is turned away from the Ascendant; because of this, and because of its meaning of death, it signifies the turning away of life. If then the Lot of Fortune chances to be there, and its lord, and the lord of the Ascendant, it makes misfortunes and irregularities. And if these [lords] are malefics, the evil is worse, for it makes unfortunate persons, and if [they are] under the Sun beams [conjunct the Sun], short-lived. Mercury being lord of the [Lot of the] Daemon, [and] being present in the 8th, [makes] unintelligent, and illiterate, and lazy persons. Mercury being lord of the 8th or of the 12th or of the 6th, being present in the 8th under the Sun beams with Saturn and Mars, [makes] deaf and dumb persons. If the Moon is there by night adding to her numbers and light, especially if Jupiter is in the 11th, [the native] will profit from matters having to do with the dead and with inheritances. If the ruler of the 4th is in the 8th, he dies abroad. The lord of the 8th makes these same things when it is cadent. Venus in the 8th makes miserable and shameful persons. If the lord of the 8th is in the 10th or the 11th or the 5th, [the native] will grow rich from matters having to do with the dead, especially if it is in its own domicile or exaltation and not under the Sun beams and if it is adding to its numbers. But if it is under the Sun beams, it makes for an inheritance, which is immediately squandered. The lord of the 3rd in the 8th destroys the brothers first. The lord of the fifth being there causes childlessness. The lord of the 12th or of the 6th being there [makes] the deaths of enemies and slaves.

And so on.

The primers and textbook compiled by the Roman astrologer Julius Firmicus Maternus and his Near Eastern colleague, Vettius Valens, were even more complete. The latter, who ran an astrological school in Alexandria in the 2nd century A.D., assembled a remarkable *Anthology* of instruction replete with case histories for his students that covered all aspects of the art. One hundred and twenty-three actual horoscopes, including that of the Emperor Nero, were included in the text. Here is Valens on the five Egyptian terms (or unequal subdivision of the sign by five) of Taurus, each having a planetary ruler of its own: "The first 8 [degrees] belong to Venus: fertile, prolific, watery, lecherous, condemned, hating strife. The next six to Mercury: intelligent, prudent, evil-doers, having few offspring, poor-sighted, causing death. The next 8 to Jupiter: high-minded, manly, fortunate, ruling and beneficent, magnanimous, temperate, loving modesty. The next 5 to Saturn: barren, childless, eunuch-like, vagabonds, reprehensible, theatrical, joyless, toilsome. And the final 3 to Mars: masculine, tyrannical, fiery, harsh, murderous, temple-robbers, utterly bad, not infertile, but destructive and not long-lived."

Elsewhere, in an example chart, Valens explained how to forecast the gain or loss of wealth by the triplicity rulers of the Moon and the Part of Fortune together with the planetary ruler of the sign the part was in. In this instance, the Moon was in Sagittarius, thus in the fiery triplicity ruled by Jupiter and the Sun. Both of those planets were accidentally dignified by being in angles, which indicated wealth; but, alas, they were also in opposing signs, which meant its loss. The Part of Fortune itself was in Scorpio, ruled by Mars, so the extent of prosperity was shown (according to his method) by the condition of Mars and state of the eleventh house *as counted from the Part of Fortune,* which in this case meant the twelfth. There he found Mars in Virgo "cadent from the Ascendant and afflicted by a square to Saturn," in dismal shape.

The *Mathesis* or *Ancient Astrology: Theory and Practice* of Firmicus Maternus commands equal interest as the most detailed work on astrology to survive from the classical world. For Firmicus, as for Manilius, astrology operated by immutable laws and allowed the true adept "to eavesdrop on cosmic secrets" if the rules of the art were followed with rigorous care. In his view, the horoscope was so fine-tuned that "a degree's movement more or less in the Ascendant, the shadow of the presence of a planet, and the native might fall from a glorious career to a life of abject toil." Cicero's colleague Nigidius Figulus had illustrated the difference a small interval could make by striking a spinning potter's wheel twice in quick succession, then pointing to the distance between the marks. Firmicus, for his part, tells us, "If, at birth, the Moon is full and moving away from Mercury toward Saturn by day, the native will have some laborious occupation, such as those who hire out their bodies to carry loads on their shoulders and back. If malefic planets are also in aspect to this combination, he may end condemned in prison, or die in squalor, unkempt and deformed." Likewise, "if the Moon is void of course, in aspect to no planet, and with no benefics on the cardines [angles], the native will be a pauper who begs for a living and is always in need of a stranger's help." Again: "Venus in Capricorn with Saturn in opposition [in a man's horoscope] makes the natives hated by their wives . . . They will be despised in sexual relations yet objects of scandal for their base desires . . . The Moon, Saturn, and Venus in the seventh house, that is, on the Descendant, makes for effeminate perverts, and for sodomites if squared by Mars." Such judgments were reached only after the most meticulous calculations in accordance with time-honored rules. Firmicus also laid down an admirable code of ethics for professional astrologers, which became, as it were, their Hippocratic oath:

> Study and pursue all the marks of virtue . . . Be modest, upright, sober, eat little, be content with few goods, so that the shameful love of money may not defile the glory of this divine science . . . See that

you give your responses publicly in a clear voice, so that nothing may be asked of you which is not allowed either to ask or to answer . . . Have a wife, a home, many sincere friends; be constantly available to the public . . . avoid plots . . . In drawing up a chart, do not show up the bad things about men too clearly, but whenever you come to such a point, delay your responses with a certain reticence, in case you seem not only to explain but also to approve what the evil course of the stars decrees for the man.

Needless to say, not all his colleagues and successors lived up to the standard he set. His standards of analysis were also extremely high. Here is Firmicus on how to assess the Part of Fortune in a chart:

It shows the quality of life, the amount of inheritance, and the course that fortune takes . . . Therefore observe who is the ruler of the sign and of the degree of the Part, that is, in whose terms it is; also in what house each ruler is, whether on first or second angles, in favorable or dejected houses, in their exaltation or fall; whether each ruler is in aspect to the Part of Fortune and if they, on the angles, are in aspect to each other.

When you have carefully investigated all this, then observe who is the [term] ruler of the degree of the Moon (in a nocturnal chart) or of the Sun (in a diurnal chart) and see if this ruler is in aspect to the Part. If one planet is found to be the ruler of all these degrees and is favorably located in signs in which he rejoices, or in his exaltation, or his own house, this indicates a lucky chart. If this planet is well located with the Sun and Moon and in aspect to the Part of Fortune, still greater good fortune is indicated by the multiple aspect. If the Part of Fortune is on one of the angles in aspect to the Moon, this even more strongly indicates the same. If there is not one [consistent] ruler of the sign of the Part and its terms and of the terms of the Sun and Moon, that which has the greatest power should figure most in the forecast; and if this one is benefic and in favorable signs, in his exaltation or in his own house, or on first angles and in aspect to the Part, that makes a great and noble chart.

If not, then otherwise. And so he proceeds at length, with more fine-tuning, until concluding with the notable admonition that one must pay very close attention to the duodecatemorion [division of the sign by twelfths] of the Part of Fortune and its ruler—"lest the meaning of the chart slip away."

With these subtle words, it may be said that the astrology of the ancient world took its leave, for the work of Firmicus was the swan-song of the era, and the last known work on the subject before the learning on which it depended was eclipsed.

★   ★   ★

AT ITS HEIGHT, and at its best, Rome—as the refuge of that learning—had enforced a kind of martial civility throughout its dominions, giving protection to all its citizens, whether native, conquered, or free-born. For much of its sway, the imperial capital had been a model of civic administration—well policed, with excellent sanitation, central heating (for the well-to-do, at least), running water, and other amenities of civilized life. The indigent had often been cared for at public expense; the arts had flourished—though none, perhaps, as much as that of war. Even so, under Antoninus and Hadrian, for example, Roman armies were seldom seen far from the boundaries of the Empire, and a relative peace and quiet reigned throughout the Roman world.

But then the night began to fall. "The decline of Rome," wrote Edward Gibbon, "was the natural and inevitable effect of immoderate greatness. Prosperity ripened the principle of decay; the cause of destruction multiplied with the extent of conquest; and as soon as time or accident had removed the artificial supports, the stupendous fabric yielded to the pressure of its own weight . . . The victorious legions, who, in distant wars, acquired the vices of strangers and mercenaries, first oppressed the freedom of the republic and afterwards violated the majesty" of the imperial crown. With Commodus, Hadrian's insane and profligate son, that decline had sensibly begun. Thereafter every emperor was either installed or ratified by the self-willed imperial guard, and few men of ability, with the notable exception of Septimius Severus, emerged at the helm. Severus conducted successful campaigns against the Parthians and Picts, but his son, Caracalla, immediately immersed himself in domestic purges while attempting to pacify the German frontier. In the 3rd century, numerous emperors of negligible stature—Elagabalus (so named from his worship of a Sun god of the same name at Emesa), Severus Alexander, Philip the Arabian, Decius, Valerian, Claudius II, Aurelian, and Diocletian, among others—succeeded each other in a bland procession that merely mark the stages of the Empire's decline.

In a fateful move, Diocletian tentatively divided the political administration of the Empire into east and west. That division took hold with Constantine the Great (Constantine I), the first Christian emperor, and was made permanent after the death of Theodosius I in 395. As the center of imperial power shifted to Constantinople in the East, Rome's own political importance was extinguished. The foreign or barbarian legions of the army that had grown up with the expanse of Roman power now turned against their masters and with a tribal, if not quite ethnic solidarity, tested their own might. The last western emperors abandoned the now shabby capital for other strongholds—Ravenna, Milan, or Trier—which were transformed into armed camps against the advance of Germanic tribes. But the tide could not be stemmed. In 410, Alaric, king of the Visigoths, swept through Italy, rejected negotiations, and besieged and sacked Rome. The whole of Italy was thereafter ravaged by a series of invasions until in 476 the last Roman emperor, Romulus Augustulus, was deposed by Odoacer of the Goths. What was left of the Empire collapsed and Europe's Dark Ages began.

Western civilization as such thereafter sought its new fortified home in the East Roman Empire of the Byzantines. In the 6th century, Emperor Justinian imposed a unified rule on the Mediterranean by fortifying the ports of Asia Minor and North Africa, and by constructing a strategic network of castles, towers, and bastions throughout Italy and Greece. The Byzantine cultural world combined Roman political tradition, Hellenic culture, and Christian doctrine, but it was ever an embattled citadel, as the Visigoths of Alaric, the Huns of Attila, and the Ostrogoths, Bulgars, Vandals, Avars, Slavs, and Persians variously challenged parts of its dominions and even besieged the capital itself. In the end, the greatest challenge came from the armies of Islam, which swept out from the Arabian Peninsula in the 7th and 8th centuries and annexed territory from Persia to Spain.

Chapter 4

I T HAS BEEN SAID that "in the general decline of learning that
overtook Western Europe during the first Christian centuries, no
science suffered a more complete eclipse than astrology." To
flourish, it had required the books, instruments, astronomical tables,
and relevant knowledge that only an environment sympathetic to
higher education could supply. The Dark Ages were conspicuously
lacking in those cultural coordinates. That was in the West. In the East,
in the Empire of the Byzantines, it continued as an object of study; and
in the Arab world it was taken up and embraced. Although the Koran
forbade the worship of the Sun and Moon, astronomy and astrology
prospered, as did the other sciences and arts.

Meanwhile, in the Near and Middle East, Islam had emerged as the
dominant faith. In A.D. 610, a forty-year-old merchant of Mecca
named Muhammad had a vision in a cave above the city that convinced him that he had been chosen by God to be the unique Arab
prophet of sacred truth. He made few converts at first, many foes, and
after an attempt on his life he fled to Medina, where he established a
theocratic city-state. He attracted an ever-larger group of adherents
and began to extend his sway. Over the next few years, Muhammad
fought several pitched battles with rivals and opponents, which he unexpectedly won—thereby gaining stature as well as ground—and before long, his missionaries began to fan out across the Arab world. As
new disciples flocked to his standard, his religious army grew, and in
630 Mecca itself fell without a fight. That proved a herald of his incon-
testable power, and, in a remarkably short time, the tribes of Arabia

were united under Islam and its creed. Part of that creed was conquest (in a parallel with the Christianity of the later Crusades), and just seven years after the death of Muhammad in 632, the Arab assault on North Africa began. Attacking simultaneously on several fronts, small bodies of mounted zealots ripped into the Byzantine Empire and from that point on the movement gathered tremendous force and speed. Palestine, Syria, Egypt, Cyprus, Tripoli, Carthage, and the whole of Persia fell in succession, and by 711 the entire coast of Africa, from Suez to the straits of Gibraltar, was under Islamic control. The Arabs then advanced into Spain, and pushed irresistibly northward into Provence. It wasn't until the victory of Charles Martel at Poitiers in 732 that their progress in the West was checked. By then the Byzantine Empire had been drastically reduced, while the once tiny militaristic sect of Islam commanded tens of millions of subjects from Gibraltar to the Aral Sea.

The administration of the whole of this territory was centered at first in Syria, at Damascus. But power struggles soon emerged among rival clans. Not long after the conquest, the ruling dynasty, known as the Umayyads, was overthrown by the Abbasids, and a decision was taken to transfer the seat of government further to the east. In 762, Abū Jafar al-Mansūr, the second caliph of the new dynastic house, traveled the length of the Tigris River looking for the ideal site. Eventually he settled on a little village on the west bank of the river where it was connected to the Euphrates by canals. The village was surrounded by palm trees, accessible from both rivers, as well as to caravans from Syria and Egypt; the Persian Gulf was a route to the Byzantine Empire and even China by the sea. Construction of the new capital then proceeded at breakneck speed. The caliph himself laid the first brick, and a huge labor force of 100,000 was drawn together from the cities of Mosul, Kufa, Wasit, and Basra, as well as from Syria, Persia, and other lands. Some of the bricks used were huge and weighed up to two hundred pounds. Within eight years, the round city, as it was called, was completed as a military stronghold, bounded by three concentric walls and surrounded with a moat. In the center stood the Golden Gate palace of the caliph surmounted by a great green dome and connected to the Great Mosque. Four main gates, set on a rectilinear axis, like the four points of a compass, opened out onto highways that connected the palace with the far-flung reaches of Arab rule.

Thus was Baghdad—known as "Dar es-Salaam," the City of

Peace—founded. It developed rapidly into a vast emporium of trade linking Asia and the Mediterranean, and by the reign of al-Mansūr's grandson, Hārūn ar-Rashīd (786–809)—in English, Aaron the Just, the caliph of the *Arabian Nights*—surpassed even Constantinople in prosperity and size. Its administration had managed to harness the Tigris and Euphrates rivers for the cultivation of grain, and a brilliant system of canals, dikes, and reservoirs drained the surrounding swamps. Immigrants of all kinds—Christians, Hindus, Persians, Zoroastrians, and so on—came from all over the Moslem world, and from lands as far away as India and Spain. For the most part, they were welcomed in an ecumenical spirit, and there was much to entice them to stay. There were many rich bazaars and covered shops along the embankments, where all sorts of artisans and craftsmen—marble workers from Antioch, papyrus makers from Cairo, potters from Basra, calligraphers from Peking—plied their trades. Food stalls sold lemon chicken, or "lamb cooked over a spit with cardamom, or small rolls dipped in honey, or flat slabs of pita bread smeared with fat." There was a large sanitation department, many fountains and public baths, and (unlike the European towns and cities of the day) streets that were regularly washed free of refuse and swept clean. Most households had water supplied by aqueducts, and some had subterranean rooms cooled by screens of wet reeds. Marble steps led down to the water's edge, where, along the wide-stretching quay, river craft of all kinds lay at anchor—from Chinese junks to Assyrian rafts resting on inflated skins. Thousands of gondolas, decked with little flags, also carried the people to and fro. On the outskirts of the city were numerous suburbs with parks, gardens, and villas, some adorned with varnished frescoes of lapis lazuli and vermilion, or faience panels and ceramic mural tiles. An immense square in front of the central palace was used for tournaments and races, military inspections and reviews.

A bustling city by day, the lamp-lit nights of Baghdad had attractions in abundance, too. There were cabarets and taverns, game rooms for backgammon and chess, shadow-theater productions, concerts in rooms cooled by punkahs, and acrobats to entertain strollers by the quays. On street corners, storytellers regaled occasional crowds with tales such as those that inspired the *Arabian Nights*.

In those glory days of Baghdad's ordered splendor, London and Paris were still grimy and chaotic little towns made up of a maze of twisting streets and lanes crammed with timbered or wattle-and-daub houses whitewashed with lime. Most of the dwellings were shabby, and a fifth of the populations lived and died in the streets. There was no real paving of any kind, and for drainage only a ditch in the middle of the road. That ditch was usually clogged with refuse—including the welter from slaughterhouses as well as human waste—and in wet weather the streets were almost marshes, awash in a depth of mud. Footpaths along the main streets were marked by posts and chains. There were some shops, of course, but most of the real commerce took place at trading stations (like the famed Six Dials in Southampton, England) where livestock and crafts were purchased or exchanged. In Paris, all that remained from its commercial development under the Romans were the vast catacombs under Montparnasse.

Yet just at this juncture the West began to recover a little of its former light and strength. This happened under Charlemagne, the eldest son of Pepin the Short, and the grandson of Charles Martel, who had checked the Saracen advance into Europe at Poitiers. Charlemagne was a king of the Franks. In 773, his authority had been confirmed in Rome by Pope Adrian I, and he subsequently enlarged his kingdom to include the Pyrénées, northern and central Italy, parts of Bavaria, and territory north of the Rhine. In 778, he crossed into Spain, seizing the area around Pamplona, Barcelona, and Navarre, then, after taking Pavia in Italy, assumed the iron crown of the Lombard kings. Over the next thirty years he also occupied Saxony, advanced west to Pomerania against the Avars and Slavs, and by 800 had consolidated all of the Germanic peoples of western continental Europe under his rule. That same year on Christmas Day he was crowned at Rome by Pope Leo III and the Holy Roman Empire was born.

As emperor of the West, Charlemagne made contact with ar-Rashīd in Baghdad, and the two exchanged several embassies over the next ten years. A contemporary account of one of these missions was left to posterity by "Notker, the Stammerer," a monk of St. Gall, whose somewhat naïve and tendentious narrative sought to exalt Charlemagne at ar-Rashīd 's expense. The Baghdad envoys, he tells us, arrived in the last week of Lent, 807, but Charlemagne delayed receiving them until Easter eve. He then donned himself in full royal attire—a brightly embroidered tunic fringed with silk and adorned with precious stones—which Notker assures us overawed the Arabs, for "he seemed to them so much more than any king or emperor they had ever seen."

Charlemagne then gave them a tour of his palace complex, during which "the Arabs were not able to refrain from laughing aloud because of the greatness of their joy." Compared to the imperial splendors of ar-Rashīd's Golden Gate Palace, with its great green dome, Charlemagne's estate at Aachen or Aix-la-Chapelle—a site chosen for the therapeutic vapors of its warm sulphur springs—was rather modest, despite the marble columns and mosaics he had obtained for its adornment from Ravenna and Rome. (As for his country estate at Asnapium, which the Arab envoys may also have seen, it featured, according to one contemporary acccount, "a royal house built of stone in the very best manner, having three rooms." No matter what those rooms might have looked like, that was not likely to overawe.)

Following Easter service on the following day, Charlemagne regaled the envoys at a banquet, but, according to Notker, they were so amazed at all they saw "that they rose from the table almost as hungry as when they sat down." (In fact, they found the food repellent but were trying to be polite.) However, one evening during their stay they got soused on barley beer and indirectly betrayed what they thought of Charlemagne's might. Notker reports it this way:

> The envoys were more merry than usual, and jokingly said to him, who as always was calm and sober: "Emperor, your power is indeed great, yet it is much less than the report of it which is spread throughout the kingdoms of the East." "Why do you say that, my children?" he replied. "How has that idea come into your head?" "All the peoples of the East fear you," they replied, "much more than we do our own ruler, Haroun. As for the Macedonians and the Greeks, what can we say of them? They dread your overwhelming greatness more than they fear the waves of the Ionian Sea. The inhabitants of all the islands through which we passed on our journey were as ready and keen to obey you as if they had been brought up in your palace and loaded by you with immense favors. On the other hand, or so it seems to us, the nobles of your own lands have little respect for you, except when they are actually in your presence. For when we entered your domains and began to look for Aachen, and explained to the nobles we met that we were trying to find you, they gave us no help at all but sent us away."

In other words, people a long way off may be impressed by rumors of your power, but up close it doesn't seem to amount to much. Notker, however, missed the point and thought Charlemagne was being praised for the awe he inspired in distant lands.

Protocol was maintained, however, and handsome gifts exchanged. The Arabs had brought him various spices and unguents, brass candelabra, ivory chessmen, a colossal tent with many-colored curtains, and a water clock that marked the hours by dropping bronze balls into a bowl. As the balls dropped, mechanical knights or horsemen—one for each hour—emerged from behind little doors which shut neatly after them as they stepped forth. Notker also mentions a white elephant (an earlier gift), which had once belonged to an Indian raja and which was later immortalized in stone in the cathedral porch of Bale. Included, too, was a beautiful astrolabe, along with a number of books on astrology "which Charlemagne, Emperor of France and Germany, commanded to be translated into Latin from Arabic." Thought Notker: "They seemed to have despoiled the East that they might offer all this to the West." Charlemagne reciprocated as best he could with some embroidered cloaks from Frisia, a few Spanish horses, and some hunting dogs "specially chosen for their ferocity and skill."

When the Arab envoys returned to Baghdad, ar-Rashīd immediately put the dogs to the test. They were released to chase a wild lion, which they managed to corner, and this (Notker tells us) so impressed the caliph that he took their prowess as emblematic of Charlemagne's superior might. As evidence, Notker cited a letter in which ar-Rashīd (with perfectly ironic generosity) affected to mourn the fact that the Holy Land was too far away for Charlemagne himself to defend, and so offered to defend it in his name.

Had war ever erupted between the two, ar-Rashīd would likely have prevailed. Like Charlemagne, he was a formidable warrior, and while still a young man had led an army of 100,000 against the Byzantines, then ruled by the Empress Irene. He had met and defeated the renowned general Nicetas, and marched thence to Chrysopolis (now Üsküdar) opposite Constantinople on the Asiatic coast. Having pitched his tents on the heights, he had threatened to sack the city if a large annual tribute were not agreed to at once. It was paid forthwith, and for several years thereafter received with great fanfare at Baghdad, where the occasion was marked by festive events and a parade. In 802, however, the Byzantine Emperor Nicephorus I had refused to oblige. His envoy brought a defiant message to Baghdad, and, threatening war, threw a bundle of swords at the caliph's feet. Drawing his own sword, or scimitar, the caliph, according to legend, cut the Roman swords in two with a single stroke without even turning the edge of his blade. Then he had dictated a letter which said: "Harun-al-Rashid, Commander of the Faithful to Nicephorus, the Roman dog: I have read thy letter. Thou shalt not hear, thou shalt see my reply." He set out at once with a large army, took and sacked the Byzantine city of Heraclea on the Black Sea, and soon forced Nicephorus to submit. After the latter then reneged, ar-Rashīd met and defeated his army in Phrygia in Asia Minor, where the Byzantines lost forty thousand men.

★   ★   ★

BY THE EARLY 9TH CENTURY, Baghdad had become the cultural and intellectual center of the Islamic world. Ar-Rashīd had done much to bring this about, but it was under his son, Abū al-Abbas al-Ma'mūn, that the great flowering of Arabic culture reached its height. He was responsible for the translation of hundreds of Greek works into Arabic, and founded the Academy of Wisdom in Baghdad, with a large library where scholars of all races and religions could mingle in fraternal pursuit.

In time, the complete medical and philosophical works of Galen, the *Physics* of Aristotle, the Old Testament (from the Septuagint Greek), Plato, Hippocrates, Ptolemy, Dorotheus, Archimedes, Euclid, and others were all translated, as the liberal arts flourished along with scientific research. Advances were made in spherical astronomy, geometry, algebra, trigonometric functions, integral calculus, and a number of other fields. The Arabs discovered algebraic equations, invented the zero, introduced Arabic (really Hindu) numerals, created the decimal system, perfected the lunar calendar, and from Spain to Samarkand built new observatories that in the end enabled them to double the level of celestial observations that the Greeks had attained. Indeed, many of the names of prominent stars, such as Betelgeuse and Rigel, and astronomical terms like nadir, azimuth, and zenith, were coined by the Arabs at this time.

If the astrological ideas that had inspired Columbus were espoused by a French cardinal, they were Arab in derivation, and the great schools of Islam were their source. Such schools flourished in Cairo and Alexandria in Egypt, in Kairwan (one of the first Islamic centers to open its doors to Coptic Jews) south of Tunis, and in Cordova and Toledo in Spain. But their central home was Baghdad, and anyone who wants to know what astrology has meant to world history must become familiar with that 9th-century intellectual capital of the world, where all the astrological ideas of antiquity—Babylonian, Greek, Roman, Persian, and Egyptian—made their vibrant home. Under the caliphs of Islam's golden age, innovations flowed: the Arabs invented the solar return as a predictive technique for natal astrology, the Aries Ingress for mundane forecasts (public events), and further developed the conjunction theory of the Persians. From 622 on, Islamic, Jewish, Persian, Greek, and Hindu scholars flourished together. What today is called "Arabic astrology" is really that body of astrological learning—heavily influenced and shaped by the Greeks—that was assembled by Arabs, Jews, and Persians from the 8th to the 12th century in Arab lands. In particular, great Arab astrologers such as Masha'allah, al-Kindi, al-Bīrūni, and Abū Ma'shar progressively joined Greek to Arab learning as Greek works became more available for them to study and assess.

Born in Basra at the mouth of the Persian Gulf, Masha'allah (known also as Messahalla to the West) was an Arab Jew who had taken part in the astrological deliberations that had led to the founding of Baghdad on July 31, 762. One of his notable colleagues in that enterprise was the great engineer Ibrahim Fazari, who also constructed the first astrolabe in the Arab world. In the course of his career, Masha'allah wrote on solar returns, ingress charts, conjunctions, and elections, but was particularly interested in how astrology could be applied to history. According to the 9th-century Christian astrologer Ibn Hibinta, his work in that field included the expert analysis of charts for the rise of Islam, the advent of Christ (which indicated that "violence befalls him from his people"), and for Muhammad (whose tenth-house Moon spared him the fate of Christ). He also wrote a commentary on Ptolemy and an astrological textbook that became a touchstone for all such studies in the Arab world.

This is part of his valuable discourse on "reception," which has to do with the affections of the planets (according to sign placement) and their inclination to act:

You have to know that reception is formed through the exaltations and the domiciles [signs], but either way, it is the same thing: for example, if any one of the seven planets is found in the exaltation of another or in its domicile, and the same thing if it unites with another by important aspects; or if they are both in one sign, and one of them is in the exaltation of the other in union with it, when it unites itself with it by body (conjunction). For example, Saturn in 20 degrees Aries and Mars in 15 degrees Aries; in this case, Mars unites with Saturn by body, and Mars receives Saturn in its domicile, but the latter is not in reception with Mars . . . Another example is of connection and reception: when Saturn was in 20 degrees Aries, and Mars in 10 degrees Capricorn, and none of the other planets is closer to Mars in union with Saturn, that is within a few degrees. If Mars is united with Saturn within a degree, in such case they are found to be in mutual reception by domicile, since Mars receives Saturn because it is in its domicile, and Saturn receives Mars because it is also in its domicile.

Also, for the same reason, the exaltation is like the domicile: but the exaltation is of greater importance in the kingdom; i.e. if the question is about the king [because] the lord of the exaltation is stronger than that of the domicile. Hence, when the Sun is in 10 degrees Aries and Mars in 10 degrees Capricorn, the Sun is united to Mars, and Mars receives the Sun because it is in its own domicile; but the Sun does not receive Mars because it is not in its domicile. Likewise, each of the other planets can be united to its companion by domicile or by exaltation, by important aspect, or by being in the same sign . . .

Meanwhile, the first of Baghdad's astrological schools had been founded in 777 by the Jewish scholar Jacob ben Tarik. That school was successively headed by al-Kindi and Abū Maʿshar. Al-Kindi was an Arab from Kufu and, like Masha'allah, was educated at Basra. Enamored of Aristotle, he was a prolific scholar of prodigious scope. Some two hundred works are attributed to him, on subjects ranging from magic, philosophy, and metaphysics to mathematics, meteorology, and optics. It was al-Kindi who translated the works of Aristotle into Arabic, and his work on great-conjunction theory contains the earliest description of the use of some ninety-seven "lots" or "parts," which, like the Part of Fortune, are arithmetically derived points on the ecliptic endowed with special force. (Many of these so-called Arabic parts, as they came to be called, were largely Greek in origin; others represent an original Arab contribution to the art.)

Abū Ma'shar (known also in the West as Albumasar) in turn became professor of astrology at Baghdad University during the caliphate of al-Ma'mūn. He had begun his career as a student of the Hadith, or sayings and traditions of the Prophet Muhammad. However, in 825 he undertook to learn mathematics in all its forms, including astrology. He studied with al-Kindi and eventually acquired immense renown as the leading astrologer of the Islamic world. After moving to Baghdad from Balkh in Khorāsān, he devoted himself to the study of Persian, Greek, and Syriac texts, his eclectic mind welding their astrological doctrines into a single system, as set forth in his principal work *The Great Introduction to the Science of Astrology,* composed about 850. Some fifty other books are also ascribed to him—including works on great-conjunction theory and solar returns—and altogether (as befits the foremost student of al-Kindi) his work had an encyclopedic range. One of his lost books, called the *Book of the Thousands,* apparently contained an outline of world history. It was Abū Ma'shar, not incidentally, who also arranged for the translation into Arabic of Ptolemy's great treatise on astronomy, thereafter known by its Arabic title as the *Almagest.*

By and large, his stature as "the teacher of the people of Islam concerning the influences of the stars" remained unchallenged during his lifetime, though his willingness to allow students to dabble in "heretical views" angered extremists, and in the caliphate of al-Musta'in—a degenerate monarch—he was flogged for not always insisting on their strict adherence to the tenets of the faith.

One of his surviving works is a little treatise entitled *The Abbreviation of the Introduction to Astrology*—the earliest such manual translated into Latin—and a good compendium of the principles of his art. That art was fastidious, as exemplified by his third chapter, where he enumerates and explains the twenty-five "conditions of the planets" that determine their health and strength. These are: domain, advance, retreat, conjunction, aspect, application, separation, void of course, wild, translation, collection, reflecting the light, prohibition, pushing nature, pushing power, pushing two natures, pushing counsel, returning, refranation, resistance, evasion, cutting the light, favor, recompense, and reception. Many of these are wholly unfamiliar to modern astrologers. For example, "pushing power" means that if one planet is in its own house or exaltation, or otherwise dignified by term or face, and applies to another planet, it pushes its own power onto it. "Refranation," on the other hand, refers to a planet that begins to apply to or move toward another but then aborts the aspect it made by reversing course. These distinctions were not insignificant, and in some cases determined the judgment of the chart.

A number of anecdotes from Abū Ma'shar's life survive. For example, "Once with some travelers," he reportedly told a pupil,

> I went to Baghdad and stayed with a friend of mine who knew a little astrology. He asked me how the Moon would be the next day, and I said, "In quartile aspect with Mars." He said, "Then you mustn't leave tomorrow"; and I said to him, "I have no intention of departing on such a day. But the other travelers won't heed my warning." "Let's test them," he said. So I said to them, "Tomorrow is an unfavorable day. Bide your time a while, and I'll even see to it that your animals are fed." But they would not listen. The next morning, they prepared to depart. As they did so, I observed that the Ascendant was in Taurus, Mars close to the Ascendant, and the Moon in Leo in quartile aspect to Mars. I said to them, "For God's sake don't go at this hour," but they laughed at me in scornful disbelief and set out. I turned to my friend and said, "I'm heartily sorry for these senseless men." We passed the morning in a leisurely repast, and were still sipping our drinks when some of the company came staggering back in. It turned out they had fallen among thieves, lost all their goods, and some had been killed. Strangely enough, some of the survivors now blamed me—"These things happened," they said, "because of your superstition"—and threatened to beat me to death. I escaped by the skin of my teeth. Then and there I swore I would never again discuss the science of astrology with such ignorant folk.

Most ancient astrologers of any standing tended to abide by that rule, mindful that their knowledge had a powerful and sacred cast. Those of note included Muhammad ibn Ahmad al-Bīrūni, Haly Rodoam, Haly Abenragel (sometimes called Albohazen), Alcabitius, and Abū Ali al-Khayyat. Abenragel was tutor to the crown prince of Tunisia in Kairwan, and wrote *The Distinguished Book on Horoscopes from the Con-*

stellations. Alcabitius, also known as al-Qabisi from his birthplace, Qabisa, a village near Mosul in Iraq, was a mid-10th-century court astronomer and astrologer in Aleppo, Syria, who wrote an *Introduction to the Art of Astrology,* mostly devoted to natal charts. Al-Khayyat, for his part, had been a student of Masha'allah and among many other works wrote a little book, *The Judgments of Nativities,* with valuable sections on everything from how to determine from a birth chart whether the native will prosper to how one's brothers will fare. This is how he begins his guide to "The Native's Wealth and its Sources, and the Things Signified by the Second House":

> Look at the second house from the Ascendant, because if the fortunes [Jupiter and Venus] are in it or in aspect to it, and the evil [planets] are not, and do not aspect it, and its lord is in a good place in the circle and from the Sun, it signifies good fortune. But if these same dispose themselves in contrary fashion, it portends loss to the native in portions of his wealth. Again, if the lord of the second applies to the lord of the Ascendant, it signifies the acquisition of wealth without much work; but if, conversely, the lord of the Ascendant applies to the lord of the second house, hard work is shown.

More formidable still was al-Bīrūni, to whom few could compare. Born near Khiva in Khwārīzm in 973, by the Aral Sea, he was already an accomplished astronomer and mathematician by his late teens, when he successfully computed the latitude of his hometown by observing the maximum altitude of the Sun. The great scientist Abū Nasr Mansūr, who developed trigonometric functions, took him under his wing, and before long considered him his peer. When he was twenty-two, al-Bīrūni wrote a treatise on the making of maps in which he explained how to accurately project a hemisphere onto a plane, but his scholarly work was soon interrupted by the pervasive unrest that had begun to plague the Islamic world. In 994, he sought refuge in a town in Persia known as Rayy, not far from modern-day Tehran, where he met and assisted the astronomer al-Khujandi, who had established a hilltop observatory, furnished with a giant sextant, from which he calculated the obliquity of the ecliptic and solstitial transits of the Sun. Over the next decade, al-Bīrūni served as court astronomer and astrologer to a number of caliphs and sultans, determined the longitude of various cities by comparing data on eclipses drawn from several locales, wrote three books on astrology, two on history, and one each on astronomy, the decimal system, and the astrolabe. In 1022, the caliph he was then serving, Mahmūd of Ghazni, invaded northern India and invited al-Bīrūni to come along. The campaign lasted for four years, as the Arab troops advanced to the Indian Ocean. During that period, al-Bīrūni took advantage of the circumstance to determine the latitudes of a dozen towns around the Punjab and the borders of Kashmir. In his exploration of the region, he also mastered Sanskrit; took copious notes on India's culture, customs, science, geography, and history; and later incorporated it all into an immense book called *India,* a work of inestimable value for a knowledge of the subcontinent at that time. In other notable works, he wrote on "shadows" in all their aspects, including "the history of the tangent and secant functions, applicable to the astrolabe"; the idea that acceleration is connected with nonuniform motion; the use of rectangular coordinates to define a point in three-dimensional space; irrational numbers; ratio theory; trisection of the angle; algebraic equations; time keeping; hydrostatics; the velocity of light; Siamese twins; even flower petals and the workings of natural springs and artesian wells. He also carried on a polemical correspondence with the great Arab scientist Ibn Sīnā (known as Avicenna to the West) about heat and light, philosophy, astronomy, and physics; and many centuries before the telescope would confirm it, described the Milky Way as a nebulous collection of innumerable stars. Avicenna today is better known in the West, because his encyclopedic compilations in science, medicine, and philosophy became primary sourcebooks of the Latin Middle Ages. But al-Bīrūni had the greater mind.

Not surprisingly, his magisterial textbook *The Book of Instruction in the Elements of the Art of Astrology*—is meticulously put together and has a strongly scientific, or mathematical, cast. It includes not only detailed explanations of the fundamentals, but algebraic and geometric demonstrations of their character. Written in 1029, it is a veritable primer of 11th-century science, with sections on geometry, astronomy, geography, arithmetic, and chronology; on the use of the astrolabe for astronomical as well as astrological purposes; on how to determine the width of a river or the depth of a well; and so on. It was al-Bīrūni's conviction that no one could call himself an astrologer without a thorough knowledge of such disciplines. His succinct discussion of the received tradition is also admirably clear. "There are certain signs," he writes,

which are described as places of exaltations of the planets, like the thrones of kings and other high positions. In such signs, the exaltation is regarded as specially related to a certain degree, but there are many differences of opinion in this matter, some saying that it extends to some degree in front of or behind the degree in question, while others hold that it extends from the first point of the sign to that degree, and again others that it is present in the whole sign without any special degree. Below are the signs and degrees according to the Persians and the Greeks: Saturn, 21 degrees of Libra; Jupiter, 15 degrees of Cancer; Mars, 28 degrees of Capricorn; the Sun, 19 degrees of Aries; Venus, 27 degrees of Pisces; Mercury, 15 degrees of Virgo; the Moon, 3 degrees of Taurus; the Dragon's Head, 3 degrees of Gemini; the Dragon's Tail, 3 degrees of Sagittarius. The opposite signs and degrees are regarded as places of dejection for the planets, where they are said to be in their fall.

Here is his comment on how one aspect of the houses may be assessed:

Prosperity is associated with the cardines, as these indicate a happy mean; adversity with the cadent houses, which point to destructiveness and excess. Being in those houses which are succedent to the angles is beyond the half-way line to prosperity, for they are the paths leading there from adversity. But this prosperity and adversity are not all alike, just as the cardines are not alike but are higher and lower in glory and dignity. And indeed the cadent houses are not alike in their destructive influences, because although the 3rd and 9th houses are cadent, the 6th and 12th are not only cadent but also inconjunct to the horoscope.

★　★　★

EVERYTHING IS BOUND to Heaven's wheel. Baghdad had been founded at a time deemed propitious by the ruling caliph, al-Mansūr. His court astrologer, a Persian by the name of Naubakt, cast for the appropriate day and hour, and then consulted with his assistant, Masha'allah, who selected the early afternoon (2:40 P.M. local time, to be exact) of July 31, 762, to lay the foundation stone. Jupiter at the time was rising and exactly conjunct the ascendant, which happened to be in Sagittarius, its own sign. The Sun, in trine to Jupiter, was then in its own sign, Leo, in the ninth house, in a sextile to Mars in the seventh, even as it moved away from a square to Saturn in the sixth. This augured well for Baghdad's early glory as the seat of learning throughout the Arab world. But seldom a chart for good or ill without its obverse side. It so happened, in this instance, that Jupiter, on the ascendant, was directly opposed to Mars in the House of Open Enemies, indicating that it was Baghdad's fate to flourish (Jupiter trine Sun) in the field of higher learning, even as it was doomed to be undone by war.

By then, the Abbasid dynasty had gone the way of palace revolutions, the death of the great al-Ma'mūn seeming to presage its course. The story is told that during his last campaign against the Byzantines he came to the River Qushairah in Asia Minor, and camped on its banks. Charmed by the clarity and purity of the stream, and by the beauty of the countryside around, he decided to pause a while to recoup his strength. So clear was the water that the inscription on a coin lying at the bottom could be read. But it was also extremely cold. One day, as a fish flashed before him "like an ingot of silver," he was inadvertently splashed by an attendant trying to catch it and caught a chill. His aides wrapped him in blankets, but he soon became delirious and fell into a deep sleep. When he awoke, he asked the name of the place where he was. He was told, "Qushairah," meaning in Persian "Stretch out thy feet." He then asked for its name in Arabic and was told "Rakkah." He sat up with a start. An astrologer in his youth had predicted that he would die in a place of that name. All along, he had thought it referred to the city of Rakkah in Syria, where he had always declined to go. Now he knew his fate. As night fell, he asked to be carried outside. He surveyed the long lines of his encampment and the light of the torches and campfires that twinkled into the distance as far as the eye could see. "O thou whose reign will never end," he cried out sadly, "have mercy on him whose reign ends now."

Of the eight succeeding caliphs, two were assassinated and two others died in exile in disgrace. Eventually, the center of Arab culture shifted from Baghdad to Cairo, under a different dynasty, and to Cordoba and Toledo in Spain.

Baghdad's horoscope played itself out. In 1258, the city was sacked by the Mongols under Hulagu Khan, the grandson of Genghis Khan, and its great library burned to the ground. Scholars, civic leaders, and a large proportion of its inhabitants were slaughtered and a mountain made of their skulls. The caliph al-Musta'sim—the last of the Abbasids—deluded himself into surrendering upon the promise that his life would be spared, but once in Mongol custody was "beaten to death in a sack." In 1401, the city was sacked again and ravaged by the Mongol Tamerlane. These devastating onslaughts were naturally followed by social disintegration and political disarray. Thereafter, in the 15th, 16th, and 17th centuries Baghdad was repeatedly riven by factional violence, and fought over by the Persians and Ottoman Turks. The Ottomans seized it in 1638, but by then it was not much of a prize. It had lost its commercial importance, while its once-extensive irrigation system had fallen into disrepair. Tribal-based pastoral nomads now drifted in to occupy the ruins. But that was not the end of it, of course. The British occupied Iraq during World War I, were driven out by the Turks, only to return and capture Baghdad in 1917. Under their aegis, modern Iraq was created as an independent state. It, too, has had its once and future trials. But that is another tale.

Chapter 5

Now, when Jesus was born in Bethlehem of Judaea in the days of Herod the king, behold, there came wise men from the east to Jerusalem, Saying, Where is he that is born King of the Jews? for we have seen his star in the east, and are come to worship him. . . .

Then Herod, when he had privily called the wise men, inquired of them diligently what time the star appeared.

And he sent them to Bethlehem, and said, Go and search diligently for the young child; and when ye have found him, bring me word again, that I may come and worship him also.

When they had heard the king, they departed; and, lo, the star, which they saw in the east, went before them, till it came and stood over where the young child was.

When they saw the star, they rejoiced with exceeding great joy.

And when they were come into the house, they saw the young child with Mary his mother, and fell down, and worshipped him . . .

And being warned of God in a dream that they should not return to Herod, they departed into their own country another way.

This narrative passage, from the second chapter of the Gospel According to St. Matthew, is the only reference in Scripture to the Star of Bethlehem, which became the universal symbol of Christ's birth. That symbol gained swift acceptance among the early Christian faithful and has enjoyed unimpeachable authority for two thousand years. There is also a little-known but coordinate legend about the star that was still current in Palestine when the 20th century began. It held that when the Magi, coming from the East, at length made their way to Bethlehem guided by the star, they lost sight of it in the daylight and wandered about the village in confusion and despair. Finally, one of them went to the well of an inn to draw water and, upon looking down, "saw the Star reflected in the water at midday." That told them they were at the place where the Christ child lay.

That well was still being shown to pilgrims as late as 1910.

Who were the Magi? And what was that star? An early bishop of Antioch, writing to the Ephesians at the start of the 2nd century A.D. (about thirty years after the Gospel According to St. Matthew was composed) tells us: "Its light was unspeakable and its novelty caused wonder." And in the 3rd century, the Church father Origen wrote that it was "a new star unlike any of the other well-known planetary bodies . . . but partaking of the nature of those celestial bodies, like comets, which appear from time to time." In the Apocryphal Gospel of James, we read: "And [Herod] questioned the Magi and said to them: 'What sign did you see concerning the new born King?' And the Magi said: 'We saw how an indescribably great star shone among these stars and dimmed them, so they no longer shone, and so we knew that a King was born for Israel' " (21:1).

In the arc of biblical history, the star appeared to fulfill the Old Testament prophecy of the soothsayer Balaam (in Numbers 24:17) that "a star shall come forth out of Jacob and a scepter shall rise out of Israel." In the two thousand years since it shone there has been endless debate as to what it was—a real or natural event? A symbolic one? Or a miracle beyond the power of science to explain? If natural, was it a comet, a nova, a meteor, planetary conjunction, or some other phenomenon? If symbolic, in what sense? If a miracle, how defined? According to the 4th-century Eastern hierarch St. John Chrysostom (ca. 347–407), "the star was not a star at all . . . but some invisible power transformed into the appearance of a star," as shown by its eccentric course. "Its divine nature," he thought, "was so powerful that its mere appearance was sufficient to bring the barbarian Magi to the feet of Christ." Or as St. Augustine put it:

The star which the Magi saw when Christ was born according to the flesh was not a lord governing his nativity but a servant bearing witness to it; it did not subject him to its power but in its service pointed the way to him. What is more, that star was not one of those which from the beginning of creation keep their regular courses under the Creator's law, but at the new birth from the Virgin a new star appeared, which performed its office by going before the faces of the Magi in their search for Christ until it led them to the place where lay the infant Word of God . . . So Christ was not born because it shone forth, but it shone forth because Christ was born; so if we must speak of it, we should say not that the star was fate for Christ, but that Christ was fate for the star.

In short, the star was both a supernatural event, and a natural event supernaturally decreed. Either way, it in no sense ruled the life Christ lived.

The first known visual representation of the star appears in a 6th-century manuscript known as the Codex Egberti; the second in Giotto's 14th-century *Adoration of the Magi,* which depicts it as a comet rather than a star. Most astrologers (and astronomers) today think it was the Jupiter-Saturn conjunction that took place in Pisces in 7 B.C., as first suggested by Johannes Kepler in 1603. The two planets, it is thought, fused into one exceptionally bright light, which became the "Star of Bethlehem" of biblical lore. That conjunction also coincided with the meeting of the two zodiacal cycles—the tropical and the sidereal—which happens only once every twenty-six thousand years. It therefore signified the dawning of a new age, and a new grand cycle of ages. As Kepler wrote: "He [God] appointed the birth of His Son Christ our Savior exactly at the time of the great conjunction in the signs of the Fishes and the Ram, near the equinoctial point"—that is, just as the vernal equinox, by precession, moved backward into the constellation before it in the zodiac. The best biblical scholarship today also places the birth of Jesus at about that time.

Other worthy theories about the star are variations on the theme. One identifies it with an occultation of Jupiter by the Moon that occurred on April 17, 6 B.C., at sunrise and (together with other factors) might have been thought to signify "the birth of a Hebrew king." Another relates the Jupiter-Saturn conjunction favored by Kepler to a series of momentous celestial events. The Jewish historian Josephus, writing in the 1st century A.D., tells us that "Herod died after a lunar eclipse visible in Jericho shortly before Passover," and we know that such an eclipse occurred on March 13, 4 B.C., a year in which Passover fell on April 11. Since the slaughter of the innocents that occurred in his reign followed hard upon Christ's birth and shortly before Herod's death, it appears likely that Christ was born the year before, in the spring of 5 B.C. In the view of astronomer Mark Kidger, who has made this subject his special field, the birth of Christ was announced by a series of heavenly omens, of which only the first was the Jupiter-Saturn conjunction that Kepler singled out. That was followed, Kidger writes, by "a massing of nearly all the planets in Pisces" (in February of 6 B.C.), in turn succeeded by "two portentous planetary pairings, also in Pisces, later that month—namely, Jupiter's occultation by the Moon, and a conjunction of Saturn with Mars." After that came a blazing nova (reported in Chinese and Korean astronomical records) that lit up the sky in February or early March, proclaiming the Christ child's birth. The Chinese chronicle the *Ch'ien-han-shu,* for example, records that a new star was sighted near Theta Aquilae in March of 5 B.C. and remained visible for seventy days. Seventy days or ten weeks was time enough, according to this theory, for the Magi to follow the star in their journey across the deserts from the East. Moreover, "during their journey, the Star, which was in the east when first seen, would have gradually swept across the sky to the south. When the Magi set out for Bethlehem, they would have seen the Star before them in the south at dawn." This may explain the peculiar phrasing in the New Testament text. They came, we are told, from the East, yet saw his star "in the east"—that is, to the west of where their journey began. What does this mean? The King James translators erred. The biblical Greek phrase *en tai anatolai* means not "in the east" but, more precisely, at "sunrise." The correct translation (now widely adopted) is, "We saw his star at its rising"—referring to the heliacal rising of a planet or star.

"All stars," explains Kidger, "except the ones that are close to one of the poles of the sky, are invisible at certain times of the year. This is because the Sun passes either in front or nearly in front of them (if they are on the ecliptic), blocking out the star completely while it passes behind the Sun. Alternatively—if a star is farther north or south of the ecliptic—at certain times of year it will rise or set at the same time as the Sun. This means the star is only above the horizon during the hours of daylight and cannot be seen at all at night. More technically, from the time the star sets at sunset, to the time, weeks or months later, when it rises at sunrise, it is at *conjunction* with the Sun." Invisible then, it eventually reappears in the morning sky, at dawn. The sighting of Sirius, not incidentally, was used in ancient times to predict the flooding of the Nile. So when the Magi said that they had seen Christ's star "in the east" they meant "in the first light of dawn." That is also why Herod was oddly unaware of what they were talking about. A comet would not have gone unnoticed. But a more occult astrological event, known to astrologers but not necessarily readily apparent to laymen? Perhaps.

The Magi who came to Bethlehem to adore the infant Christ were neither "wise men" nor "kings," as later story had it, but astrologers. The biblical Greek term *Magos* referred to a specific Persian caste of astrological seers. The story of the three kings was introduced later, in the 6th century; and their description as "Wise Men" was also a later, somewhat embarrassed attempt to acknowledge their sagacity while glossing over who they were. By the 9th century they had begun to appear under the names of Gaspar, Melchior, and Balthasar, representing three different continents and races—Europe, Africa, and Asia—thus appearing to be the first gentiles to certify that Jesus was the Christ. (Modern biblical scholarship has caught up with the truth, and most translations today identify the Magi as astrologers, either in the text itself or an accompanying note.)

If the birth of Christ was proclaimed by a celestial and astrological event, so, too was his death. The Bible tells us that his crucifixion was marked by a solar eclipse—"about noon," according to Luke (23:44) when "darkness came over the whole land until three in the afternoon, while the sun's light failed." This eclipse, as seen from Jerusalem, must have been total or nearly so, as corroborated by the Apocryphal Gospel of Peter, which tells us: "And many went about with lamps, supposing it was night, and fell down."

★   ★   ★

THE EARLY CHURCH had declined to take an official position on astrology, although certain theologians—Tertullian, for example—were suspicious of its "pagan" origin. Tertullian argued that astrology had been valid until the birth of Christ, when it was superseded by Christian revelation. That, in his allegorical view, was the "hidden" meaning of the dream "sent to the Magi telling them to return home by a different route"—that is, to change their ways. But others (including Origen, as well as the Neoplatonic philosopher Plotinus) held that insofar as the stars were set (according to Genesis) in the heavens for signs and for seasons, their artfully choreographed ballet and configurations were the alphabet of God. Astrology was therefore a divine science, and those who mastered its celestial letters could only grow in piety and faith.

Some of the later Church fathers, including St. Augustine, formally condemned astrology, although others accepted it as compatible with faith. St. Thomas Aquinas, for example, acknowledged the influence of the planets upon human affairs, and attempted to reconcile astrology with the Christian doctrine of free will. Roger Bacon, a Franciscan and the greatest scientist the Middle Ages produced, was expert in judicial or mundane astrology, which seeks to correlate planetary cycles and patterns with world events, and devoted to the study of "elections." Another medieval adept was Giovanni di Fidanza, better known as Bonaventura, one of the greatest of the medieval saints. Indeed, for all its misgiving, the Church could never quite escape the enormous fact that astrological events had figured so prominently in the life of Christ. How was one to condemn the study of the stars when the Holy Spirit itself had used a star to announce the savior's advent? Or to ignore the fact that it had required the skill of three astrologers to locate the manger where he lay? In truth, not much in Scripture was against it. After all, the Sun, Moon, and Stars were created in the beginning as "signs," Christ's birth signified by a star, his death marked by an eclipse, and the Second Coming, according to Luke 21:25, to be announced by "signs in the sun, and in the moon, and in the stars." Almost all the passages which seem to look askance at astral divination (Isaiah 47:13; Jeremiah 10:2–3; 2 Kings 23:54; Zephenia 1, 4–5) really condemn the idolatry of nature, that is, worship of the planets or stars as gods. But they do not condemn them as signs, or as an expression of God's will. And so, according to God's will, in the Book of Judges (6:19–20) the prophet Deborah tells us that "the stars in their courses fought against Sisera," who had oppressed the Israelites, and so working through her brought him low.

Nevertheless, the Church eventually saw the authority of astrology as competing with its own, and condemned it: the Council of Laodicea forbade priests to practice it, and the Council of Toledo threatened with a curse anyone who believed in astrology or divination. This position was subsequently confirmed by the councils of Braga, Agda, Orleans, Auxerre, Narbonne, and Rheims. Meanwhile, the Constitution of the Apostles in the 4th century had refused the rite of baptism to all who "pretended to divine"; Salic law (the penal code of the Franks) condemned astrologers as "casters of spells"; and the first Christian emperor, Constantine the Great (despite his own use of astrology on occasion), had threatened them with death. His successor, Constantine II, swore to have them ripped apart with iron claws.

But Church authority could not root the subject out. It sustained an underground existence in the West; flourished in the East throughout the Byzantine Empire; and in India, the Arab world, China, and Persia, assumed a collateral importance that was no less marked. Marco Polo tells us in his *Voyages* that there were five thousand astrologers at the court of Kublai Khan. Hindu practitioners thrived from the Indus River to Ceylon, and the caliphs of Islam had maintained its study in their great institutions of learning, such as the Academy of Wisdom founded (with the help of Jewish adept Jacob ben Tarik) in Baghdad by al-Ma'mūn. Writing about 1136, Geoffrey of Monmouth tells us that the 7th-century Northumbrian king Edwin consulted an astrologer named Pellitus from Spain, and that the legendary King Arthur maintained a school of adepts at Caerleon, men "learned in *astronomia* and the other arts, who diligently observed the movements of the stars and who at that time foretold to king Arthur by accurate calculations the coming of any remarkable thing." Monastics read, copied, and covertly studied the *Mathesis* of Firmicus Maternus, despite Church disapproval; Gerard, archbishop of York under Henry I, was refused Christian burial by his canons after a copy of Maternus was found under his pillow at his death.

During the Carolingian Renaissance an interest in astrology had also been revived, led by Alcuin, the Anglo-Saxon poet and scholar whom Charlemagne made head of his great Palatine School at Aachen, and who had formerly directed the Cathedral School at York. By the reign of Charlemagne's successor, Louis the Pious (reigned 814–40), it was said that "every Carolingian lord had his own astrological adviser," as did William the Conqueror, who relied on Gilbert Maminot, bishop of Lisieux.

By the 12th century, astrology had openly returned to the West enriched by a number of Arab concepts, including conjunction theory, dozens of new lots or parts—such as the Part of Friends or the Part of Fate—and by a deep study of the lunar calendar, with its examination of the fixed stars in relation to the degrees or "mansions" of the Moon. Latin translations of Arabic works (derived in part from the Greek) enthralled the learned community, and for five centuries thereafter, astrology pervaded European culture, just as it once pervaded that of imperial Rome. Before long, many cities (such as Florence) maintained a city astrologer in much the same way that a modern community maintains a health officer, while scarcely a figure of importance—pope, general, or king—could be found without his court astrologer to advise him: Henry II and Charles IX of France; Catherine de Médicis (herself proficient in the art, which she practiced at an observatory near Paris); the Holy Roman Emperor Charles IV; Charles V of France; and so on. When Charles V of France died in 1380, almost a tenth of his vast library was found to consist of astrological works. At least twelve popes were also votaries of the art—among them Julius II, Paul III, Sixtus IV, Innocent VIII, Leo X, and Urban VIII. Julius II used astrology to set the day of his coronation; Paul III, to determine the proper hour for every Consistory; Sixtus IV fixed all important dates and receptions according to the planetary hours, and enlisted the help of the German astrologer Regiomontanus in his calendar reform. Innocent VIII consulted horary charts in shaping his foreign relations; Leo X founded a chair of astrology at the Sapienza, and not incidentally, helped to provoke the Reformation when he sought to finance St. Peter's Basilica through the sale of indulgences. Leo's own favorite astrologer was Franciscus Priulus, "who," we are told, "was so dismayed by his own ability to read and predict the events and secrets of people's lives that he became inconsolably depressed and committed suicide." Some of the Reformation popes who followed looked to astrologers to tell them whether Luther would prevail. In the Rome of Urban VIII, "astrology was even more popular among the common people than cards and dice. Along the banks of the Tiber and along the Corso, in the warrens of the Borgo and Trastavere, street vendors hawked sheets called *Avvisi,*" which not unlike the tabloids of today "purveyed a juicy mix of news, gossip, and prognostication," with an emphasis on what the stars foretold about events great and small.

With respect to Luther, Protestants and Catholics alike pored over his chart in an effort to discover whether he was a prophet or the Anti-Christ. They could scarcely know, since the exact year and time of his birth were in doubt. Luther himself wasn't quite sure of when he was born, and his mother couldn't remember. That left room for a good deal of speculative play. Protestants preferred a time in 1483 that coincided with a configuration of planets that indicated "a theologian of great commitment to the cause of religion and unshakeable firmness of purpose." Catholics guessed that he had been born at 1:10 A.M. on October 22, 1484, which allied him with a fearful conjunction of five planets in Scorpio in the ninth house, to accord with his identity as "a sacrilegious heretic, a bitter enemy of the Christian religion, and profane." Neither was right. But Luther himself might not have cared much either way. Although he accepted the prevailing belief in celestial portents ("whatever moves in the heavens in an unusual way is certainly a sign of God's wrath," he wrote), and expected the Second Coming to be announced by a profusion of comets in the form of the Cross, he looked askance at astrology as a whole. Philipp Melanchthon, however, his chief lieutenant in the Reformation, was a firm believer in the art and a friend of Joachim Camerarius, who published a Greek edition of Ptolemy's *Tetrabiblos* in 1535. Melanchthon read it and decided that a Latin translation was warranted, which he completed in 1553.

To fully understand this tremendous efflorescence of astrology in the West, one must go back to the 12th century and the reconquest of Moslem Spain.

It is said that St. James the Apostle, the brother of St. John, once preached in Galicia, where, according to his wishes, his body was later laid to rest. In time, much to the dismay of the faithful, the place of his burial was forgotten or lost until one night he appeared to the Emperor Charlemagne in a dream and told him where the body lay. He urged him to lead an army across the Pyrenees to reclaim it, along a path marked out by the stars. A star later guided a bishop to the spot. And so the place of his tomb became known as Compostela—from *campus stellae,* field of the star.

Yet more than that saint's tomb would lure Crusaders to the prize. Two centuries after its founding, Arabic Spain had become the wealthiest and most advanced state in Europe, with the greatest of its cities—Cordoba, Toledo, and Seville—comparable in grandeur to Baghdad at its height. At a time when Rome was in ruins and London, Paris,

Venice, and Antwerp were still untidy medieval towns, Cordova—"the cupola of Islam," "the minaret of piety"—shone like a golden bowl among vessels of clay. The greatest of her palaces, the Azahara, was almost a city in itself, and her great mosque, the Mezquita, built from demolished churches, was sheathed in glittering gold. Within, colorful mosaics gleamed and glistened in the radiance of ten thousand lamps. Many of her splendid homes were furnished with marble balconies to catch and hold the summer breezes and hot-air ducts beneath mosaic floors for winter warmth. Over three hundred public baths served a population of half a million, who at the hours of prayer flocked to seven hundred mosques. The streets were also paved and lit. In the library of her great academy of learning, founded in 948, some 600,000 manuscripts were stored.

A number of Western scholars were aware of this cultural richness, and came to taste it, such as Gerbert d'Aurillac (later Pope Sylvester II), who studied at Cordova and Seville before directing the cathedral school at Reims. William of Malmesbury tells us that when in Spain among the Saracens, Gerbert had studied "the astrolabe of Ptolemy and Firmicus on Fate." Aside from being a convinced student of astrology, he possessed a Renaissance breadth of learning; constructed globes, astrolabes, a famed abacus, and "observation tubes" (precursors of the telescope); and invented the pendulum clock. He also introduced Arabic numerals to the West.

★   ★   ★

SINCE ITS OCCUPATION by the Saracens, Spain had been an independent emirate, formed soon after the Umayyads of Damascus were overthrown. It had consistently been blessed by administrators of exceptional ability and wisdom, including the famed grand vizier Almanzor (Abū 'Amir al-Mansūr), who had enlarged and completed the Mezquita before he died in 1002. Almanzor had also managed to hold Moslem Spain together by his own unrivaled skill. But thereafter it broke apart into little fiefdoms, which allowed the Christians to begin to recover ground. From the mountains of Asturias and Galicia, crusading armies steadily pushed their way south, taking Toledo, Cordova, Valencia, Seville, and other towns. The Moors (as the Moslems of Spain were called) regrouped and recovered some of their losses, but by 1248, only the kingdom of Granada remained in Saracen hands.

And that would eventually fall before the combined armies of Aragon and Castile.

Though the Christians came as lords, they soon proved captive to what they found. This proved true in every field, but none so much as in learning, which revolutionized the West. In a sense, Western culture now regained its own lost footing with an Arab staff, for the learning of the Arabs had kept Western civilization almost from extinguishing itself. As the Christian armies advanced, they discovered that the huge Arab libraries in the enclaves they vanquished contained many "lost" works from classical times, as well as advanced works in philosophy and mathematics that the Arabs themselves had composed. From Spain came the philosophy and natural science of Aristotle and his Arabic commentators, in the form which was to transform European thought; the ancient medical classics of Hippocrates and Galen; Euclid and the new algebra; the work of Arab physicians like Avicenna; treatises on perspective and optics; and the new planetary tables of King Alfonso X, the Wise. New and remarkable information was to be had in astrology, astronomy, pharmacology, psychology, physiology, zoology, biology, botany, mineralogy, chemistry, physics, mathematics, geometry, trigonometry, music, meteorology, geography, mechanics, hydrostatics, navigation, and history—just to give one list. Arabic terms also now entered the language of Western science and commerce: *zero, algebra, cipher, algorism, almanac, nadir, alchemy, alcohol, elixir, tariff, arsenal, admiral,* and *alembic.* The reclaimed riches of classical and Arab culture alike were stupendous, as a tremendous translation effort ensued. Almost from the moment the reconquest began, European students and scholars had begun crossing the Pyrenées to Toledo, Cordova, and other centers, where they collaborated in rendering the treasured texts.

As in Alexandria, in Hellenistic times, Jews, Christians, and Moslems worked closely together and engaged in a universal exchange of ideas. Gerard of Cremona translated the works of Hippocrates and Galen, Ptolemy's *Almagest,* Aristotle's *Posterior Analytics,* the *Spherics* of Theodosius, and some sixty other works; Herman of Carinthia took up Aristotle's *Ethics, Poetics,* and *Rhetoric;* Adelard of Bath (who also wrote a treatise on the astrolabe) translated Euclid's *Elements* and the astronomical tables of Muhammad ibn Mūsā al-Khwārizmī. Theodore of Antioch translated various works on zoology; Plato of Tivoli, Rudolf of Bruges, Hugh of Santalla, Robert of Chester and others translated Aristotle's *Physics* and *Metaphysics,* Proclus's *On Motion,* the *Pneumatics* of Hero of Alexandria, Euclid's *Catoptrics,* and a hundred other seminal texts. Bartholomew of Messina, Burgundio of Pisa, Egidius de Trebaldis of Parma, Arnold of Barcelona, Blasius Armegandus of Montpellier, Herman of Palmatia, and Robert of Retines likewise lent their talents to the task. Moreover, by the middle of the 12th century, most of the important works on astrology had also been made available to the Latin West. Abū Ma'shar, Alcabitius, and Masha'allah were translated by John of Seville; Haly Abenragel by Judah ben Moses, who rendered the text from Arabic into Old Castilian first, from which it was later translated into Latin; the *Centiloquium*—a collection of one hundred aphorisms ascribed to Ptolemy—by John de Luna Hispalensis, who also wrote a textbook on the art of elections, which Chaucer consulted when writing *The Canterbury Tales.* Abraham Ibn Ezra, a Sephardic Jew, explained the use of the decimal system with respect to the calendar and also wrote his own brief introduction to astrology called *The Beginning of Wisdom* in 1148. All this formed part of the great European renaissance of the 12th century, which not only anticipated the more famous efflorescence of the 15th, but saw the culmination of Romanesque art, the beginnings of the Gothic, the emergence of a vernacular literature, the recovery of Greek science, and the revival of Latin classics, poetry, and Roman law. Because of this flood of knowledge, the first universities began to appear, and college and university degrees were developed to grade the level of knowledge obtained. A number of scholars versed in Arabic learning came to England from Spain, hence the phrase "Oxford Don" that came to be bestowed upon a professor who "held a chair"—that is, on one who had earned the right to occupy an academic throne of authority in some field.

Since Aristotle was the lord of the new learning, astrology now basked in his scholastic prestige. His scientific thought had first been transmitted to the West, in fact, through a translation of a work by Abū Ma'shar; and there was little in Aristotelian science with which astrology was not in basic accord. Aristotle himself had said, in *On Generation and Corruption,* that "the earth is bound up in some necessary way with the local motions of the heavens, so that all power that resides in this world is governed by that above." Within the sciences themselves, the three judged most important—astrology, astronomy, and mathematics—were scarcely told apart. Peter of Blois, who had studied theology under John of Salisbury (secretary to the martyred archbishop Thomas à Becket, afterward a saint) tells us: "Mathematicians are those who, from the positions of the stars, the aspect of the firmament, and the motion of the planets, discover things that are to come." And astronomy, in the view of Adelard of Bath, "describes the whole form of the world, the courses of the planets, the number and size of their orbits, the position of the signs . . . By this science, a man acquires knowledge, not only of the present condition of the world, but of the past and future. For the beings of the superior world, endowed with divine souls, are the principle and cause of the inferior world here below." In effect, astronomy, astrology, and mathematics were one.

★   ★   ★

IN THE EAST, the Byzantines were touched by the same flame. Although Eastern emperors, in the early days at least, had condemned astrology, others fell back upon it, or allowed its cultivation at court. The first Christian Emperor, Constantine the Great, had been prepared to ascribe astrological import to natural disasters, and reportedly asked an astrologer to elect the most propitious time for the founding of Constantinople to ensure that the city would endure. The historian Ammianus Marcellinus tells us that the emperor Valens, upon being warned by the astrologer Heliodorus of a plot against him, appointed him to his own inner cabinet, while the Emperor Zeno enlisted astrologers to "analyze the prospects of political rivals" in holding on to power. There was also a considerable exchange of ideas, including astrological ones, between the Arab and Byzantine worlds. At the end of the 8th century, for example, the Byzantine scholar Stephanus the Philosopher brought a treasury of astrological manuscripts to Constantinople from Baghdad, where he had studied with Theophilus of Edessa, the Greek-speaking military astrologer of the Caliph al-Mahdi. Some (like Rhetorius the Egyptian) were also Arab-trained. Byzantine anthologies of astrological material were eventually compiled, and by the early 11th century there was a proliferation of Greek translations of works by, among others, Masha'allah, al-Kindi, and Abū Maʿshar.

Astrology gained still more ground and stature during the Byzantine renaissance of the 9th century (which paralleled the Carolingian renaissance in the West) along with Greek astronomy, which was recovered in part from Arabic translations of Greek texts. The Byzantine Emperor Manuel Comnenus (reigned 1143–80) relied on astrology and defended it on the basis of natural science, Scripture, and Church teaching. Under his rule, many Greek translations of Arabic and Persian treatises were made, and the time of his own coronation at St. Sophia on March 31, 1143, was astrologically elected (the chart for the event survives), which may or may not account for his splendid reign. His constant companion, John Camaterus, archbishop of Bulgaria and later patriarch of Constantinople, was also an adept. He wrote two astrological poems dedicated to his prince, and cast to know the fate of everything from the fall harvest to military campaigns. The Emperor Andronicus II had an expert astrologer in Theodorus Metochites, who served as "prime minister" for most of his reign. Metochites wrote with authority on philology, history, philosophy, rhetoric, poetry, and the sciences; a work on astrology prized by Chaucer; yet somehow managed to balance his private studies with his conduct of affairs of state. His contemporary, Nicephorus Gregoras, said of him: "From morning to evening he was wholly and most eagerly devoted to public affairs, as if scholarship meant nothing to him; but in the evening, after having left the palace, he became absorbed so completely in his studies that you would think his whole life were scholarship." Another emperor, Andronicus IV (who deposed his father, John V Paleologus), was advised by the astrologer John Abramius on his siege of Constantinople and entered the city at the "elected" time of 8:30 A.M. on August 12, 1376.

By the 13th century, astrology east and west was absorbing the greatest and most influential figures of the age—Michael Scot (ca. 1175–1234), one of the translators of Aristotle from Arabic into Latin and counselor to the Holy Roman Emperor Frederick II; Guido Bonatti, adviser to the Ghibelline condottiere Guido da Montefeltro, "who never fought a battle without consulting the stars"; Lutbert Hautschild, abbot of St. Bartholomew of Eeckhout and protégé of the Duc de Berry, whose celebrated *Très Riches Heures* were decorated with astrological symbols; Jacques Coeur, also an alchemist; the Yorkshire-born John Sacrobosco (also known as John of Holywood, by an English translation of his name), whose *De Sphaera Mundi* was one of the source books used by Chaucer for his *Treatise on the Astrolabe;* and Arnaud de Villeneuve, an authority on the Hermetic sciences who became rector of the University of Montpellier before running afoul of the Church.

Scarcely an authority of any magnitude demurred. Robert Grosseteste, the 12th-century bishop of Lincoln, made a special study of comets; Roger Bacon, whose encyclopedic knowledge embraced theology, mathematics, geography, astronomy, perspective, physics, alchemy, and the experimental method, thought the coming of the Anti-Christ might be divined by astrological means; Albertus Magnus (Albert the Great), a Dominican friar and the foremost scholar of his age, upheld "the influence of the planets over global affairs," as did his pupil, St. Thomas Aquinas, who reconciled astrology to Church faith.

Aquinas, for his part, insisted on the role of grace. Opposing a rigidly deterministic interpretation of astrology as inconsistent with the doctrine of free will, he argued in the *Summa Theologica* that the stars exert a direct and essential influence on men's bodies but an indirect and accidental influence on their souls. He was convinced that astrology "worked" but reasoned that it did so insofar as a person was bound up with his corporeal nature—with his physical body, and its appetites and needs. Then, like any other "body," he was ruled by the stars. But when his soul was in communion with God, his will was freed from this bondage, and he had the capacity to act above and beyond the compulsion because his spirit was in touch with a higher power. For that reason, mundane astrology, which dealt with societies, was often right, because people in groups, like a "mob," cannot possess an independent volition, but by nature move as a herd. In short, the stars might incline, but could not compel, because the will was free to resist. The wise man who could master himself could also therefore master the stars. This solution to the problem of fate and free will became the classic one for later writers and crystallized the orthodox attitude toward astrology in the medieval Church. Yet it remained a matter of dispute whether free will could alter or affect events, or only the posture of the soul. As St. John of the Cross later put it: "I am made and unmade not by the things that happen to me, but by my reactions to them. And that is all God cares about."

In the late Middle Ages, astrology was incorporated into the curricula of a number of universities in Europe and was considered an indispensable part of the training of any physician in the diagnosis and treatment of disease. Chairs of astrology were established at the universities of Paris, Padua, Milan, Bologna, Florence, Vienna, and Oxford, for example, and occupied by some of the great scientists of the age, such as Pietro d'Abano, Regiomontanus, and Giorgio Peuerbach. Regiomontanus, a professor of astronomy and astrology at Vienna, whose work was valued by Columbus, wrote the most advanced books at the time on trigonometry, compiled some of the earliest ephemerides, and translated Ptolemy's *Almagest.*

Yet a certain ambivalence remained. This was provoked by errant dabblings in magic, by attempts to apply astrology to the life of Christ (by casting his horoscope), and by other presumptive use of the art. Cecco d'Ascoli, one celebrated Italian professor of astronomy and astrology, was burned as a heretic in Florence in 1327 after he claimed that from the chart he cast "one could have predicted Jesus' great wisdom, his birth in a stable, and his death on the Cross." He atoned for the sins of many who before and after him did such castings, including, not incidentally, Cardinal Pierre d'Ailly, who somehow escaped reproof.

There were also some skeptics. Rabbi Moses ben Maimon, or Maimonides, born in Cordova, Spain, but raised in Cairo, Egypt, thought the verse in Leviticus 19:26, "Ye shall not observe times," seemed to forbid astrology, and he condemned it as "a tree under the shadow of which all sorts of superstitions thrive, and which must be uprooted in order to give way to the tree of knowledge and the tree of life." When some rabbis in southern France sought his advice on whether the art could provide any guidance to prayer, he replied: "The real science of the stars is knowledge of the form of the spheres, their size, their motion, the time of their revolution, their northern or southern inclination, their rotation to the east or to the west, the orbit of each star and its way"—in short, astronomy. "From this science eclipses of the luminaries can be known; when and where they will occur; from which degree the [new] moon appears as a crescent that grows to fullness and then gradually decreases; when the moon is visible or not visible; why there is a long day and a short day; why two stars will co-rise but not co-set; why the days differ in length at various locations whereas at one location the day equals night." Yet he also observed, "For as much as God hath created these stars and spheres to govern the world, and hath set them on high and hath imparted honor unto them, and they are ministers that minister before him, it is meet that men should laud and glorify and give them honor."

The notion that astrology was inherently fatalistic also weighed against it, despite the scholastic solution to the problem of free will. But elections, of course, were not fatalistic at all. Albert the Great, indeed, went so far as to say that it was "rash and working against the freedom of the will not to elect astrologically propitious times for one's actions." This doctrine was also related to astral magic—an attempt to channel, harness, or direct the power of the stars into objects by creating them at times when the sky was best aligned for their use. This was commonplace in the compounding of medicines, which were thought to have more efficacy if made at the proper time. Certain plants were also believed to have the "virtues" or powers of the planets that ruled them or to which they seemed to correspond. Modern notions to the contrary, the idea that heavenly power could be drawn down into images was not, strictly speaking, idolatrous or superstitious but had to do with a particular concept of time. That time was astrological, as in elections. For at the right moment, it was supposed, the power or virtue of the planets could be harnessed by talismans fashioned of apt materials to receive their power—as solar panes today are fashioned to absorb the energy of the Sun. By capturing these astral emanations, astrologers, in theory, could divert the power of the heavens to their own ends. Thus "love charms" made of copper (the metal of Venus) were created in the hour Venus ruled, when the planet was in its dignity or exaltation, and engraved with some appropriate image or symbol that expressed its special force. But it was not always so simply done.

One of the great exponents of astrological magic was the astrologer Thebit ben Corat (or Thābit ibn Qurrah), who served the caliph of Baghdad and died in 901. Here is an example of his advanced technique. To promote a political career, he writes,

When you want to make an image of a man who wishes to become the head of a city or province, or judge of a prefecture or a town . . . carve the head of the image when the Dragon's Head is in the Ascendant, and let the lord of the ascendant be a benefic, free from aspect of the malefics. Carve the body of the image under whatever rising sign the

Moon shall be in, carve the shoulders and breasts with Venus in the ascendant, the haunches with the Sun rising in one of its dignities, the thighs with Mercury in the ascendant, but not retrograde or combust, or afflicted, and in one of his fortunate [dignified] places; and the feet under the ascendant of the Moon in conjunction with Venus . . . And see that the ascendant be fortunate, and its lord, and the tenth house, and that the malefics be remote from the ascendant and its lord, and let the lord of the eleventh be one of the benefics, in aspect of the ascendant and its lord; and let the lord of the tenth be in conjunction with the lord of the ascendant in a friendly conjunction or with complete mutual reception. When you have done this and made the image in this manner, he will obtain what he desires from his king and be given the post he seeks. Preserve the image as I have told you, and it will do the work, if God wills.

Albert the Great, not incidentally, approved of "engraving astrological images on gems and minerals to work marvels." Roger Bacon likewise subscribed to the belief that images and verbal charms, if made under the proper astrological conditions, were endowed with "heavenly" power. Indeed, he was convinced that many of the miracles of saints had been performed by such means, including magic words spoken at the proper time.

Yet the influence of St. Augustine was not easily overcome. His is a curious case for so mighty a thinker because of awkward contradictions in what he wrote. He claimed to have been drawn to astrology in his youth, only to reject it with age; but in his later critique of it, he seemed unacquainted with most of its traditions and borrowed some of his objections from Cicero's works. He considered astrology untenable, for example, based on the divergent fate of twins, and pointed to Jacob and Esau as a case in point. How, he asked, could the heavens account for the dramatic differences in the destinies of two such children—one a desert wanderer, the other the father of a mighty tribe—born at nearly the same time? An astrologer would say that "nearly" is the answer. For, to allow the art to defend itself on its own terms, no adept would have supposed their charts identical in the first place. In the course of just four minutes, a new degree (and therefore possibly a new decan, subrulership, fixed star, and so on) can rise above the horizon, and give a wholly different cast to a child's fate. Radio experts, as one writer points out, "operate in fractions of seconds," makers of glass lenses "in millionths of an inch." Again, "if someone were to forget one figure in the combination of a safe, could they unlock it?" So it is with twins, where the slightest difference in the time of birth can make for "a very appreciable difference in character and the timing of events." Yet this was something Augustine somehow failed to grasp.

In Book 5 of *The City of God,* he tells us:

> As to what they [the astrologers] attempt to make out from that very small interval of time elapsing between the births of twins, on account of that point in the heavens where the mark of the natal hour is placed, and which they call the "horoscope," it is either disproportionately small to the diversity which is found in the dispositions, actions, habits, and fortunes of twins, or it is disproportionately great when compared with the estate of twins, whether low or high, which is the same for both of them, the cause whose greatest difference they place, in every case, in the hour on which one is born.
>
> It is to no purpose, therefore, that that famous fiction about the potter's wheel is brought forward, which tells of the answer which Nigidius [Figulus, the Roman astrologer] is said to have given when he was perplexed with this question . . . For, having whirled round the potter's wheel with all his strength he marked it with ink, striking it twice with the utmost rapidity, so that the strokes seemed to fall on the very same part of it. Then, when the rotation had ceased, the marks which he made were found on the rim of the wheel at no small distance apart. Thus, said he, considering the great rapidity with which the celestial sphere revolves, even though twins were born with as short an interval between their births as there was between the strokes which I gave this wheel, that brief interval of time is equivalent to a very great distance in the celestial sphere. Hence, said he, come whatever dissimilitudes may be remarked in the habits and fortunes of twins.

In the physics, or spherical geometry of his example, Figulus was absolutely right. And in the end, not unlike Cicero, Augustine kept a foot in both camps. He acknowledged that astrologers sometimes demonstrated an amazing power to predict, which he could only ascribe to the help of demons, though he later admitted in *The City of*

God that the stars hold some obvious sway over the physical life (if not the soul) of man. If the stars affect terrestrial change, he added, "it does not follow that the wills of men are subject to the configurations of what the stars might ordain." (That had given Aquinas his own starting point for reasoning out the matter.) Even so, he was dolefully convinced that it led men to resign themselves to fate instead of engaging their own free will to strive toward grace. He was therefore repelled by any abject dependence on the art, and tells the story of a man who was so anxious that his son be born at an auspicious moment, when the planets, as he supposed, were well aligned, that when his wife went into labor beforehand, he stood at her bedside exhorting her in a panic to somehow delay the birth.

★   ★   ★

AS THE HEIRS of Greek and Arab thought, most medieval thinkers, like Pierre d'Ailly, were also convinced that historical change was regulated by planetary cycles, "beginning with the shortest term, the monthly lunar cycle, which became especially important if it culminated in an eclipse, and moving up to the long-term cycle of the Jupiter-Saturn conjunctions." In d'Ailly's view, such great conjunctions had heralded or coincided with the Great Flood, the fall of Troy, the death of Moses, the foundation of Rome, and the advent of Christ. The "element" in which they occurred was some guide to the grief they would inflict. "So if you want to know the kind of misfortune and torment [associated with each]," wrote Masha'allah, "look at the shift of the conjunctions from one triplicity to another unto the lot of the transit, in whatever sign it takes place, and if it takes place in a fiery sign, the misfortunes will be from fire. If in a watery sign, from water; if in an airy sign, from wind; and if in an earthly sign, from a landslide, earthquake or falling rocks."

D'Ailly's own apocalyptic views had been prompted in part by the Great Church Schism, which endured for forty years, led to the establishment of rival popes, and split Western Christendom in two. Europe as a whole was also variously afflicted by plague, war, famine, rebellion, and economic chaos, all of which, "contributed to an atmosphere of apocalyptic frenzy." That seemed to d'Ailly, according to a common interpretation of 2 Thessalonians, a foreboding of the world's last days, and so he believed the next great conjunction would

mark the beginning of the end of the world. According to d'Ailly, however, the full effects of that fateful conjunction would not come until
1789, when the Anti-Christ would appear and the old order, which he
took to mean the end of the world, would be completely overturned.
He was not entirely wrong, of course, since in that year the French
Revolution turned the political world of Europe on its head. "If the
world shall last until that time, which only God knows," D'Ailly concluded, "then there will be many great and amazing alterations and
changes in the world, chiefly with respect to laws and sects [reli-
gions]." Those who see the French Revolution as the beginning of the
triumph of secularism might be tempted to think him right.

Following Abū Ma'shar, d'Ailly also believed that Jupiter was the
planet that signified religion, worship, and faith and that since there
were six other planets to which it could be joined by conjunction, "it
followed that the world would see six major religions," each with the
characteristics of its planetary mate. According to this scheme, Judaism
was signified by the conjunction of Jupiter and Saturn; idol worship
by Jupiter and Mars; the worship of the stars by Jupiter and the Sun;
Islam by Jupiter and Venus; Christianity by Jupiter and Mercury; and
the sect of the Anti-Christ by Jupiter and the Moon.

Other conjunctions had their own import. In 1186, for example,
Europe was thrown into a panic because of a supposed conjunction of
Mars and Saturn in Libra (an air sign), which was thought to portend a
windstorm of cataclysmic force. Special services were held in churches
to prepare for the worst, and people built caves and other shelters for
themselves in the sides of mountains and even tunneled underground.
It took the superior astronomy of an Arab astrologer to help calm fears.
In correcting the prediction, he pointed out to the bishop of Toledo
that on the projected day of the conjunction, September 16, Mars
would in fact not be in Libra, but Virgo (an earth sign), and since
Venus would also be in Scorpio, which is the house of Mars, it would
soothe or countermand the latter's malevolent force.

Along the same lines, the Black Plague that swept through Europe
from 1345 to 1348 and killed nearly a third of its population was
also thought to have an astrological cause. In 1348, King Philip VI of
France asked the medical faculty of the University of Paris to account
for the calamity, and the formal reply they submitted that October
ascribed it to the conjunction of Saturn, Mars, and Jupiter in Aquarius,

as exacerbated by a lunar eclipse. "For when the sun is directly opposite the moon," wrote the astrologer Geoffrey de Meaux, "then the power of each of them reaches the earth in a straight line, and the mingling of influences of sun and moon with that of the superior planets creates a single celestial force." To make things worse, Mars was in Leo, a fire sign, together with the Dragon's Head, in square to Jupiter—which corrupted the whole atmosphere and made it ripe for pestilence.

Again, in advance of 1524, the impending conjunction of all seven planets in Pisces was thought to "portend certain changes and transformations for the whole world . . . such as we have hardly heard of for centuries before our time." Two preceding lunar eclipses of 1523 contributed to the general fears. The fact that it was a water sign seemed to portend torrential rains, and perhaps a flood on the scale of the biblical deluge. In anticipation of the inundation, many people sold off their lands and other possessions for cash, pitched tents on hilltops and mountains, and built little arks to carry them over the waves. A certain Prior Bolton of St. Bartholomew's, Smithfield—a kind of early "survivalist"—built himself a fortified house on Harrow Hill and stocked it with provisions to last him a year. The flood did not materialize, but there was a great Peasants' Revolt in Germany that seemed almost as fearsome to some and shook central Europe to the core.

★   ★   ★

IF THE MIDDLE AGES ACCEPTED the conjunction theory of the Arabs, it also embraced the ancient lore of comets and eclipses that belonged to the astrology of the Chaldeans and the Greeks. In ancient Babylon, comets had been associated with instability in government and the death of kings, and were portents of disaster in the world of Greece and Rome. Homer in the *Iliad* speaks of a comet unloading disaster from its "burning" locks, and we have seen its calamitous role in the politics of imperial Rome. In China, as well, war, executions, the deaths of generals and royalty, and any great natural shock, such as an earthquake—"all rode in on the comet's tail."

Meaning was detected in their shapes—variously described as "bearded," "hairy" (the meaning of the Greek *kometes*), or like a spear or sword—a dread precursor and weapon in the sky. According to Josephus, the fall of Jerusalem in A.D. 70 had been preceded by a comet, which appeared above the city "like a spear." Pliny in his *Natural History* associated comets with civil unrest, and in the 1st century A.D. Seneca remarked that some comets are "blood-stained and threatening, bringing prognostication of bloodshed to follow in their train." Not unlike conjunctions, comets in Earth signs were associated with drought; in water, with torrential rains and floods; in air, with mighty winds and pestilence; in fire, with war. Marcus Manilius linked them to war, treachery, insurrection, and natural disaster, and thought a comet accounted for the devastating plague that struck Athens at the beginning of the Peloponnesian War. Comets were also said to have heralded the battles of Philippi (42 B.C.) and Actium (31 B.C.), and the annihilation of three Roman legions under Varus in Germany in A.D. 9. When a comet appeared in A.D. 11, it was popularly believed to signal the impending death of Augustus, who published his horoscope to demonstrate that his demise was not yet due. Yet his end was near, and a blood-red comet was seen overhead four years later when he died.

Comets were also said to have coincided in A.D. 337 with the death of Constantine the Great; in 453 with the death of Attila the Hun; in 455 with the death of the Emperor Valentinian III; in 729 with the inroads of the Saracens into Gaul and the deaths of the Anglo-Saxon ruler Osric and St. Egbert; in 1066 with the Norman Conquest; in 1199 with the death of Richard I of England; in 1477 with the death of Burgundy's Charles the Bold; in 1515 with the victory of Francis I of France over the Duke of Milan's Swiss troops at the battle of Marignano, as seen in the Low Countries on September 15, the day the battle raged; and in 1560 with the death of Francis II. They would also be associated in 1618 with the beginning of the Thirty Years' War. In signifying the demise of the high and mighty, comets were nothing if not ecumenical and applied indifferently to Arabs, Christians, and Jews. The death of Muhammad was said to have been so signified; and a number of popes perished under their light. The early Church fathers (Origen, Synesius of Cyrene, and St. Jerome, among them) had accepted comets as fatal omens, as did Isidore of Seville, the Venerable Bede, John of Damascus, Peter Abelard, and Robert Grosseteste. St. Jerome, indeed, thought comets would be among the fifteen signs to precede the Day of Judgment when the world came to an end. The custom of ringing church bells at noon originated in a papal edict meant to propitiate the comet of 1456, which coincided with the invasion of Constantinople by the Turks: "Lord, save us," intoned Pope Callistus III, "from the devil, the comet, and the Turk." A comet rising before the Sun was said "to accelerate the manifestation of the events it portends, whereas a vespertine, or evening, comet delays it." The interpretive colors were related to the planets, according to the Babylonian scheme: white, Venus; orange, Jupiter; dusky or grey, Saturn; red, Mars; variable, Mercury; yellow, Sun; blue, Moon.

Solar and lunar eclipses were also closely watched. Of importance were the date and time of occurrence (position of the eclipsed body in the sky), duration, magnitude, direction of the shadow, color, shape, its sign, decanate, and celestial house, parts of the disc eclipsed, and its aspect to planets and stars. Babylonian omen texts tied eclipses to the deaths of monarchs—for example: "On the 16th day an eclipse takes place. The king dies" (1700 B.C.). Or, slightly later: "On the 20th day an eclipse happens. The king on his throne is slain." Pliny noted that a solar eclipse had occurred after the murder of Julius Caesar. The Anglo-Saxon Chronicle tied such an eclipse in 664 to the death of a king. Ptolemy ascribed great importance to eclipses and predicted the regions they would affect from the area of the zodiac in which they occurred. So Casca in Shakespeare's *Julius Caesar* tells us: "For I believe, they are portentous things / Unto the climate that they point upon." Just as the tail of a comet pointed to the area of the world affected, so, according to tradition, an eclipse impacted those countries from which it was "visible or total, or nearly so." When falling in a fixed sign as well as a royal one, its effects were said to be lasting upon those lands and dynasties concerned. Also, if a comet was first seen in Leo, for example, it boded ill for monarchs; if in Virgo, for the harvest. The duration of an eclipse indicated how long its effects would last: for a lunar eclipse the hours were converted into months; in the case of a solar eclipse, into years.

Far to the east, astrologers thronged the Mongol and Chinese courts. Anyone planning to undertake a trip, or business venture, would consult an astrologer to ascertain its prospects, and scan the skies for some indication of heaven's intent. Comets, conjunctions, occultations, "guest stars" or nova, "odd appearances of the Sun and Moon," solar and lunar halos, and so on, E. C. Krupp writes, were "unwelcome postings on the celestial bulletin board" and "greeted with interest and concern."

In China, as in the days of imperial Rome, the palace astrologer occupied a precarious niche. Knowledge of the heavens was privileged, and any ruler had a vested interest in controlling what an astrologer might say. In the twilight of the Sui dynasty (A.D. 581–618), for example, the astrologer Geng Xun told the emperor Yang Di that a military campaign he had just undertaken would fail. The emperor, enraged, condemned him to death. But before the sentence could be carried out, news from the battlefront proved him right. Geng was freed, rewarded for his skill, and promoted; but a few years later the emperor was assassinated, and a struggle for the throne ensued. Geng consulted the stars to ascertain its outcome and promptly concluded that he was on the wrong side. But before he could announce his change of allegiance, he was slain on the palace grounds. Two hundred years later, in 840, in an effort to reserve astrological knowledge to the state, the emperor issued an edict prohibiting imperial adepts from talking to other officials or even members of the population about their work. In 1583, when Matteo Ricci, the famed Jesuit astronomer, traveled to Beijing, he was told that it was a capital crime to study mathematics as used in astrological computations without the emperor's leave.

In natal astrology, eclipses have traditionally been regarded as malefic, especially when they fall on an angle or a planet in the chart. Some astrologers believe that "the last eclipse prior to birth is of importance to the native and that its path may mark areas on the Earth that will be of significance in the life." For example, the eclipse of February 29, 357 B.C., prior to the birth of Alexander the Great, "was on the Midheaven at his birthplace, Pella in Macedonia. Its path of totality swept through the very lands which he was later to conquer: Egypt, Mesopotamia, and Persia." Similarly, according to the *Larousse Encyclopedia of Astrology*, "the path of the eclipse that occurred on the day of Karl Marx's birth (May 5, 1818) swept directly across the Russian Empire." In such ways do natal and mundane astrology sometimes intersect. According to a kindred scheme, those born when a comet appeared were often thereby tied years hence to epochal events. "The year that Mithradates was born" notes Justin, a writer of the 3rd century A.D., in his *Philippic History*, "as well as the year that he ascended the throne [120 B.C.], there appeared for seventy days a comet of such great brilliance that the sky seemed on fire. It was so big that it occupied a quarter of the heavens, and so bright that it eclipsed the light of the sun. Four hours elapsed from its rising to its setting." A comet also flashed across the sky at the birth of Alexander the Great.

Some, of course, imagined that the Star of Bethlehem had been a comet, and indeed the periodicity of comets would later figure in attempts to determine the date of Christ's birth. Meanwhile, the great cathedrals of medieval Europe had been transformed into solar observatories in a more general attempt to bring the Church calendar into accord with the celestial map. Astronomers and astrologers alike (there being no clear difference between them) collaborated on the task. Tradition had fixed the date of Christmas for December 25, because it coincided with the pagan midwinter festival of Sol Invictus, "the undefeated Sun." In an early bid to attract its "sun-worshipping" adherents to the faith, the Church had coopted that day as its own. In A.D. 525, Dionysius Exiguus, a Scythian monk and Church scholar, had decided to call the first year of Christ's life A.D. 1—though we know now that Christ was born five to seven years before. Meanwhile, the date for Easter (with its roots in the Jewish lunar calendar) had been set two hundred years earlier by the Council of Nicaea, in A.D. 325, which proposed that it fall on the first Sunday after the first full moon after the vernal equinox. Yet determining a *universal* date for Easter bedeviled the Church.

Since the full Moon and equinox might both occur at different times at different places on Earth, the Church, for the sake of unity of worship, tried to determine an "averaged" date that it could reliably calculate and announce in advance. "Everything depended," writes J. L. Heilbron, "on exact average values of the periods between successive vernal equinoxes and between successive full moons." At the same time, adherence to the outdated Julian year had skewed the normal calculations since it undermeasured the length of a solar year by about eleven minutes—an error that threw the calendar out of sync with the seasons one complete day every 125 years. So by the 12th century, the supposed date of the vernal equinox—and therefore Easter—was no longer in harmony with the celestial clock. To correct the discrepancy, and rightly calculate the time of the Sun's return, cathedrals all over Europe—in Rome, Milan, Florence, Bologna, Chartres, Antwerp, and elsewhere—were subtly converted into solar observatories, by tracing a "meridian line" from south to north (in effect, from solstice to solstice) on the cathedral floor, which sunlight, let in through a hole in the roof, would trace in turn throughout the year. The time it took for the Sun to make its way to the same spot—usually marked by a brass marker or a metal rod—measured the length of the year.

One of those who had an ingenious hand in this, and helped to refine the method at a later date, was Paolo Toscanelli, the astrologer who had supplied Columbus with his map. Toscanelli was not only a great cosmographer, but, in his own day, an astronomer almost without peer. His painstaking and exact observations and calculations of the orbits of various comets (including Halley's) that appeared between 1433 and 1472 survive in manuscript, as does his astrological commentary on their meaning. In the left transept of the Cathedral of Santa Maria del Fiore in Florence may still be seen the famed gnomon he constructed in 1468, consisting of a marble slab in the dome with a hole in it, which enabled him, by the shadow it cast from the Sun, to determine the altitude of each solstice and, to within half a second, the time of each midday.

★ ★ ★

MANY ASTROLOGICAL ADEPTS were colorful characters. Galeotti Martius, astrologer to Louis XI, for example, was a tall, stately man with a long, sweeping beard who dressed in clothes of the richest velvet, was renowned as a wrestler, and also captain of a legion of horse. A formidable if ostentatious scholar, he entertained his powerful clients in a lavishly decorated study, which included a silver astrolabe (a gift of the German Emperor) and a Jacob's staff of ebony jointed with gold (received from the reigning pope).

No career in the early revival, however, was more distinguished than that of Michael Scot. Born in Balwearie near Kirkcaldy in Fife, Scotland, about 1175, he studied first at the Cathedral School of Durham, then at the universities of Oxford and Paris before crossing the Pyrenées to Toledo, where he learned enough Arabic to help translate Aristotle into Latin for the glory of the West. From Toledo he went to Sicily, became a priest, an aide in succession to two popes, Honorius III and Gregory IX, then court astrologer to the Holy Roman Emperor Frederick II. Both popes offered him preferment, including the exalted posts of archbishop of Cashel in Ireland, and archbishop of Canterbury in England, but it was at Frederick's court that he thrived. The emperor, who carried on a regular correspondence of his own with Arab scholars, treasured Scot's overall knowledge of Arabic learning and was said to have especially valued a book, now lost, that Scot wrote on algebra. Other works ascribed to him are doubtful. One was a so-called *Magic Book,* meticulously written in yellow, green, red, and black ink on vellum, in mysterious characters, with a Latin gloss. The characters are incredibly beautiful, of no known language though of an Arabic cast. The Latin gloss also suggests that it was a guide to summoning or dismissing spirits and demons from within a circle with a wand, mitre, habit, sigil, and the incantation of the proper spells.

Scot's posthumous reputation as a fearful wizard probably began when he was placed by Dante in his imagined Hell. In later years, it spread through Europe, and eventually entered the myths and legends of the Scottish Border region, where he became a figure of romance. In Sir Walter Scott's *Lay of the Last Minstrel,* he is portrayed as a mighty warlock in command of nature herself. With a mere gesture of his arm, he can cleave "the Eildon hills in three" and "bridle the [river] Tweed with a curb of stone." The *Lay* also makes him an earl, and gives him a coal-black stallion whose pounding hoofs can shake cathedral towers, and who feasts his friends on dishes of exceptional relish brought by spirits from the royal kitchens of France and Spain.

No examples survive of Scot's astrological work, but he is said to have foretold the place of the Emperor Frederick II's death, which Frederick supposed to be Florence but turned out to be Florentiola in Apulia, in 1250. Scot also seems to have known he would one day be killed by a stone falling on his head. To protect himself, he wore an iron skullcap beneath the hood of his robe; but just once, as he was about to enter a church for the elevation of the host, he removed it, and the fatal stone fell from the roof.

If Scot remains elusive, his real life almost eclipsed by the legend to which it gave rise, the opposite is true of Guido Bonatti, the outstanding European astrologer of the Middle Ages. His written legacy is substantial, though comparatively little is known of his life. Born in Forli around 1200, he evidently studied astrology at Bologna, was a landowner, traveled widely, and won the confidence of a number of monarchs and princes, who relied on his advice. For most of his career, however, he was caught up in Italian politics and in the chronic Guelph-Ghibelline strife. In 1246 he warned Frederick II of a plot against him, which was thereby foiled; joined the entourage of Guido Novello, the Ghibelline leader in Tuscany, whom he advised at the battles of Montaperti and Lucca in 1260 and 1261, and as the official astrologer of Florence, concluded a treaty with Siena that averted an impending war. Four years later, he was back at Forli, which under Guido da Montefeltro became the Ghibelline stronghold in the Romagna from 1275 until 1283. In the Annales of Forli it is recorded that in 1282 the town was besieged by an army commanded by Pope Martin IV, an ally of the Guelphs. Bonatti suggested that his master could entrap the papal army by withdrawing his own forces and allowing the pope to occupy the town. This was done, and while the papal forces were celebrating their illusory triumph, Montefeltro, who knew every weak point in the town's defenses, attacked and retook the town. This confrontation apparently followed an enlightened but futile attempt made by Bonatti a few years before to reconcile the two camps. Having wearied of the factional strife, which had long wracked Italy to no purpose and beleaguered his home town, he had proposed that the two sides join together to build a new town wall. This was to be done at a time elected by him as propitious, and a tentative agreement to do this was forged. At the time set, a representative from each party was to cement a new foundation stone in place. As the solemn moment approached, workmen stood poised with their implements and the two chosen partisans stood holding their stones. Bonatti gave the signal— and the Ghibelline at once put his stone in place. But the Guelph balked, suspecting a ruse. Moments passed; the planets shifted, then were realigned. "God damn you," Bonatti shouted. "You ignorant fool. Do you know what you've done? This was more than just the chance of a lifetime, for a moment like this won't come again for another five hundred years!" Shortly thereafter, the annals tell us, the Guelph faction in the town was crushed.

"All things," wrote Bonatti, "are known to the astrologer. All that has taken place in the past, all that will happen in the future—everything is revealed to him, since he knows the effects of the heavenly motions which have been, those which are, and those which will be, and since he knows at what time they will act, and what effects they ought to produce." In a stunning demonstration of his own expertise, Bonatti devoted his last years to the completion of his masterpiece, the *Liber Astronomiae,* a comprehensive treatise on astrology based on Arab sources, which has justly been described by Robert Hand as "probably the most important single astrological work written in the Western tradition between Ptolemy and the Renaissance." Within its pages, the whole substance of the art as it was understood was gathered up.

He is ever thorough. Here are a few of his 145 famed "Considerations," as seen through the hourglass of time:

The 111th is, To consider in Nativities and Questions especially of Law suits and controversies, whether the Dragon's Tail be in the Seventh? For that signifies damage or overthrow to the Native's enemies and prosperity to the Native or Querent, because the Dragon's Head will then be in the Ascendant. If it be in the eighth, it denotes the decay and loss of their estate or substance, and increase of the native's. In the third, prejudice to the Native's Brethren. In the fourth, to his Parents. In the fifth, to his children. In the sixth, to his servants. In the ninth, to his journeys. In the tenth, to his preferment. In the eleventh, to his Friends. In the twelfth, to his cattle of greater sort, etc. And so to all other things signified by each house respectively: so do Saturn and Mars also, but not so much. Likewise, 'tis observable that other ill positions may make void the said significations, but no so much as Saturn and Mars, unless they themselves are significators of the mischief, and then much of their malice is abated.

The 120th Consideration is, to observe whether the Lords of any of these Eight Houses, viz., the third, fourth, fifth, sixth, ninth, tenth, eleventh, or twelfth, be in the Seventh. For which soever of them is there, the person by him signified will prove the Native's enemy, unless a perfect reception, with some good aspect as Trine or Sextile intervene. Yet a Square or Opposition with Reception will abate the enmity, but not wholly prevent it. Thus if it be the Lord of the Third, his Brethren will prove him enemies; if of the Fourth, his Parents; if of the fifth, his children, etc., nor shall he gain of or by them so much as he shall lose another time; or if any of them sometimes appear kindly, it will be but from the teeth outward, and for their own ends.

By such careful and intricate considerations, predictions about life and fate could sometimes be made. In general, they required a thorough knowledge of mathematics and astronomy as well as astrology's own compendious body of received doctrine; to this one had to add years of study and a certain impalpable gift for interpretation that, beyond the principles and rules, enabled the astrologer to untangle the strands of fate.

Though Bonatti could be dauntingly detailed, he could also be pithy and succinct. Here is his guidance for discovering, again by horary means, whether a threatened battle between two armies will take place:

Look at the first house and its ruler and the Moon and the seventh house and its ruler, and see whether they are joined together by body in any one of the angle, since that signifies that there will be a battle between them. But if they are not joined by body, see whether they are joined by opposition or square aspect, since that similarly signifies that there will be a battle. And if neither of these configurations exists, then see whether any planet transfers the light between them by opposition or square aspect, since that signifies that there will be a battle if there is no reception between them. But if the heavier one of them receives the one which transfers their light between them, it signifies that there will not be a battle, or, if there is, that it will not last long . . . If there is no planet to transfer the light between, there will be no battle at all.

Here, pertaining to a birth chart, is his judgment as to whether someone is destined for the crown: "Note in the Nativities of Kings and rich men, and such grandees as are fit to bear rule, whether both Luminaries are in the Degrees of their Exaltations, or in their own Houses, in the same degree one with the other, and free from affliction? For this signifies that the native shall obtain great honors, for he shall be made Emperor or something like it; so that he shall be as it were monarch of the world."

Every presidential aspirant might want to learn this by heart.

✶   ✶   ✶

AMONG CONTINENTAL MONARCHS, none was more conspicuous for his cultivation of occult learning than Charles V of France. Charles assembled a library of more than one thousand books, a collection huge for its time, with a rather large assortment on astrology and divination. Among them were the works of Ptolemy, Ibn Ezra, Abū Maʿshar, Masha'allah, Guido Bonatti, and Haly Abenragel. He also commissioned translations of Arab commentaries on many of these texts, though it is not known to what extent he may have been guided by their advice. Christine de Pizan, the daughter of one of his Italian astrologers, wrote in *The Book of the Body Politics* that Charles was captivated by the subject, and we know that "horary questions and elections were common at his court." We also know that the principal general in his employ—Bertrand du Guesclin—sought tactical advantage in battle from his study of the art and pored over the horoscopes of at least two of the English commanders he faced in the Hundred Years' War. Charles himself endowed a college of astrology and medicine at the University of Paris in 1371, and evidently "regarded himself, and wished himself to be regarded by others, as a learned man, skilled in the natural and occult sciences among other things."

✶   ✶   ✶

THOUGH THE ENGLISH were not as quick to take up astrology as their Italian and French counterparts (the earliest surviving English birth chart is that of Edward II, born on April 25, 1284), astrology had planted itself firmly at court, in the Church, and at the great universities of the realm by the reign of Henry VI (1422–61). In between, the art had enjoyed a steady rise. Though earlier kings had shown only a limited interest in the occult, some nobles, particularly those with contacts in Italy and France, had already sought out the advice of astrologers on various matters, while "an ever-increasing number of astrologer-physicians," Hilary Carey tells us, "most educated in the medical schools of the great Italian universities"—at a time when astrology was an intrinsic part of medical diagnosis and treatment— "enjoyed the patronage of English kings."

Pedro Alphonso, physician to England's King Henry I, wrote: "It has been proved by experimental argument that we can truly affirm that the Sun and Moon and other planets exert their influences in earthly affairs . . . And indeed many other innumerable things happen on earth in accordance with the course of the stars, and pass unnoticed by the senses of most men, but are discovered and understood by the subtle acumen of learned men who are skilled in this art."

England's Henry II (reigned 1154–89), the son of Geoffrey Planta-genet, Count of Anjou, was a student of the art, and at his invitation the Arab-Jewish astrologer and physician Abraham Ibn Ezra came from Toledo to lecture in England in 1158. The king also took note of the astrological advice of Adelard of Bath, who had translated Euclid, al-Khwārizmī's astronomical tables, and Abū Ma'shar. Edward III may not have embraced astrology completely, but he owned a person-alized copy of the *Secretum Secretorum* or *Book of Secrets,* an apocryphal compendium of magical, political, and medical lore supposed to have been written by Aristotle for Alexander the Great. It contained pre-cepts on the science of government, and featured all sorts of occult lore ranging from astrology to the magical properties of numbers, gems, and plants. It also contained an onomantic table (a form of div-ination that used letters in names) for predicting the outcome of bat-tles, though whether Edward consulted it is not known. Nevertheless, after the Battle of Crécy in 1346, in which English long-bowmen demonstrated their supremacy over armored French knights, the king's chaplain, Thomas Bradwardine, evidently felt obliged in his vic-tory sermon preached at Neville's Cross, near Durham, to warn the king not to ascribe his triumph directly to the stars. Taking 2 Corinthi-ans 2:14 as his text ("Give thanks to God who always leads us in tri-umph") he declared: "What astrologer could predict this? Or foresee such a thing? There is one prediction which will never be proved false: whatever God wishes to happen or to be done, that is done; whomso-ever God wishes to be victorious, he is victorious; and whomsoever God wishes to reign, he will reign. Although therefore the heavens and the earth, and all things under the heavens should be against you, if God is for you, what can harm you? And although the heavens and the earth and all things under the heavens should be for you, if God is against you, what can help you?" He would hardly have put it that way if astrology in court circles had not been gaining ground.

As indeed it was. A calendar commissioned by John of Gaunt, one of Edward's sons, included astrological tables for medical use, and one of the horoscopes to survive from the period was clearly cast in re-sponse to a question asked by a courtier as to whether the future

Richard II would gain the throne. At the time, the expected heir, Edward the Black Prince, had just died, and the reigning King Edward III was in decline. The chart indicated, in fact, that Richard would be crowned. Once crowned, Richard moved to incorporate the occult sciences into the rich and eclectic life of his fashionable court. In this, his taste was allied with that of Charles V of France, King Wenceslas of Bohemia (the "Good King Wenceslas" of Christmas song), and other monarchs. Richard's beautifully decorated book of divination, much of it consisting of extracts from the *Secretum Secretorum,* included a set of geomantic tables, based on the treatise of an Arab astrologer, in which 3,200 possible answers were given to twenty-five representative questions. Some of these were typical of what the client of an astrologer might ask: "Should I marry?" "Will I become rich?" "Is my wife (or mistress) pregnant?" "Will the pregnancy go well?" "Who will win the battle?" "Will I be promoted at court?"

Astrologers making such predictions sometimes lived on the razor's edge. In 1441, Richard Bolingbroke and Thomas Southwell, both "reputable physicians and senior Oxford scholars," were tried for treason after they were found to have cast a horoscope that predicted the death of Henry VI. As later dramatized in Part 2 of Shakespeare's *King Henry VI,* this episode also involved Eleanor Cobham, wife of the Lord Protector of the realm, who together with a woman known as "the Witch of Eye" (Eye being a village in Suffolk) "had consulted Bolingbroke and Southwell for astrological help in conceiving a child." In the aftermath of the trial, Bolingbroke was hanged, drawn, and quartered at Smithfield; Southwell died in prison; Eleanor Cobham was banished; "the Witch of Eye" burned. The king, meanwhile, understandably spooked, had "commissioned two [other] astrologers to produce an alternative interpretation of the horoscope" but the first would prove more nearly correct, at least with respect to the manner of his end.

Yet many astrologers at this time, such as John Holbroke and John Argentine, also enjoyed considerable repute. Holbroke, a doctor of theology, was a master of Peterhouse College, Cambridge, chancellor of the university, and a mathematician and astronomer of renown. His chief work was an ephemeris which, by more accurately recording the true and mean motions of the Sun, Moon, and planets, with a table of the ascensions of the signs, revised and corrected the Alfonsine tables then in use. John Argentine, no less distinguished, was provost of

Kings College, Cambridge, as well as royal physician and astrologer to both Edward IV and Edward V. Henry V was also a student of the art and commissioned splendid astrolabes of gold and silver for his use. Henry VII cherished Bonatti's work and had at least two astrologers, Baptista Boerio and William Parron, at his court. Henry VIII consulted an astrologer by the name of John Robyns and insisted that Thomas Cranmer keep him informed about any comets that appeared. Sir Thomas More's son-in-law, William Roper, relates that Henry also had More from time to time "up into the leads there to consider with him the diversities, courses, motions, and operations of the planets and stars." Meanwhile, Cardinal Wolsey, Henry's first chancellor, hoped that by studying the king's horoscope he could successfully anticipate his actions and "pander to his whims." Wolsey also tried to elect auspicious times for some of his diplomatic missions, but not all of them worked out. Sir William Paget, state secretary to Edward VI, could scarcely get enough of what Bonatti taught, and his successor under Elizabeth, Sir Thomas Smith, "could scarcely sleep at night" for his study of the art. Smith's pupil Gabriel Harvey, "friend and counselor to Philip Sidney and Edmund Spenser," was similarly possessed.

If Mary Tudor's tenure on the throne was "tainted by popish superstition," that of Elizabeth did not shy from the occult. When Elizabeth came to power in November 1558, few would have predicted a long and prosperous reign. Fear and uncertainty were rampant after a tumultuous succession of ill-fated rulers and the bitterness and hatred of religious strife had led to conditions akin to civil war. Even Ivan the Terrible in Moscow, enmeshed in intrigue and presiding over a sundered realm, could regard the English crown as no more stable than his own. The queen faced plots both at home and abroad, and feared that Spain and France might overcome their own mutual enmity and unite against her rule. Astrologers belonging to contending factions were also at work. Those who favored Mary, Queen of Scots urged revolt on the grounds that Elizabeth would not survive her first year— and, indeed, a lunar eclipse on April 2, 1558, together with a conjunction of Saturn and Mars, seemed to augur the worst.

But the cryptic prophecies that caused the greatest alarm among the English were those of Nostradamus, whose opaque, encoded forecasts seemed to apply to Elizabeth's reign. According to the course of public events, they were thought to have predicted the marriage of Mary,

Queen of Scots to the dauphin of France; the recent death of Queen Mary of England; the death of Henri II of France in a jousting match; the threatened war with France over Scotland; and the accession of Elizabeth herself. Elizabeth's new archbishop of Canterbury, Matthew Parker, scoffed at his pronouncements as a "fantastical hodge-podge," yet "the people did so waver, and the whole realm was troubled," and seized with a terrible doubt.

Elizabeth herself was prepared to give astrology its due. She accepted the notion that, as she remarked in a letter to Mary Stuart, "our dispositions are caused in part by supernatural signs, which change every day," but at the same time discouraged her subjects from exploring the matter on their own. From 1581 on it was a felony to cast her horoscope, or even to possess it, and a capital crime to predict her death. Even so, it was the mischief of prediction that she feared, not the stars themselves. She had a stalwart heart. In May 1582, when all England was alarmed by a comet, her courtiers "went about to dissuade her majesty (lying then at Richmond) from looking on [it]," but she "with a courage answerable to the greatness of her state, caused the window to be set open, and declared, '*Iacta est alia,* The dice are thrown.' " When her anxious attendants asked her, "Should the comet not be feared?" she replied, with an exemplary piety, that "her steadfast hope and confidence was too firmly planted in the providence of God, to be blasted or affrighted with those beams."

That comet was but one of several celestial events that seemed to augur ill. Most were thought to point toward 1588, when a great new Saturn-Jupiter conjunction would take place. "I am astrologically induced to conjecture," wrote Richard Harvey, "that we are most like to have a new world, by some sudden, violent, & wonderful strange alteration, which even heretofore hath always occurred." Few doubted this was so. In 1475, the year before his death, Regiomontanus had published an almanac in which he predicted that 1588 would be momentous for European politics. He thought that it would be ushered in by an eclipse of the Sun in February and marked by two total eclipses of the Moon, one in March and one in August, when Saturn, Jupiter, and Mars would also hang in ominous conjunction "in the Moon's own house." He was not far off. In that notable year the Spanish Armada was launched and defeated, the king of Denmark (Frederick II) died, the king of France (Henri III) was driven out of Paris, and there was a terrific struggle for the Polish throne.

★ ★ ★

IN WAYS THAT WERE SOMETIMES SUBTLE, sometimes obvious, astrology touched nearly every aspect of secular and religious life. Despite the official position of the Church, the persecution of astrologers was also fairly rare. The Church found no fault with medical astrology, and openly embraced "elections" since its practice made "a calculated effort to modify the future favorably" by a strategic act of free will. In the Church's own heavenly hierarchy, the nine orders of angels were also astrologically assigned. "Thrones were of the realm of Saturn, since Saturn is static," as one writer tells us. "Dominions were of the realm of Jupiter, since Jupiter is a planet of rulers; Powers were assigned to Mars, the planet of power; Principalities were assigned to the Sun, symbol of Kings; Virtues were assigned to Venus; Archangels were assigned to Mercury; and Angels to the Moon. Above the Thrones were the Cherubim and Seraphim, who controlled the Fixed Stars and the Primum Mobile, making the Nine." That stately system, based on the astrological—not astronomical—signification of each planet, may be viewed as the cabalistic counterpart of fortune-telling with playing cards, which is an astral method of divination. The four suits represent the four fixed signs, as well as the four seasons, and the thirteen cards in each suit correspond to the thirteen lunar months. The four suits of 13 cards add up to 52 (the weeks of the year). The numbers 1–13 are added to make 91, multiplied by 4 to give 364, or the approximate number of days in a year. The Joker was added to make 365.

The art of the Middle Ages and the Renaissance found inspiration in astrological motifs. Down to the 18th century, many literary works, buildings, and works of art are nearly unintelligible without a knowledge of such ideas. We find symbolic figures representing the zodiac in many of the great cathedrals—in Notre Dame, Amiens, Reims, Cremona, and St. Mark's, for example—in sculpture, mosaics, paintings, frescoes, illuminated manuscripts, even household wares. A famous astrological frieze adorns Merton College at Oxford; Oxford itself is named for a zodiac sign (Taurus, the celestial Ox or Bull). Many of the paintings of

Raphael, Botticelli, Michelangelo, and Tintoretto, Dürer, Rubens, Hieronymus Bosch, and others, can scarcely be understood without reference to astrological theme. Raphael's famed *Vision of Ezekiel,* for example, depicts God surrounded by the four fixed zodiac signs, and his decorations for the cupola of the Chigi Palace show the planets and constellations as gods. Guido Reni portrayed St. Michael the Archangel with the Balance of Libra in his hand; Dürer's *Melancholia* is a complete portrait of Saturn exalted in Libra, in which the figure of melancholia, winged, and crowned with the myrtle leaves of Venus, "counsels a patience that will mean the triumph of the soul over sorrow through a penetration of the meaning of life and death." A number of Botticelli's paintings—including *Venus and Mars*—are astrological allegories; Holbein's *The Ambassador's Secret* incorporates a complete horoscope in encoded form.

The work of Chaucer, Dante, Rabelais, Shakespeare, and other masters is rich with related themes. For example, the famous speech in Shakespeare's *As You Like It* about the "Seven Ages of Man" is based on a planetary scheme. In the *Divine Comedy* Dante begins his wanderings when the Sun is in Aries—the beginning of the zodiac, but also (as Dante tells us) the sign ruling when God created the world: ". . . The hour was morning's prime and on his way / Aloft the Sun ascended with those stars, / That with him rose when Love Divine first moved / Those its fair works." Pisces, representing Christ, is also rising over the horizon; Virgo—the stars of his mother, the Virgin— is opposite on the descendant, the two together "cradling the Earth." Venus exalted in Pisces represents divine love in the first Canto of the *Purgatorio;* in the *Paradiso* Dante boldly attributes his own poetic gift to the constellation of Gemini, the sign that dominated his natal chart. "O glorious constellation!" he exclaims, "O mighty stars pregnant with holy power." Though he placed two famed astrologers in his poetic Hell, their sin was not so much astrology as pride. Dante himself, in any case, had no doubt about the influence of the stars upon human life. Were it not for the influences of the stars, he wrote, children would be exactly like their parents—which turned Augustine's argument about twins on its head. "Our life, and also the life of every living thing here below," he wrote, "is caused by the heavens," which, as he put it in the *Purgatorio,* "according to its stars direct every seed to some end." Marco's speech to the poet in the sixteenth canto of the *Purgatorio* is also a pure statement of orthodox doctrine on the art.

Chaucer, a high court official under Edward III, was familiar with Ptolemy, Alcabitius, al-Kindi, Masha'allah, and others, and deeply versed in the work of Bonatti and Michael Scot. His *Treatise on the Astrolabe,* written about 1391 for "Little Lewis," a boy of ten who may have been his own son, set out to deal with astrology in all its aspects, as well as the technicalities of the astrolabe itself. His work, which remained unfinished, explained the characteristics of the signs, the correspondence between them and the parts of the human body, and the planetary hours. In considering how to judge whether an ascendant is fortunate or not, he tells us, it must not contain an evil planet (Mars or Saturn), or be in aspect to them, or conjunct the Dragon's Tail. However, should it contain a fortunate planet (Venus, for example), that would be a cause for "joy." (A simplistic view, of course, but then his work was meant to introduce the subject to a child of ten.)

The Canterbury Tales is full of astrological lore. The bawdy Wife of Bath ascribes her lust to Venus, her hardiness to Mars, and her general sensuality to her Taurus ascendant. "Alas! alas!" she cries, "that ever love was sin! / I followed ay [always] mine inclination / By vertu of my constellacion." We learn that those with Mercury in Pisces (the sign in which the planet has its "fall") fare poorly in love because Pisces is the sign where Venus is exalted; Venus, on the other hand, suffers in either Gemini or Virgo, where Mercury reigns. In "The Knight's Tale," the advice of Arcite to Palamon accepts the rule of the stars over man's destinies as an unavoidable fact, and counsels Stoic resignation:

> For Goddes love, tak al in pacience
> Our prisoun, for it may non other be;
> Fortune hath yeven [given] us this adversitee.
> Som wikke [evil] aspect or disposicioun
> Of Saturne, by sum constellacioun,
> Hath yeven us this, although we hadde it sworn;
> So stood the heven whan that we were born;
> We moste [must] endure it: this is the short and pleyn.

The tale also explains the planetary hours and the procedure for calculating a fortunate ascendant for an election chart. In "The Man of

Law's Tale," the Emperor's daughter, Lady Constance, is betrothed to a Turkish Sultan, but her father fails to elect a propitious time for her voyage, so that the unfortunate ascendant under which it begins brings a string of misfortunes in its train. All this, we are told, a competent astrologer could have helped her to avoid. The lawyer asks, "Was there no philosopher [astrologer] in all thy town [to elect a better time]?" Aries was rising with Mars, its ruling planet, in Cancer, the sign of its "fall," just as Constance embarked, and a "feeble Moon" in "disastrous," fashion was moving away from Aries—its place of mutual reception—toward a conjunction with Mars. As a result, the hoped-for marriage was "slain." One cannot always know when Chaucer is speaking in his own voice, although astrology in the *Tales* is disparaged only by such disreputable characters as the Franklin, who rails against the mansions of the Moon. Chaucer dolefully concludes: "For in the sterres, clerer than is glas, / Is written, God wot, whoso coude it rede, / The death of every man, withouten drede." ["For in the stars, clearer than in glass, is written, God knows, whoever can read it, the death of every man, without doubt."]

Some of the tales, such as "The Nun's Priest's Tale," would seem to be celestial allegories in which the characters themselves represent planets or stars. Thus an apparently "unsophisticated, farmyard story of a vain cockerel" who narrowly escapes becoming the feast of a fox turns out to mirror an actual stellar event. The tale takes place on May 3, as Chaucer tells us, with the Sun in Taurus. The exact date was May 3, 1392. The cockerel (representing the Sun) announces the break of day to his seven hens (the Pleiades). On that day, the fox (Saturn) neared conjunction with the Sun (thus placing the cockerel at risk), but the Sun is saved by "the timely arrival of a widow and her two daughters, who correspond to the rising of the Moon and the twin scales of Libra." Again, in "The Franklin's Tale," Chaucer allows us to fix the date of the events precisely—as Christmas Day, December 25, 1387—when exceptionally high tides temporarily submerged a rocky stretch of the Brittany coast. In the tale, a magician pretends to make the coast appear to disappear, by understanding in advance the Moon's effect on the tides.

In *Troilus and Criseyde,* Chaucer's other poetic masterpiece, set in the Trojan War, the Trojan astrologer Calkas defects when by "calculating" he discovers the Greeks will win. Troilus fares well in love when

Venus is in Criseyde's seventh house; Pandarus chooses to convey a message to Criseyde when the Moon is "in good plyt"; the two lovers consummate their desire when Jupiter, Saturn, and the Moon conjoin in Cancer to cause a torrential rain; and that night Troilus can only hope that the "badde aspects" made by the malefics in his chart, as well as his "Venus combust" will somehow be overcome by Jupiter's benignant rays. Jupiter fails him, Criseyde is obliged to leave Troy as part of a prisoner exchange, and as she departs, she bemoans "the cursed constellation" under which she was born.

★ ★ ★

ALMOST ALL THOSE who knew anything about astrology were learned, as the science of it required. But some of this learning also trickled down in simplified form. Most of the popular lore was lunar, and had to do with interpreting the phases of the Moon. Planting by the Moon was time-honored, of course. A fast or waxing moon, "increasing in light, well placed and aspected," was usually favored for most ventures or occasions—unless one was being bled; mental effort was best made when the Moon and Mercury were in friendly aspect or both in the signs that Mercury ruled. A haircut was thought to last longer when the Moon was on the wane.

PART TWO

"There are in Astrologie (I confess) shallow Brooks, through which young Tyroes may wade; but withal there are deep Fords, over which the Giants themselves must swim."

—ELIAS ASHMOLE

Chapter 6

THE 15TH AND 16TH CENTURIES have justly been called "the golden age of astrology" in the West. By the end of the Renaissance, astrology had permeated all classes of society—from the farmer who consulted his almanac for planting according to the phases of the Moon, to the popes and monarchs who relied for guidance on the complex castings of occult seers at court.

One scholar has nicely described the ambience of a typical consultation during the art's more reputable years. "Like the patient who visits the office of a modern dentist or ophthalmologist," he wrote, "the client who stepped across the threshold of an astrologer's chambers in the fifteenth or sixteenth century entered a realm that seemed both strange and familiar, both laden with powerful symbols and equipped with impressive tools." The consulting rooms contained reference works and instruments to establish the astrologer's professional authority—a globe of the heavens, for example, or an astrolabe and armillary sphere, as well as an almanac with planetary and ascension tables (the "essential desk reference") lying open on a stand. The technical information he assembled with their help was "recorded on a form divided into categories, which enabled him to select, organize, and record the significant facts" according to a time-honored scheme that incorporated the data into a coherent map—noting the signs, houses, planets and their aspects, nodes of the Moon, fixed stars, decans (or subdivisions of each sign), the paranatellonta (stars that rise alongside a given zodiac degree), the angles of the chart, and so on.

That map of course was the horoscope, meaning a view of the

hour—in particular, that moment of the hour when the degree of a sign rose over the horizon for a particular place on earth. In a precise way, it might be likened to a clock of heaven, giving an overall picture of the person, his life, and fate. The incidents or developments of that life were ascertained through the transits and progressions of the planets, lunar nodes, angles, and other sensitive points, together with the directions of the stars. Before any astrologer would risk some grand pronouncement, he would consider the chart as a whole, the temperament of the person it revealed, and the manner in which that temperament was expressed. The ascendant and its Lord (or ruling planet) was judged in those days of more general importance than the Sun or its sign, and the essential dignities of the planets (their strength by sign, exaltation, triplicity, term, and face) more important than house or aspect. Astrologers recognized that heredity, environment, and education all played their part but that the birth time ultimately ruled. The reward was therefore not only a penetrating analysis of one's innermost attributes, including one's virtues and vices, but also a forecast of the life that would be lived. How you lived it (if you believed in free will) was up to you.

All of this was unique. Technically speaking, according to the house system of the horoscope, no two people except those born within four minutes of each other in the very same place could lay claim to the same fate. Even among twins such identity was rare.

Two other kinds of consultations, as we have seen—called "horary" and "elections"—were also common when the art was still proficient and esteemed. Horary charts gave answers to questions—of any kind, from the outcome of a battle to the prospect for a trip—based on the position of the celestial bodies at the time the question was asked or received. Elections provided a means of selecting propitious times for any action or enterprise.

All things being equal, the calculations required an expert knowledge of mathematics, spherical geometry, astronomy, trigonometry, geometry, and so on, and were a learned and time-consuming process, even as they possessed something of the magical aura of the acts and incantations of a priest.

That does not mean that astrologers always knew what they were talking about. "In all the better families," noted Jacob Burckhardt, in his great work *The Civilization of the Renaissance in Italy,* "the horoscope of the children was drawn up as a matter of course, and it sometimes happened that for half a lifetime men were haunted by the idle expectation of events which never occurred." No doubt many other predictions failed to come true, though these, to bedevil any history, have not been as faithfully preserved as those that did. Astrology had its opponents, of course, though they tended, even if brilliant, to speak with an unconvincing voice. The Italian love poet Francesco Petrarch, for example, had a personal grudge against it, which subtracts from his critique. In 1350 he had been given the honor, it seems, of delivering the formal oration at the installation of the new duke of Milan, and, as was his custom, "he labored mightily to produce a perfect Latin speech." He was halfway through it when the official astrologer—whose task it was to announce the exact moment when the new duke took charge—attempted to interrupt him, since he had gone on too long. Before the shocked gaze of the assembled multitude, the two took part in an undignified wrangle, but of course the astrologer won: no matter how much the duke might esteem Petrarch's eloquence, he wanted to be sure he assumed his office at a propitious time. As a result, "Petrarch was left with the mangled remnants of his speech."

A more formidable antagonist was Pico della Mirandola, a Renaissance humanist and polymath who after wandering through the great universities of Europe—where he studied Greek, Latin, Hebrew, Syriac, Arabic, philosophy, science, and the Kabbalah—undertook to maintain nine hundred theses against all comers in a public debate in Rome. After thirteen of his theses were condemned as heretical by Pope Innocent VIII, he bowed his chastened head, "destroyed his poetical works, gave up profane science, and determined to devote his old age to a defense of Christianity against Jews, Moslems, and astrologers." It had been predicted to Pico (by Lucius Bellantius, an astrologer of Siena) that he would die in his thirty-first year. Like Cicero, Pico found such a prediction unwelcome, and it turned him against the art. He also levied a torrent of abuse against it, using all the old arguments, some of them culled from Augustine, but again like Cicero, that did not prevent the prediction from coming true. Niccolò Machiavelli, on the other hand, was one of those hard-eyed realists who accepted astrology without being quite sure how it worked. "Whence it comes I know not," he wrote, "but both ancient and modern instances prove that no great events ever occur in any city or country that have not been predicted by soothsayers, revelations, or by portents and other celestial signs."

Astrology at the time was on the side of "science," and its opponents were often allied with a superstitious prejudice that confused mathematics and magic, and a study of the heavens with an idolatrous worship of the stars. "Some of the greatest scientific minds of the age believed in the art," notes one abashed historian. "And as we look over their books, we notice to our amazement what intelligent writers most of the astrologers are; on the other hand, we are only too often confronted with an anti-astrologer who is both ignorant and dull. There are various reasons for this. To be a ranking member of the astrologer's profession in the sixteenth century required a mastery of astronomy and mathematics . . . To be an opponent of astrology, one needed only enough Latin to read Pico and abridge his arguments." But as one 16th-century astrologer sensibly put it, the astrologer was also "not a divine oracle, but simply a learned and erudite man, who can be deceived like other men. 'Who will not bump his head among so many shadows; who will not stray among the windings of such a twisted labyrinth?' "

One who seldom strayed was Marsilio Ficino, one of the greatest figures in the Italian Renaissance. Born at Figline, Italy, on October 19, 1433, Ficino had trained in law, medicine, theology, and music, and achieved early renown in the recovery of classical learning under the patronage of Cosimo de'Medici (Cosimo the Elder), who had wisely selected him to head his new Platonic Academy of Florence, which became the very center of contemporary science and art. With a largess that might make a modern scholar green with envy, Cosimo established him in a villa at Careggi on the city's outskirts and subsidized his scholarly endeavors with stipends generous enough to relieve him of all material care. From 1459 on, Ficino was able to pursue his invaluable work in comfortable and idyllic surroundings, and devote himself to translating many newly recovered texts.

At Cosimo's behest, Ficino undertook a Latin translation of the whole of Plato's work; translated a number of the Hermetic texts (deeply imbued with astrology) ascribed to Hermes Trismegistus, a legendary Egyptian sage; the works of Synesius of Cyrene, Psellus, Iamblichus, Porphyry, Proclus, Plotinus, Theon of Smyrna, and Dionysius the Areopagite, the *Homeric Hymns,* Hesiod's *Theogony,* and

Pythagoras's *Golden Thoughts.* His own original work included *The Three Books of Life, The Christian Religion,* and an unfinished commentary on St. Paul's Epistle to the Romans, aside from various editions of his voluminous correspondence published after his death. An ordained priest and a canon of the cathedral of Florence, Ficino insisted that there was no contradiction between classical teaching and Christian revelation, and in his own writings combined his religious convictions with astrology, alchemy, and magic. He kept an "eternal flame" burning before a bust of Plato in his study, thought Plato should be read in churches, and regarded Socrates as a forerunner of Christ. The whole intellectual life of Florence in a sense came under his spell, and many of the scholars and artists who seem to embody Renaissance art and thought at its height—Lorenzo de Medici (whom he tutored), Alberti, Poliziano, Landino, Botticelli, Michelangelo, Raphael, Titian, Dürer (in Germany), and others—were inspired by Ficino's work. He was their center, and they were the center of the Renaissance.

Among his foreign correspondents were John Colet, the dean of St. Paul's Cathedral in London; the chancellor of the French parliament; King Matthias Corvinus of Hungary; and the German humanist Johann Reuchlin. Though Ficino was not the first to revive the study of Plato, it was principally through him that Europe was fully awakened to Platonic thought.

A short, slender man with blond curly hair, a high forehead, and a smooth but florid complexion, Ficino by all accounts was a refined and mild-mannered man of simple tastes, with a palate for fine wine. Though naturally eloquent, he stuttered slightly and lisped on the letter *s* and over time developed a slight hunch at the shoulders from his scholarly toil. Though "cheerful and festive in company," he was sometimes "benumbed" in solitude with melancholy thoughts, which he endeavored to dispel by the energy he devoted to his work. Most of his intimate friends were men, but in his private life he seems to have yearned for the sublimation of his passions, which, it is said, had a homoerotic cast. It was Ficino who "coined the phrase 'Platonic love' " to express the essential chastity of the union to which he aspired.

Ficino's thought was fundamentally shaped by the work he had helped to recover and revive. His life, in that sense, was all of a piece. In the *Timaeus,* Plato had spoken of the planets as incarnations of spiritual intelligences, which endowed them with divine authority and power. The "untroubled course" of their motion, he wrote, was set as a sign and pattern in the heavens "to guide the troubled revolutions in our own understanding, which are akin to them," so that we would have an idea of how we should be. That formed part of Ficino's creed. In Plotinus, in turn, he found: "We may think of the stars as letters perpetually being inscribed on the heavens. Those who know how to read this sort of writing . . . can read the future from their patterns, discovering what is signified." And so Ficino believed. And in the creation myth of the *Corpus Hermeticum,* he beheld an image of man, as created or conceived by the Divine Mind, imprinted with the qualities of the seven planets, which governed his destiny on earth. Yet by the very nature of this creation, man also partook of the absolute freedom of the divine, and by acting from the immortal part of his soul, strove to achieve a more direct knowledge of God. That was the sacred hope of Ficino's life.

Ficino, an able astrologer, accurately predicted that Giovanni de' Medici, the great-grandson of his first great patron, would one day be pope (in 1513, he was crowned Pope Leo X), and regularly gave astrological advice to his friends. He was also preoccupied with the strength of Saturn in his own chart, and ascribed the whole tone and tenor of his life to its influence, for better or for worse. He once complained to a friend in a letter: "Saturn seems to have impressed the seal of melancholy on me from the start." He struggled under its weight, but found ultimately that surrendering to it deeply gave him inspiration and freed him from its malevolent power. Indeed, it endowed him with a serious disposition, with its strong sense of obligation, and led him toward the scholarly and contemplative life he chose. In that specific sense, Ficino believed one should "follow one's star," or go with one's celestial strength: "The heavens will promote your undertakings," he wrote a friend, "and will favor your life to the extent that you follow the auspices of the lord of your geniture [birth], especially if that Platonic doctrine is true . . . that every person has at birth one certain daemon, the guardian of his life, assigned by his own personal star, which helps him to that very task to which the celestials summoned him when he was born."

In the end, astrology itself "could . . . only be justified if it provided some means or guidance for the human soul to begin to know itself as an image of God." Nothing about that was incompatible with astral magic, which Ficino also embraced. For it was not possible, in his view, for "a material action, motion, or event to obtain full or perfect efficacy except when celestial harmony conduces to it from all sides." Through rituals that channeled astral forces, the human spirit could align itself with the spirit of the heavens and "receive the gifts of its special planet as it vibrates in sympathy, like two strings of a lute."

★ ★ ★

IT TAKES A KIND OF INGENUITY to interpret a horoscope with coherent rightness, but plausibility and coherence do not always make a reading right. This is so in all the sciences, of course, including medicine—as anyone knows from the misdiagnosis of disease. Some astrologers, such as Abū Ma'shar, Guido Bonatti, and Luca Guarico, seem to have known what they were doing, at least on their own terms. Others were merely, or mostly, ingenious—lacking the requisite knowledge and skill. Yet the literature abounds with striking tales.

In 1514, for example, five years before he prepared to circumnavigate the globe with five ships, Ferdinand Magellan met the astrologer Ruy Faleiro in northern Portugal, where together they attempted to determine where the Spice Islands might lie. Their astronomical and geographical speculations placed them within the sphere of the discoverable world that the pope had awarded to Spain in the Treaty of Torsedillas, and so it was to Spain they went for support. Having received it, they fitted out their ships, and in September 1519, on the verge of embarkation, Magellan invited Faleiro to join the expedition. Faleiro demurred. Not long before, he had cast a horary chart for the voyage and by it had foreseen not only its ultimate glory but a slaughter that would befall the men at a landfall, somewhere among the islands of the sea. On September 20, the ships set sail, and over the next eighteen months their stalwart captain discovered the straits that would bear his name, christened the Pacific Ocean, weathered mutiny and tempests, ate raw ox hides and sawdust to avoid starvation, and, after reprovisioning on the island of Guam, made landfall in mid-March in the Philippines. There Magellan unwisely involved himself in rivalries among the different tribes, and on April 27, 1521, was killed with many of his confreres on the island of Mactan.

Beyond such accounts of historical interest, there was an irrepressible fund of cautionary tales (some fabulous, of course) that lent credence to the ineluctability of fate.

Everyone knows, or used to know, the story about the merchant in Baghdad who one day sent his servant to market where he learned his impending doom. The servant had hurried off to purchase some item or other but soon returned trembling and white as a sheet. "What happened?" his master asked. The servant replied: "When I was at the bazaar, I was jostled by a woman. But when I turned and looked I saw the face of Death. She looked hard into my eyes and made a threatening gesture. Master, please lend me your horse, for I must hurry away to escape her." "Where will you go?" he asked. "To Samarra, where she will not think to look." So he took the horse and rode away as fast as he could. Later that day, the master himself went down to the market and saw Death standing in the crowd. "Why did you threaten my servant?" he asked her boldly. She looked at him and smiled: "Oh, that was not a threat, but a gesture of surprise. I was startled to see him in Baghdad, when we have an appointment in Samarra tonight."

There is an ironic fascination about such stories that one cannot escape, because it is both gratifying and appalling to imagine that the apparent chaos of life is governed by an exact and intricate design. Tommaso Campanella, in a work published at Lyons in 1629, tells us that the notable astrologer Valentin Naibod, foreseeing his own violent death, tried in vain to escape it by shutting himself up in his house. Robbers, however, seeing the house completely sealed, thought no one was at home and broke in. When they encountered Naibod, they stabbed him before absconding with his goods. Another adept of the time, Bartolommeo della Rocca, was likewise said to have foretold he would be killed by a blow to his head. As the day approached, he naturally grew less sanguine of his skill, and in fear of the event, which he somehow hoped to avoid, began (as Michael Scot had done) to wear a metal plate concealed in his cap. That did not save him, for on the appointed day, a vendor of kindling, whom he thought he knew, showed up at his door and for some unknown reason struck him on the temple with a stick. That strange story is akin to yet another told of the scholar Johannes Stoeffler, who had foreseen that on a certain day he might be crushed. He decided for safety's sake to spend the day indoors and, to pass the time, called in some learned friends for a conversation over a jug of wine. In the course of their discussion he reached up for a heavy tome to settle some point, and the whole wall of books gave way and crushed him to the floor.

The lore associated with Nostradamus—physician in turn to Henry II, Francis II, and Charles IX of France—is of questionable interest, in part because his claimed powers of prediction have been overplayed for five hundred years. In opaquely phrased doggerel verse, he is supposed to have prophesied the rise to power of Oliver Cromwell in England, the Great Fire of London of 1666, the birth of Napoleon, and a number of our modern wars. With respect to the Great Fire, his prophecy reads: "The blood of the just requires, / Which out of London reeks, / That it must be razed by fires. / In year three score and six . . ." Another bit of verse, impenetrable at the time, was later taken to be a prophecy of the Crimean War.

> In thrice one hundred years the Bear
> The Crescent will assail;
> But if the Cock and Bull unite
> The Bear will not prevail;
> In twice ten years again
> Let Islam know and fear;
> The Cross shall stand, the Crescent wane,
> Dissolve and disappear.

In 1853, Russia (the Bear) attacked Turkey (the Crescent) and the Cock and the Bull (France and England) combined to save the Turks. Nevertheless, the Crescent waned. Such prophecies, however, are impossible to credit on their own terms. Riddling enough to mean anything, or nothing, their divination was not unlike construing the entrails of an animal, as was done by the augurs of republican Rome.

Nostradamus also made predictions for Catherine de Médicis when she was queen of France, but she preferred the advice of Ruggieri the Elder, who predicted the principal events of her life, "with an accuracy," wrote the French novelist, Honoré de Balzac, "that is enough to drive disbelievers to despair." Ruggieri forecast the disasters that, during the siege of Florence, affected her early life, as well as her marriage with a prince of France, the latter's unexpected accession to the throne, and the number and gender of the children she would bear. Three of her sons, she was told, were to reign in succession, with two of her daughters also becoming queens. Yet none would have children of their own. "And this proved out so exactly," wrote Balzac, that "many historians have regarded it as a prophecy after the event."

Another astrologer of great repute at the time was Luca Guarico, who more demonstrably proved his skill on a number of occasions, and served as court adviser to Catherine de Médicis as well as several popes. He predicted the defeat of Francis I at the battle of Pavia; the death of the Duc de Bourbon on the walls of Rome during its sack in 1527; the fall of Giovanni Bentivoglio (Signore Giovanni II), the tyrant of Bologna; the election of Alessandro Farnese as pope; and the promiscuous papacy of Giulio de'Medici, who, as Clement VII, would father a remarkable number of bastards. To date, historians have identified twenty-nine. Meanwhile, after Farnese became Pope Paul III, he made Guarico bishop of Giffoni, and in April 1543 commissioned him to determine an auspicious moment for laying the cornerstone of the new Farnese wing of the Vatican complex in Rome. The moment was set, and as soon as the astrologer shouted for construction to begin, "straightway Ennio Verulano, cardinal of Albano most reverend, clad in a white stole with a red tiara on his head, set in the foundation a huge marble block beautifully polished and engraved with the papal arms."

Gaurico's perspicacity was not always appreciated, however, and his candor put him at risk. One unhappy client, Giovanni Bentivoglio, being told that his power would soon be usurped, dangled him by a rope down a winding tower staircase and allowed him to swing for two days against the walls. Even so, in 1552 he rather daringly published his interpretation of the horoscopes of a number of prominent people, including popes, cardinals, princes, and notable men in the arts and learning—some of them still alive. This book of celebrity horoscopes helped foster a trend and was soon followed by similar books by Francisco Giuntini of Florence and Girolamo Cardano of Milan. The latter analyzed the horoscopes of a hundred distinguished individuals, including Erasmus, Albrecht Dürer, and Henry VIII.

Perhaps Gaurico's most famous prediction was made in 1555. At queen Catherine de Médicis's behest, he warned King Henri II of France, then thirty-seven years old, that in or around the forty-second year of his life, he ran the risk of being killed "during single combat in an enclosed space through an injury to the head." It so happened that in that same year, the French astrologer Nostradamus also seemed to predict the manner in which that death would come. This is of interest because it is perhaps the one verifiable prediction he made. Yet it was specific enough to be startling. In 1555, the following quatrain had appeared in his famed book of prophetical verse: "Le Lyon jeune le vieux surmontera, / En champ bellique par singulier duelle, / Dans cage d'or les yeux lui crevera, / Deux classes une puis mourir mort cruelle." ("The young lion will overcome the old on the tournament ground in single combat. Through the cage of gold his eyes will be pierced. Two wounds become one, followed by an awful death." Henri was not mentioned by name, but he "was instantly recognizable not only by his gold jousting helmet, but by the golden lion which formed its crest."

The fatal year came round, and in June 1559 the king celebrated the marriage of his daughter, Elizabeth of Valois, to King Philip II of Spain. As part of the festivities, he decided to hold a three-day jousting tournament in which he was determined to take part. For the occasion, the rue de St. Antoine, in front of the Tournelles, was cleared, its paving stones dug up, and a great wooden amphitheater with raised boxes erected for the crowd. July 1 was the last day of the revels, and after a series of contests leading up to the big event, Henri himself emerged encased in his royal armor, with its distinctive gold helmet, and was hoisted upon a Turkish stallion, which he had just received as a gift from the duke of Savoy. Joining several other great nobles in the lists, Henri ran the prescribed three courses, and in the first two prevailed. But in the third, against Gabriel Montgomery, captain of his own Scottish guard, he was almost unhorsed. Resolved on making a better showing, he insisted on a second try, even though the queen begged him not to and Montgomery himself offered to withdraw. But the king would not yield.

Once more the two took up their stations, reared, and spurred their mounts. As they met in a furious charge, both shattered their lances, but as they did so, the king's vizor flew open, and splinters from Montgomery's lance were driven through his eyes into his brain. In agony and screaming, he was carried off the field and the renowned anatomist Andreas Vesalius was summoned at once by courier from Brussels in an effort to save his life. Vesalius arrived on the 3rd, but was at a loss as to what to do. In a trial operation, he was allowed to open the skull of a recent murder victim, probing with a silver needle through the dead man's eye. But the king's case was clearly hopeless. In a semidelirious state, he beat his chest in contrition, swore to extirpate heresy, vowed to go on pilgrimage to a shrine of the Virgin near Orléans, and called for his mistress, Diane de Poitiers, whom the queen firmly barred from the room. On the morning of the 10th, he received the last sacraments and at one o'clock that afternoon he died.

In the year Gaurico predicted, and in the manner Nostradamus had described, the king of France was slain. What was it the two had seen in the king's chart? In part basing her analysis on Firmicus Maternus, one modern astrologer has valiantly attempted to sort out the clues. Henri's Sun and Moon, she tells us, were both in a square to Saturn, indicating a violent death, and Jupiter was under Mars in the sixth house (where illness is described). In the year of his death, the ascendant by primary direction, or progression, had been brought into conjunction with Mars, which in turn was in a square to his Sun in Aries and opposite Saturn, which ruled his eighth house. Transiting Saturn was also precisely conjunct the Part of Fortune and the Dragon's Head, the ascending north node of the Moon. Finally, the king's ascendant was near the Pleiades, and Mars on a nebula called the Eyes of the Crab in Cancer, both, by tradition, said to endanger sight. In sum, by sign and aspect these factors indicated a head wound entailing blindness; moreover, there had been an eclipse earlier in the year, around the time of the king's birthday, which fell on his Sun-Moon-Venus conjunction in the twelfth house.

It must be left to those who may know, to judge if this analysis is right.

Yet if a second opinion could have been had, none at the time would have been more qualified to provide it than Girolamo Cardano of Milan. Astrologers were often men of great ability beyond their special field, but in general genius, Cardano perhaps excelled all of his peers. Aside from his expertise at divination, he helped lay the foundations of modern algebra, expounded the rules for solving cubic equations, mastered subjects from philology to perspective, and invented the universal joint. As a doctor he also enjoyed considerable renown. Among his friends and patrons, he could count cardinals, archbishops, generals, and princes; Andrea Alciati, the premier jurist of his day; Vesalius, the foremost anatomist; Protestant nobles in the English court of Edward VI; the French ambassador to England; and John Hamilton, the last Catholic archbishop of St. Andrews in Edinburgh. He also wrote a number of popular books, and some of his own students went on to have distinguished careers, including Ludovico Ferrari, professor of mathematics at the University of Milan. Various European courts—Denmark, Scotland, Mantua, and France, among them—tried to lure him to their service with lucrative fees, and opportunities for preferment also came from Pope Paul III and Pope Pius IV. From his autobiography we also learn that he was an expert swordsman, a fine angler, a good part-singer, and overcame a number of physical deficiencies through a dedicated exercise routine. He was also careful with his diet, and ahead of his time in emphasizing fruits, vegetables, and fish.

Yet the cup of his sorrow was always full. Born in Pavia out of wedlock, Cardano was an unwanted child. His mother tried to abort him, and when ripped just in time from her womb after three days labor, he had come forth "almost dead." Doctors plunged him at once into a basin of warm wine to revive him, but as he wailed out with his first breath—at 6:40 A.M. on September 24, 1501—"both luminaries [in his horoscope] were cadent," he tells us, "and neither applying to the ascendant, being in the sixth and twelfth house." To make matters worse, Mars was also square to his Moon. However, "the Sun, both malefics, and Venus and Mercury were in the human signs," he added, "so at least I was not born a monster." However, like Ficino, he stuttered (a trait inherited from his father), had odd-shaped hands and feet, and some congenital disability that precluded a normal sex life before his thirty-first year.

His infancy was hectic and from the first his health was frail. His parents passed him from one wet nurse to another, and after one nurse died of the plague, he was drenched in vinegar as a disinfectant before another would take him on. When he was eight, he nearly died of dysentery, and, being accident-prone, fell down a flight of stairs and cracked his skull. A few months later he was struck on the head by a stone. Something of a hypochondriac, he was variously afflicted throughout his life with indigestion, bronchitis, pyorrhea, gout, hemorrhoids, heart palpitations, insomnia, bladder infections, eczema, erysipelas, and was twice struck by the plague. He was also, he tells us, "tormented by a tragic passion so heroic" that it inclined him to suicide. His mother was somewhat unloving, if conspicuously devout; his father, a prominent jurist, more attentive and had tried to pass on to his son what he knew. From the age of ten, Cardano had served as his page, learned Latin, Greek, classical astrology, and the first six books of Euclid under his guidance, and at the universities of Pavia and Padua took degrees in medicine and the arts.

No student could have shown more promise. Nevertheless, barred by his illegitimate birth from the College of Physicians in Milan, he became an acid-tongued critic of the medical elite, and was obliged to make a living for many years as an impoverished country doctor. In time, however, his talents made their way, but as he gained in stature his irascible nature also lined his path with foes. He had a number of close calls with death, was arrested by the Inquisition for rashly casting a horoscope of Christ, and had two wayward sons, one of whom scraped by as a petty thief, while the other was condemned to death and beheaded for poisoning his wife with cake laced with white orpiment. Cardano was sensitive to omens, and recalled that the day his son was born he had noted with alarm that the child had "small, white, restless eyes," two toes joined together on his left foot, and was deaf in one ear. The moment he was lifted from the baptismal font, a huge wasp had also flown into the room, circled the baby's bed twice, and thrummed loudly "like the beating of a drum." Cardano recalled, too, that he had had a sickening premonition that his son would come to a violent end. Years later, the day his son was arrested, a red mark in the shape of a dagger had appeared on one of the fingers of his hand.

On the whole, Cardano was inclined to a solitary life, insofar as a public career allowed it, and reveled in a home overrun with household pets that included rabbits, goats, and storks. A master at chess, he was also drawn to games of chance, and was, by his own admission, a compulsive gambler—though no one alive was better, perhaps, at calculating the probabilities at cards and dice. Indeed, though he once gambled away his wife's jewels, he won as often as he lost and even supported himself for a time by his winnings. In losing, he was not to be crossed. On one occasion at age twenty-eight, in Venice "at the festival of the birth of the Virgin," he tells us, "I cut a man who had cheated me at cards across the face with a sword. I was in his house, and he had two retainers. Two spears were hanging from the rafters, and the house-door was fastened with a key. I pointed my sword at the two [retainers] and threatened to kill them if they did not unlock the door." A little book Cardano wrote on the science of games is full of curious information about gambling tricks (such as the use of soaped cards) and describes all the various card games—Spanish, French, German, and Italian—current at the time. For various reasons, he loved the world to which gambling belonged. At the gambling table, he wrote, a man can not only forget his cares and refresh his spirits, but "friends and acquaintances tend to open their souls unwittingly, as their passions and propensities break out over changes of the game. One then gets to see them as they are." The gambling table was also, he wrote, "as true a leveler as death," where princes and paupers might meet.

★ ★ ★

BY CARDANO'S DAY, medical astrology had been developed to a high degree. Hippocrates, the great Greek physician and source of every doctor's oath, had insisted that his students master astrology for its help in determining when patients were most at risk. This seemed to him of such importance that "the doctor who does not understand astrology," he declared, "is a fool." A doctor classified his patient according to the four humors—choleric, sanguine, melancholic, phlegmatic—as determined by the general quality of his sign, considered the position of his Sun and Moon, and tried to identify any afflicting planet that might help explain the disease. For afflictions of the throat, shoulders, or neck, he might look for Mars in Taurus; for heart trouble, Saturn in Leo. A conjunction of Jupiter and Saturn in Scorpio pointed to venereal disease. Even such minor things as a bee sting might be treated—as in one famous case, effectively—with the help of an astrological chart. At the University of Bologna, the motto of the School of Medicine was: "A doctor without astrology is like an eye that cannot see." In those days, when a student went up for his medical exams, the first question asked was: "Where do you look for the disease?" The correct answer was: "In the sixth house." In some respects, medicine was also allied to alchemy and its common processes associated with signs of the zodiac: Aries-Calcination, Taurus-Congelation, Gemini-Fixation, Cancer-Solution, Leo-Digestion, Virgo-Distillation, Libra-Sublimation, Scorpio-Separation, Sagittarius-Ceration, Capricorn-Fermentation, Aquarius-Multiplication, Pisces-Projection.

Cardano's skill as a physician was second to none, and in the years he taught, the University of Bologna was viewed with awe. He is said to have restored to health more than a hundred patients who had been given up as hopeless, not only at Bologna but Rome and Milan. His first breakthrough seems to have come when he was still in his early thirties and estranged from the medical establishment, which he had infuriated with a little treatise on medical malpractice. On this occasion, three doctors—Cardano being one—had been called to the bedside of a child (the son of a senator, Francisco Sfondrato) whose feverish convulsions had left him partly paralyzed. The other two doctors had more seniority, but neither could diagnose the cause. When it came time for Cardano to speak, he said, without hesitation, "This is a case of opisthotonos," meaning the excessive action of one class of muscles by which limbs are curved back. One of the other two physicians stared blankly, for he had never even heard the word before. The other asked, "How can you know?" Cardano then showed how the child's head was forcibly held back, and could not be pulled forward into a natural position. His colleagues then "lauded his discernment" and concurred. The therapy Cardano prescribed also apparently worked, for the grateful senator (afterward a cardinal, almost a pope) thereafter did all he could to assist his career.

Sfondrato was in a position to do a lot. After becoming a cardinal, he was appointed to the Secret Council of Holy Roman Emperor Charles V, and upon the death of Paul III missed becoming pope by just three votes. He was a big man, in one description, "tall, frank-looking, fat and rubicund, genial, elegant, joyously disposed, not without wisdom and erudition. In business, he was cautious, prudent, prompt, and successful. He delighted in gambling, and that, too, for large sums. He believed in fate, and in the *Sortes Virgilianae* [a divinatory use of the *Aeneid* akin to the *I Ching*], of which he testified that he had often found them true." In 1550, the year after he stood for pope, he was poisoned and died at the age of fifty-seven. By then, Cardano was one of the five or six most famous doctors in the world.

Cardano's medical astrology was too subtle to abide by the art's more simplistic forms. "Whether or not it is true," he wrote once, "that the head is threatened when the moon is in Aries, and the neck when it is in Taurus, and the chest when it is in Cancer, and the heart when it is in Leo, and the viscera when it is in Virgo, I . . . can neither affirm nor deny." But "medicine," he explained, "is a many-sided art," practiced by a variety of specialists—surgeons, "oculists" (eye doctors), "bone specialists" (osteopaths), herbalists (pharmacists), and so on, and "each of these special fields is shared by many groups devoted to some particular phase" of their craft. The general practitioner had better be versed in all of them, he tells us, and know how to deal effectively "with nurses, assistants, and druggists"; and even see that the sick room is clean, well-heated (in winter), with plenty of water, and that the patient's food is properly prepared. In his conscientious attention to all such matters, Cardano was ahead of his time.

Astrology, of course, was also part of his art. Here is Cardano's account of the onset and cause of an illness that befell one Giovanni Antonio de Campioni on May 23, 1553, at 3 P.M.

From the start he had the moon quartile to Venus. Lack of temperance in food and drink made him ill, and his condition quickly worsened because of the rapid motion of the moon. On the seventh day he felt considerably worse, for the moon was with the dragon's tail, and devoid of any aspect with Jupiter, and moving towards opposition to Venus and the sun was afflicted by the square of Saturn. On the eighth he seemed to be relieved by a flow of blood from the nostrils, but his strength declined because of the moon's opposition to Venus. On the ninth he seemed to breathe a little because the moon was trine to the sun. The tenth took the place of the eleventh, since the moon had reached the angle, that is, the beginning of Pisces, in opposition to Jupiter and Mars. On the eleventh day it was reasonable for him to die, since the moon had come into conjunction with Saturn at the tenth hour, and into quartile to the sun at the eighteenth hour. He died on June 5, three hours before noon, and it was the beginning of the fourteenth day, and the moon had arrived at the point exactly opposed to its [original] position.

To a modern reader this might seem a little involved. But it is about as straightforward as traditional astrology gets. Like the pulse line on a hospital monitor, the course or journey of the Moon along the ecliptic described a rhythmic therapeutic line.

In 1552, Cardano accepted an offer to go to Scotland to attend John Hamilton, the archbishop of St. Andrews, who was suffering from a baffling disease. After erecting the prelate's chart, he discovered the ailment (a form of asthma) and cured it, for which the grateful archbishop gave him a Thoroughbred horse, jewelry of great value, and nine hundred gold crowns. Yet Cardano had also been asked to look into Hamilton's future and, with obliging candor, took his leave with these words: "I have been able to cure you of your sickness, but I cannot change your destiny—nor prevent you from being hanged." Eighteen years later this churchman was arraigned for treason by the Queen Regent of Scotland and hanged by her palace gate. Meanwhile, from Scotland Cardano proceeded to England to the court of Edward VI, where he was a guest at the London home of Sir John Cheke. (While there, he made one of his few predictive mistakes, and failed to see that the delicate king would soon expire.) Cardano's return through Europe—by way of Bruges, Ghent, Brussels, Louvain, Antwerp, Liege, Aix-la-Chapelle, Cologne, Basle, and Zurich—was a triumphal progress, as from all sides came abundant recognition of his abilities with offers, invitations, and gifts.

In his practice, Cardano tried to follow the code of ethics laid down by Firmicus Maternus, emphasized clarity and concision, and advised against making judgments "about insignificant things . . . [And] let the astrologer himself be prudent, gentle, of few words, elegant, well-dressed, grave, faithful, and honest, and exemplary in every way. For often the artist can adorn the art." He also sensibly counseled against "firmly predicting" some "great evil" to a prince or other personage of power, unless pressed, since the astrologer was likely to provoke the client's wrath if he did not allow for some reprieve. At the same time, Cardano was justly proud of his skill, and happily recounted instances when he had outshone some colleague or rival. In his book of *Aphorisms,* we get a glimpse of how sharp he could be. On March 21, 1546, for example, a fellow astrologer had brought him a test horoscope to read: "Looking at it, I said: 'This man is Saturnine and melancholic.' He replied: 'Where do you get that?' I answered, 'Because Saturn rules over the ascendant, and holds the degree opposed to it, and looks at it. And Saturn is in Leo, which adds to the sorrow. But,' I added, 'he has a deceptively calm and gentle demeanor and is capable of smooth and easy speech.' 'And where do you get that?' 'Because,' I answered, 'Aquarius [on the ascendant] is a human sign, and Saturn [which rules it] produces men who talk that way, and the head of the dragon—which is very important—is also in the ascendant, and gives the demeanor I've

described.' " His colleague, apparently stunned, exclaimed: " 'You captured the man perfectly . . . But please go through the rest.' To which I
replied: 'Well, he will certainly die a bad death.' 'But how do you know
that?' 'He has Saturn condemned with the dragon's tail in the seventh
house.' 'How then?' he asked. 'By hanging.' 'How do you know that?'
'Because Saturn and the dragon's tail, in the seventh, show that he will
be hanged.' " And so he was. The horoscope belonged to a counterfeiter
by the name of Francesco Marsili—who met the fate described.

On another occasion, a colleague showed him the chart of a prosti-
tute and trickster. He looked at it, saw the Moon in Scorpio, Mars with
Saturn in Leo, opposite Venus, but trine the ascendant, which, again
(as he explained in similar detail), told him what she was. Then there
was Paolo Sforza, a prince who hoped to become ruler of Milan.
Sforza had asked another astrologer about his prospects; the latter, in
turn, asked Cardano what he thought. "When he showed me the figure," Cardano recalled, "I said that Sforza would die that year. For the
moon was among the Pleiades in the sixth house, in quartile to Mars
and Jupiter, which were moving through the fixed house of Saturn."
And die he did.

Some of Cardano's *Aphorisms* are pungent and telling. Here are four
"Relating to Nativities": "When Jupiter shall be in the Tenth House in
trine of Mars and strong; and the Sun with the Dragon's Head, and the
Moon with Cor Leonis [one of the fixed stars]; such Native, though
the son of the meanest peasant, shall be wonderfully exalted." Or:
"When Venus and Jupiter are in the Seventh House, the Moon be-
holding them in her own dignities [the five essential dignities of triplicity, sign, exaltation, term, and face], and the Dragon's Head joined
with them or with Mercury, the native shall get a great estate by means
of his wives." Again: "When Venus is with Saturn and beholds the
Lord of the Ascendant, the Native is inclined to Sodomy, or at least
shall love old hard favored women, or poor dirty wenches." And finally: "Mars is seldom joined with Mercury for Good, for he makes
people naughty and Imprudent, yet industrious in Art, whence it
comes to pass that the best Artists are too often the worst men."

Cardano wrote over one hundred books and tracts on various
topics—from geometry to palmistry to poisons to dreams, as well as a
life of St. Martin and some seven thousand pages of learned commen-

tary on Galen and Hippocrates. He also wrote extensively on music—on its structure, rhythm, instruments, and variety of forms. Still another work was on mechanical inventions, on levers and pumps, an original contrivance for raising sunken vessels, and a device that anticipated the telegraph. He also produced a heavily illustrated, multivolume study of metoposcopy, the art of interpreting the lines on the forehead by astrological means. The whole brow, in his scheme, was intricately mapped out like a graph, with seven horizontal lines drawn at equal distances, one above the other, beginning just above the eyes, into planetary zones. Other works did not survive. Two—one on saliva, the other on venereal disease—were spoiled by cat piss ("amba hi libris corrupti urina felis"), and when he was thirty-seven, he burned nine others himself in manuscript "because I knew they were empty and would never be of use." In 1552 a pet dog left at home too long alone jumped on his desk and tore his public lectures to shreds.

One of Cardano's better-known books, the philosophical treatise *On Consolation,* was later translated into English as *Cardano's Comforte* at the behest of the Earl of Oxford in 1573. It had an odd introduction written by one Thomas Churchyard, who exhorted its prospective readers not to nod off "with slobbering hands or head to blot or blemish the beauty of this book," which made it seem uninviting. In fact it had an eloquence that even in translation came through. "Men in this worlde are like trees," he wrote,

some slender, some great, some flourishing, some bearing fruit, some withering, some growing, some blown down, which in one harvest time are brought together and laid upon one stack . . . Only honesty and virtue of mind doth make a man happy, and only a cowardly and corrupt conscience do cause thine unhappiness. Because the worst that a good man can fear, is the best that the evil can wish for: which is the destruction of the soul in death. But as he ought not to hope thereof, so should not the other fear it. For God the eternal father hath sent us into this world as children and heirs of his kingdom, and secretly beholdeth how we fight and defend ourselves, against our senses, the world and the Devil. And who so in the battle, valiantly fighteth, shall be called and placed among the Princes of heavenly kingdom. And who so slothfully or cowardly behaveth himself, as a slave in features shall forever more be bound.

Cardano today is perhaps best known for his pioneering writings on mathematics, which are still of interest, for his aptitude was very great. At odd hours, he loved to work through such abstruse problems as: "Make me of ten four quantities in continued proportion whose squares added shall make sixty." Or: "Two persons were in company, and possessed I know not how many ducats. They gained a cube of the tenth part of their capital, and if they had gained three less than they did gain, they would have gained an amount equal to their capital. How many ducats had they?" On one occasion, he scolded a fellow mathematician this way:

> Where did you ever find [how could you have ever supposed] that the discovery of the root pronica media, which lies at the bottom of the solution of all the thirty questions [proposed by a colleague] which is founded on the eighth problem of the sixth book of Euclid, could resolve a question of cube and number equal to the census [the quantity represented by x squared], under which section is to be ranked the proportion which says, "Find me four quantities, in continuous proportion, of which the second shall be two, and the first and fourth shall make ten." I speak in the same way of the others, so that while you wished to show yourself a miracle of science to a bookseller, you have shown yourself a great ignoramus to those who understand such matters; not that I myself esteem you ignorant, but too presumptuous; as was Messer Zuanne da Coi [another mathematician], who thinking to get credit for knowing what he did not know, lost credit for knowing what he did.

The recipient of this arcane diatribe was Niccolò Tartalea, a mathematician of immortal fame. A few years before, he had discovered the solution to cubic equations, but had refused to disclose it, and Cardano was probing his mind. Tartalea was furious and absurdly derided Cardano's own technical observations as "so weak and ill-conditioned, that an infirm woman could beat them to the ground."

Suffice it to say that Cardano eventually obtained Tartalea's secret, which he was on the verge of grasping himself, and in his great book *Algebra,* published in 1545, gave full credit where credit was due. But he also went beyond anything Tartalea had done. For he was the first to lay out the rules for cubic equations in all their forms, "having all their terms or wanting any of them, and having all possible varieties of signs." He then proceeded to demonstrate every rule geometrically, and "in a manner before wholly unknown," to apply algebra to the solution of geometrical problems. In this way, the whole doctrine of cubic equations was first published to the world.

Cardano once observed that "mathematics is its own explanation; . . . for the recognition that a fact is so, is the cause upon which we base the proof." Elsewhere, he gave an example: "Considering the proposition that the exterior angle is equal to the sum of the two opposite interior angles—there is no reason why this should be, but that it is, is simply a fact." Astrologers, both ancient and modern, have occasionally made the same point about their art. One of the moderns, John Frawley, deeply versed in the tradition, observes: "There is no apparent reason why the position of the planets thirty-six days after the birth should show the same events as the Solar Return taking place that many years later and the Lunar Returns taking place each month during that same year. Yet they do, for in His infinite wisdom, the Almighty has shaped a universe that fits together in coherent fashion. Not the least of the many delights of the study of astrology is the chance to marvel at the precision and intricacy of this construction as it turns."

It was said of Cardano that "no other man of his day could have left behind him works showing an intimate acquaintance with so many subjects," though their peculiar variety led one writer to grotesquely compare his "half-medieval, half-modern" mind to that of "a magnificent moth half-released from the state of chrysalis, its head and feet and front wings working out towards free space and upper air, but all the rest bound by some morbid adhesion to its dusky shell." That is a modern prejudice. To Cardano, the world and everything in it was connected and all of a piece.

He died just four days shy of his seventy-fifth birthday on September 20, 1576, and was buried in St. Mark's Church in Milan.

★   ★   ★

CARDANO WAS ONE of those who took to heart the admonition of Ptolemy that "the science [of astrology] demands the greatest study and a constant attention to a multitude of different points." He also shared the latter's view of the skeptic. "It has been supposed," Ptolemy wrote,

> that even such events as have been truly predicted have taken place by chance only, and not from any operative cause in nature. But it should be remembered that these mistakes arise, not from any deficiency or want of power in the science itself, but from the incompetency of unqualified persons who pretend to exercise it. . . . The reproach . . . thus brought upon the science is wholly unmerited; for it would be equally just to condemn all other branches of philosophy [scientific disciplines], because each numbers among its professors some mischievous pretenders.

As then (in classical times), so in Cardano's day—and ours.

Cardano's counterpart in England for learning was John Dee. Born in 1527, Dee was the son of Rowland Dee, a gentleman server to Henry VIII, who traced his ancestry back to Roderick the Great, the ancient prince of Wales. Little is known of his childhood, but in 1542, he entered St. John's College, Cambridge, where, by his own account, he was "so vehemently bent to study" that he slept only four hours a night. He excelled in all subjects; went on to study mathematics on a fellowship at Trinity College, Cambridge, under the great humanist Sir John Cheke; and in May 1547, traveled abroad to "speak and confer with learned men." His precocious quest took him to the Low Countries, Italy, and France, where he collected manuscripts and books, and established a collegial relationship with some of the most accomplished men of the age. He met Gerard Mercator, the renowned mapmaker; Gemma Frisius, the great cosmographer and adviser to Emperor Charles V; Pedro Nunes, the cosmographer royal of Portugal and professor of mathematics at Coimbra; and Oronce Fine, professor of mathematics at the Collège Royal. In a subsequent journey, he visited Antwerp, where he was received by the famed cartographer Abraham Ortelius, and Paris, where he lectured on Euclid at the university to standing-room-only crowds. Dee's continental tours were abundantly fruitful. From Frisius, who had recently pioneered the use of triangulation in land surveying, Dee learned the use of the cross-staff, astronomical rings, and other instruments; from Mercator, the science of mapmaking in its most advanced form. Mercator at the time was working on a series of globes and maps that were revolutionary in design. Whereas medieval charts had depicted the world as a disc or semicircle with three continents "divided by the Mediterranean, with Asia at the top, Europe to the left, Africa to the right, and Jerusalem in the center," Mercator incorporated most of the recent discoveries made by explorers, and mathematically projected the curved surface of the Earth into a rectangular plane.

Dee returned to England determined (in a way that was still possible in those days) to master all that was then known about astronomy, astrology, mathematics, algebra, geography, navigation, alchemy, and the kabbalah after the fashion of his idols, Roger Bacon and Albert the Great. But it was the stars that he worshipped most of all. Each clear night, he tells us, under the canopy of heaven, he set up his quadrant or cross-staff, and made "observations (very many to the hour and minute) of the heavenly influences and operations actual . . . Of which sort I made some thousands in the following years." For a time, he served as tutor in the household of the Earl of Pembroke, "a mad fighting fellow" who "could neither read nor write, and used a stamp to make his mark." Subsequently, he joined the household of the Duke of Northumberland, who then held power in young King Edward VI's name.

In those hectic times, Dee was a rare resource, and so it was inevitable that his skills would be coveted by the revolving powers of the throne. He was a presence at court during the brief reign of Edward, was retained by Mary Tudor (despite a charge of sorcery against him) after Edward's death and went on to serve Elizabeth I. Under her Dee enjoyed considerable royal favor, and could count among his patrons a number of powerful men—William Cecil, Lord Burghley; the Earl of Leicester; Sir Christopher Hatton; and Sir Francis Walsingham.

He was deeply grateful. Though Dee's father had prospered as a customs official in London under Henry VIII, Queen Mary had stripped him of his worldly gains. That had left Dee without much of an inheritance—or "due maintenance," as he put it in a letter to Lord Burghley—but Elizabeth came swiftly to his aid. She had known him before her accession, when for three months they were prisoners together at Hampton Court, and remained grateful for his efforts then to give her heart. He had cast her horoscope in those dark days, and had assured her that all would come right. When she came to power, it was Dee who determined the most propitious date and time for her coronation, and predicted, against the common expectation, that she would have a long and glorious reign. He also acted as her "intelligencer" or secret agent, undertook missions for her abroad, assembled descriptions of newly discovered lands for her edification, worked on codes for the use of her secret service, and in numerous other ways made himself useful to her rule. It was Dee's curious custom, not incidentally, to sign his dispatches "007," with "the top of the 7 extended over the zeros to represent the idea that he was the queen's eyes." (The novelist Ian Fleming later borrowed this cipher for his hero James Bond.) Once, when an image of the queen was found under a tree in Lincoln's Inn Fields pierced through with pig's bristles like a voodoo doll, Dee was asked by the Privy Council "to counteract any harm intended against her" by astrological or magical means.

In an astonishing demonstration of his range, in other work Dee explained the appearance of a new star in 1572; made calculations to facilitate the adoption of the Gregorian calendar in England; probed magnetic theory, which (in anticipation of Newton) he joined to the science of optics; backed the heliocentric theory of Copernicus and helped to make it "acceptable science" in England; wrote on the ebb and flow of the tides; hypothesized (before Galileo) that two bodies of unequal weight fall to the ground at the same speed; and introduced Euclidian geometry and classical architectural doctrine to England. He taught chemistry to Sir Philip Sidney and Sir Edward Dyer, and seamanship to Sir Humphrey Gilbert; gave indispensable advice to such great navigators as Richard Chancellor, John Hawkins, Stephen Borough, Martin Frobisher, and Sebastian Cabot, in their explorations of new lands and routes to China or Cathay. He also coined the phrase "the British Empire" to embody the hope and expectation of the queen's far-flung commercial aims. In 1575, in a work drafted for her benefit, Dee proposed that England develop a substantial navy as the "Master Key" to her expansion, and foresaw that she would one day challenge Spain, colonize North America, and found an empire overseas.

Dee was also a devoted bibliophile. Under Queen Mary he had advocated the founding of a royal or national library and when she died, and Protestant fanatics ransacked monastic libraries and other archives to expunge any trace of "popery" and root out all "Catholic" texts, Dee, though a Protestant, intervened and salvaged what he could. In time his library at his house at Mortlake in Surrey—a rambling place to which he kept adding rooms—became one of the largest and most diverse in the world. It included works on scientific and nonscientific subjects, religion, philosophy, anatomy, medicine, music, architecture, geography, navigation, literature, history, and the occult. He had works by Ficino, Cardano (whom Dee had met), by Pico and other Renaissance humanists, as well as writings of the Church fathers and Hebrew scholars, and owned tracts by Luther and Calvin, numerous copies of the Bible, and even a copy of the Koran.

In addition to his library, Dee assembled a notable collection of scientific instruments—celestial and terrestrial globes, quadrants, crossstaffs, a sea compass, and even a "watch-clock," or portable timepiece, "accurate to the second"; maintained an alchemical laboratory "with five or six bubbling stills"; archives related to his antiquarian studies; and a number of special collections, including Welsh and Irish records, genealogies, and ancient seals. There was also a dark room where he once entertained a neighbor's child by showing her "the image of a solar eclipse projected through a pinhole."

The queen and her Privy Council visited Dee at Mortlake on more than one occasion, as did other distinguished visitors, including most of the leading figures of the English Renaissance. As his reputation grew, Dee "enjoyed almost universal esteem," and in the course of his life a number of potentates, including Ivan the Terrible, Boris Godunov, and Emperor Charles V, offered him large annuities to come to their courts. Years later, Dee's son, Arthur, also an astrologer, enjoyed a successful career in Moscow as physician to Tsar Mikhail I.

Yet the notion that Dee was some kind of sorcerer also attached to his name. This had been so from the time he was a student at Trinity College and had devised a flying machine that had baffled the audience at a production of Aristophanes' *Peace*. Later, he "performed 'marveilous Actes and Feates' that popular opinion ascribed to diabolical powers, even though they were 'Naturally, Mathematically, and Mechanically, wrought and contrived.' " Yet this was a time when magic and mathematics were confused—the very word "calculing" being used to describe them both. Indeed, during the iconoclastic rampage of the Protestant reformers under Edward VI, a mathematical diagram had been enough to condemn a manuscript as a "diabolical" text. John Aubrey, many years later, wrote of "those dark times [when] Astrologer, Mathematician, and Conjurer were accounted the same." He recalled that Thomas Allen, the Oxford mathematician and astrologer to the Earl of Leicester, had been thought a fiend simply because he had some mathematical instruments and glass vials in his room. Allen's personal servant didn't help matters much, since he used to say (to bait the credulous) that in his master's house he sometimes "met the Spirits coming up the stairs like Bees." Dee, however, enjoyed the queen's special protection—"great security," as she told him, "against any of her kingdome"—to pursue his studies as he chose.

At times he tested the limits of what she might allow. For six years in Europe (from 1583) he dabbled in alchemy and scrying, and with a self-professed medium by the name of Edward Kelley, an Aztec obsidian mirror, and a crystal ball, tried to summon spirits and subject them to his power. For such dark doings he was expelled from Prague by the Emperor Rudolph II (an alchemist himself) and in 1589 summoned home by the queen.

Dee's own astrological theory was eclectic, and drawn from all sources, Arab as well as Roman and Greek. He regarded astrological symbols as a universal language—"characters imbued with immortal life"—and astrology as "an Art Mathematical which demonstrateth reasonably the operations and effects of the natural beams of light and secret influence of the Stars and Planets in every element and elemental body at all times in any Horizon assigned." In some ways he straddled two ages. A Pythagorean by inclination, perhaps, he was an empirical scientist at heart, and seems never to have been sure himself where the heart of astrology lay. The 1st-century B.C. Roman stoic Geminus of Rhodes had observed that "a star is often a sign of certain kind of weather and not its cause, just as a beacon is a symbol but not the cause of war. People think," he wrote, "that Sirius produces the heat of the dog days, whereas Sirius merely marks the season" when the Sun's heat is at its height. The positions of the planets at the time of birth, in this view, merely indicate the play of the various forces that are at work at that moment, so that the child born then is an image of an image—a part of the cosmic forces of the moment rather than a being shaped directly by their power. That was perhaps the more traditional (or classic) idea of what a celestial pattern meant, though many had also held that a direct influence of some sort—by rays of light, and so on—was involved. Still others were convinced that astrology was really "a Pythagorean art." Its basis, as Robert Zoller explains, "is not the physical movements of the planets, or some kind of 'radiation' from the stars, but the esoteric nature of geometry and number." In that sense, it worked fundamentally with degrees and "arithmetically derived points" on the ecliptic, as well as the so-called Arabic Parts.

An impressive figure by any standard, Dee in his person was tall and slender, wore a robe "like an artist's gown," with hanging sleeves and a slit, and had a rosy complexion and a beard as "white as milk." His own horoscope, drawn in his hand, survives and gives his birth as July 13, 1527, at 4:02 P.M., in or near London, 51 degrees, 32 minutes north latitude. That placed Sagittarius (4 degrees 54 minutes) on his ascendant conjunct Antares, a powerful fixed star. Despite his prominence, Dee lived out his last years under a cloud. Under James I, he was forced from his modest living as warden of Manchester College (bestowed upon him by Queen Elizabeth in 1596), and returned to the family home at Mortlake, where he died in poverty in December 1608.

James I was a somewhat superstitious monarch, and his belief in the enmity of the devil was uncommonly fierce. He sometimes blamed evil spirits for the misfortunes of his life, and was convinced, for example, after a ship he was on was nearly wrecked in the North Sea, that witches had cast black cats into the waves to create the storm. He didn't trouble his head much with astrology, but in other respects, he was not unlike Pope Urban VIII, who did. Although Urban's own election as pontiff had occurred at an auspicious hour—when the Sun was conjunct Jupiter in Leo, the strongest of the leadership signs—heavenly omens still haunted his daily rule. Any notice or sign of divine displeasure—in the form of comets, occultations, "guest stars," or odd appearances of the Sun or Moon—alarmed him greatly; and on one occasion, terrified of an approaching solar eclipse, he released a Dominican magus, Tommaso Campanella, from a Neapolitan jail, and in the papal apartments in the Vatican created with his help an artificial universe out of crystal in which they attempted to reproduce favorable configurations of the planets and stars. This sympathetic magic saved him, as he thought, and gave him such assurance at the helm that, ironically enough, when an errant astrologer predicted his death in 1631, he condemned the art as a whole. Even so, the prediction was accepted widely, and "a considerable number of foreign cardinals journeyed to Rome in the confident expectation of electing a new Pope."

Chapter 7

I N AN ORDINARY SENSE, the sky is a celestial clock, and we tell ordinary time by it—measuring out our lives in years, months, and days by the motions of the Sun and Moon. "What purpose, then," asked Tycho Brahe, "do the five planets going round in other orbits serve? . . . If the celestial bodies are placed by God where they stand in their signs, they must necessarily have a meaning especially for mankind, on whose behalf they were chiefly created."

Astrology and astronomy were originally one. All of the great astrologers of antiquity (Hipparchus, Ptolemy, Manilius, Firmicus Maternus, and so on) were also capable astronomers, while many of those responsible for the greatest advances in astronomical knowledge—including Nicolaus Copernicus, Brahe, Johannes Kepler, Galileo Galilei, and Isaac Newton—had some contact with astrology. Copernicus established the theory that the Earth rotates daily on its axis and the planets revolve in orbits around the Sun; Brahe vastly improved the art of celestial observation; Kepler discovered the laws governing the elliptical orbits of the planets; and Newton identified the law of universal gravitation, based in part on Kepler's work. Some of the pioneering work done by Brahe and Kepler—for example, on planetary orbits—was undertaken to supply accurate data for astrological charts. None of the foremost lights of early modern astronomy, in truth, was immune from astrology's allure. On the contrary, the occult and mystical yearnings of Copernicus, Brahe, Kepler, and Galileo helped to inspire their scientific work.

We do not know exactly what Copernicus thought of astrology, but he was a sometime student of the art, as we know from annotations in his own hand in his copy of Ptolemy's *Tetrabiblos*. He also owned the works of the Arabic astrologer Haly Abenragel (or Albohazen), which he prized; pored over the work of Regiomontanus, court astrologer to the king of Hungary and to Cardinal Bessarion in Rome; and entrusted the publication of his revolutionary work to the astrologer Georg Joachim Rheticus, whose *Narratio Prima* of 1540 was the earliest printed account of the new astronomy. In his *Narratio,* Rheticus included observations on the course of empires as linked to the stars, attacked Pico della Mirandola for attacking astrology, and held that the overall account of celestial phenomena in Copernicus bolstered astrological ideas. Thus was the new astronomy brought forth into the world.

Copernicus himself was many things—a doctor, canon lawyer, painter, numismatist, and cathedral official, as well as an astronomer. But it was Neoplatonic Sun worship, not mathematical calculation, that inspired his conception of the universe as having a heliocentric design. "In the middle of all sits the Sun enthroned," he wrote. "In this most beautiful temple could we place this luminary in any better position from which he can illuminate the whole at once? He is rightly called the Lamp, the Mind, the Ruler of the Universe; Hermes Trismegistus names him the Visible God, Sophocles' Electra calls him the All-Seeing. So the Sun sits as upon a royal throne ruling his children, the planets, which circle around him." Copernicus was completely captivated by what he called "the choral dance of the stars," and it was from reading the astrologer Manilius that he first encountered the idea that the Earth turned on an axis and revolved about the Sun.

Copernicus spent most of his life as a reclusive canon at the cathedral in the East Prussian town of Frauenburg, now Frombork, Poland. Brahe was quite a different personality and the greatest observational astronomer since Hipparchus, the 2nd-century B.C. Greek who developed trigonometry and catalogued over one thousand stars. A Falstaffian "giant of a man with a metal-alloy nose" (the original having been sliced off in a duel), Brahe was born at Knudstrup, Denmark, a child of wealth and rank. His father was the governor of Hälsingborg Castle and a privy councilor of the realm; his mother, a countess and lady-in-waiting to the queen. But it was his uncle, the stern lord of Tostrup Castle, who took him under his wing. Under his guidance, Brahe studied mathematics at the University of Copenhagen, law at

Leipzig, and prepared for a government career. Astronomy and astrology also drew him in. In 1560, when he was fourteen, a total eclipse of the Sun first awakened his interest in the stars, for the prediction of the event, as well as the eclipse itself, seemed to him "something divine." Three years later, for the same reason, a Jupiter/Saturn conjunction held him in thrall. His uncle tried to quash his new enthusiasm, and assigned him a tutor who doubled as a spy. But Brahe managed to obtain some books and even a few astronomical instruments on the sly, as well as a tiny globe the size of an orange on which he plotted the major constellations, using an ephemeris as a guide. Brahe's uncle eventually made a virtue of necessity, acquiesced, gave him support, and Brahe's great career began.

Brahe studied astronomy with Bartholomew Schultz at Leipzig; read Ptolemy, Copernicus, Regiomontanus, and Sacrobosco; developed new and improved instruments to assist his observations, including a huge quadrant, which he perched outside Leipzig on a hill; and began constructing an observatory in Denmark's Herrevad Abbey, complete with an alchemical still. In 1572, a new star or nova appeared in the constellation of Cassiopeia, and Brahe was the first to catch sight of it, one evening while on his way to catch some fish. The sun had just set when, looking upward, with his incomparably trained eye, he saw a "bright star which appeared as distinct as Venus" in the sky. "It surpassed all the other stars in its brilliance," he wrote, "and was shining almost directly above my head; and since I had, from boyhood, known all the stars of the heavens perfectly, it was quite evident to me that there had never been any star in that place of the sky, even the smallest, to say nothing of a star so conspicuous and bright as this." The nova was so distinct that for sixteen months it was visible to the naked eye and for two weeks at the beginning could even be seen at midday. At length, it began to fade and change color, from bright white to yellow to a faintly reddish hue, until it vanished forever into eternal night. But from his careful study of it, Brahe proved that the nova belonged to the supposedly immutable realm of fixed stars. This upset accepted doctrine, based on Aristotelian theory, and coincided with the speculations of Copernicus shaking the old astronomy to its roots.

The nova inspired Brahe to dedicate the rest of his life to astronomy, while in a related undertaking he resolved to revise and improve the planetary tables then in general use. These were so woefully inadequate that by their outdated calculations the Jupiter-Saturn conjunction he had witnessed in 1563 had been off by a month. That annoyed him immensely, and he had since embarked on his own meticulous observations to correct them. At about this time, he also began to set forth his own, unique cosmological scheme—an ingenious compromise between the Copernican and Ptolemaic systems—which held that while the Sun and Moon revolved around the Earth, the other planets revolved around the Sun. This seemed perfectly plausible to contemporary astronomers and was generally embraced.

By 1574, at the age of twenty-eight, Brahe's reputation was established. His revolutionary findings on the nova had been published, his theories were widely discussed, and his work on the planetary tables had begun to generate great expectations among astronomers and astrologers alike. The Danish king Frederick II considered him a national treasure and declared that no facility was too expensive or magnificent to house his work. Brahe agreed and built Uraniborg, a semi-underground observatory on the island of Hven (now Ven) in the Øresund strait, and christened it Stellberg, or the City of the Stars. On August 8, 1576, when the Sun and Jupiter were rising, he laid its foundations, presaging fame to the project; combed Europe for the finest astronomical instruments; and on rotating platforms atop the turrets of a magnificent fortress (furnished with an intercom system, flush toilets, and other advanced facilities) mounted quadrants and armillaries of his own construction and design. The crown jewel of the observatory was "a gleaming brass celestial globe, five feet in diameter, on which a thousand stars were inscribed, one by one, as Tycho remapped the visible sky."

At Uraniborg—named for Urania, the muse of astronomy—Brahe and his battery of assistants kept to a grueling schedule, working from dusk to dawn, with lengthy afternoon consults in which Brahe often sternly critiqued their work. As the days, months, and years passed, they compiled "a matchless series of numbers, dates, and times marching up and down unlined foolscap, the stiff pages piling up on splintery boards in observatory rooms overlooking the sea." It was unremitting and exhausting work. Kepler would later marvel that the observations—all made before the invention of the telescope—were accurate to within a minute of arc.

But in other respects, Brahe lived in splendor like a Renaissance prince. The observatory itself resembled a castle, and his private apartments were luxuriously arrayed, with hand-carved furniture, rich tapestries, and other emblems of his wealth. He set up his own printing establishment to produce and bind his manuscripts, and enlisted Dutch and Italian artists to embellish the décor of his astronomy rooms. Massive stone gates guarded the east and west portals of the ramparts, where mastiffs were kenneled to warn of arriving guests; and the spacious grounds included "private game preserves, sixty artificial fishponds, extensive gardens and herbariums, and an arboretum with three hundred species of trees." Brahe's evenings were given over to hard-drinking banquets, during which, surrounded by a retinue of retainers, he liked to hold forth on the importance of his work. He also maintained (in cap and bells) a dwarf named Jeppe, his own court jester or fool. Not unlike a prince, Brahe was surrounded by admiring scholars. He received from across Europe, among them curious nobles such as James VI of Scotland, later England's James I.

During all this time, Brahe also served as court astrologer to the Danish king. Brahe's devotion to the art was ardent. He was convinced of its medical value; held that the Bible did not condemn it but only idolatry masquerading as a reverence for the stars; rejected the argument Augustine had made against it based on the divergent fate of twins (Brahe was a twin himself, having lost his brother at birth); and in a famous lecture on astronomy delivered at the University of Copenhagen in September 1574, he memorably declared, "We cannot deny the influence of the stars without disbelieving in the wisdom of God." He also had no doubt that "new stars" (novas), comets, eclipses, conjunctions, and other heavenly developments signified momentous terrestrial events. When the nova of 1572 appeared, "all Europe," recalled one writer, "stood at gaze, vehemently expecting more strange and terrible alterations than ever happened since the world began." Brahe, according to his own Protestant bias as an astrologer, thought it signified the end of the hegemony of the Roman Catholic Church.

He was also convinced that comets foretold tumult, war, and the fall of those on high. "Such unnatural births in the heavens," he wrote, "always have had something great to deliver to this lower world." The "blazing star" or comet that suddenly materialized in mid-November 1577 he viewed as an oracle of God. It first appeared "just above the

Tropic of Capricorn," but "within three days had moved above the head of Sagittarius, glowing with a clear, white light. It was surrounded by a discolored mane, and had a long tail curved like a Turkish sword." Its scimitar shape fueled fears of a Turkish invasion, though its head was to the west, "suggesting that its influence might be in that direction, toward the New World." Brahe, however, noted that "its tail pointed to the northeast, where it would 'spew its venom' over the Muscovites and Tartars." (This, not incidentally, was the same comet that had thrown the English court into a state of such confusion and from which Queen Elizabeth had been urged to avert her eyes.)

When it came to eclipses, Brahe also followed the interpretive rules that Ptolemy had laid down long before. Shakespeare would write: "When beggars die, there are no comets seen." So too, as Brahe explained, with eclipses: For "generally speaking, kings and princes are more affected by eclipses than private people (as I have observed myself), because the sun and the moon are the princes among the planets. The effect will last as many months as the eclipse lasted hours, and the beginning of it depends on the moon's distance from the horizon at the commencement of the eclipse." He cited several recent examples: the lunar eclipse of April 3, 1558, after which Charles V had died; the solar eclipse of April 18, 1558, which did not begin to take effect till the end of the year, at which point Denmark's Christian III died; and shortly afterward, the lunar eclipse of November 7, 1565, near the Pleiades, that had augured torrential rain. The lunar eclipse of October 28, 1566, close to Orion, had also produced an extremely wet winter, as Brahe predicted it would.

Weather prediction in fact was of great interest to him, as it was to Kepler, and he believed, as farmers down through the ages have believed, that the heliacal rising and setting of certain stars produced rain, wind, and other atmospheric changes, particularly when the planets joined their effect to that of the stars. He thought that the Jupiter-Saturn conjunction of 1563, which took place near Praesaepe, a coarse star cluster in Cancer that Ptolemy considered pestilential, had signified the coming of the plague.

Yet Brahe also invariably got the *astronomy* of the phenomena right. He had showed that the nova was a real star by its lack of parallax, or displacement, when viewed from two different angles or points; and that contrary to tradition or Aristotelian belief, the comet of 1577 was not a sublunary exhalation from the Earth's atmosphere, but more distant from the Earth than the Moon. In his day, at least, it was still possible to be an empirical astronomer and a convinced astrologer at the same time.

As court astrologer, Brahe made annual predictions for the Danish royal family, which he recorded in large quarto volumes handsomely bound in green velvet with gilt edging along the spine. When Frederick II's son Christian IV was born, Brahe was called upon to do a character analysis and life forecast based on his chart. In proceeding, he considered the two principal and time-honored systems for predicting future conditions and timing future events. One, known as primary directions, considered a degree of right ascension as equivalent to a year of life; the other, known as secondary directions (or progressions), considered a day in the life as equal to a year. In the first, the number of degrees by which a planet (or a point, like the ascendant) advanced through the horoscope corresponded to the native's age; in the second, the calculation was made by days. For example, the position of the planets on the thirtieth day after birth, as well as the aspects they might form with natal planets, represented conditions during the thirtieth year of the life. The first was based on the apparent motions of the planets (moving at the same rate of speed, in the same direction) as a result of the Earth's rotation on its axis; the second, on the actual, and varied, orbital motions of the planets along the ecliptic. (In short, secondary progressions assumed a meaningful relationship between the Earth's daily rotation on its axis and its annual orbit around the Sun.) In practice, the two systems were closely related, and often yielded the same result.

Kepler favored secondary progressions. Brahe followed primary directions, and therefore cautioned the king that an error of four minutes in the stated time of birth would make a difference of one degree, or an error of one year. His forecast for the newborn prince was this: that the years of his infancy would pass without danger, since Venus was favorably placed in the ninth house, although an opposition between Mercury and the ascendant in the second year indicated an illness, but it wouldn't amount to much. When the boy was twelve, however, the ascendant would be in a square to Saturn, indicating some risk; and there would be danger again at the age of twenty-nine, when the Sun would be in a square to Saturn, and Saturn and Venus opposed. At fifty-six, the prince might die, Brahe thought, when Venus, the ruler of the chart, was in the eighth house, and the Sun, by direction, overtook Mars. But if he survived, he would have a happy old age.

Having first determined the probable length of life, he proceeded to describe the nature of the man. By planetary position, strength, and aspect, he concluded that Venus would make the prince "pleasant, comely and voluptuous, fond of music and the arts"; Mars, "brave and warlike"; Mercury, "clever and astute." He would be naturally cheerful, "generous and ambitious," have "good luck in his undertakings," but "indulge too much in sensual pleasure" and engage in struggles with the Church. A number of favorable fixed stars of importance, such as the Twins in the tenth house, Spica in the first, and the Southern Crown on the cusp of the fourth, would give the prince honors and dignities in abundance; wealth, in turn, was assured by the conjunction of the Part of Fortune with his Sun. However, Scorpio on the cusp of the second house showed that his wealth would be gained by war.

As for the prince's marriage, he could look forward to a good deal of strife, as indicated by afflicted aspects between Venus and Mars and Venus and Saturn, especially with his sixth-house Moon; he was also not likely to have many children, since Saturn was lord of the fifth house and in a sterile sign. On the other hand, he would have a number of extramarital affairs. As for his friends, they would be "solar people," such as kings and princes, because the Sun ruled the eleventh house of friends; his enemies, on the other hand, would be churchmen and men of war. How so? Because Jupiter was unluckily placed in the twelfth house, and Mercury, ruling the twelfth, was in the seventh, where by mutual reception it assumed the nature of Mars. Mercury, in square to Saturn, made captivity or exile a risk, but the prince would ultimately defeat his enemies, because Venus, his ruling planet, was at the apogee of its eccentric and much higher than Mars in the sky. There was also nothing to indicate a violent death, so Brahe predicted the prince would die from some illness brought on by self-indulgence, according to his pleasure-loving ways.

This is not the place to rehearse Christian's life, but anyone acquainted with its course can judge the aptness with which Brahe framed its career.

Frederick II died in 1588, and under his son, Christian IV, the very one whose chart this was, Brahe's influence waned. When the new king came of age, Brahe—who for two decades had been lord of his own little fiefdom—was soon stripped of royal support. At first the two had seemed to get along. In 1592, the future king, then fifteen, had come to Uraniborg to pay his respects, and "after a pleasant day of sightseeing and amiable, wide-ranging conversation," Brahe had presented him with a mechanical brass globe, worked by cogwheels, that simulated the daily rotation of the sky. The young prince immediately returned the favor by presenting Brahe with a gold chain, engraved with his own likeness, from around his neck. But Christian had scant appreciation for Brahe's work, and other things in time drove a wedge between the two. Brahe, who was tempestuous by nature and not always deferential to the king, had quarreled with the tenants he subtended land to on his estate and had neglected to properly maintain a lighthouse entrusted to his care. More notably, perhaps, he had also failed in the upkeep of the Chapel of the Magi at Roskilde, where Christian's father was interred. The king seized on these lapses to reduce the flow of royal funds, which Brahe's self-importance simply could not brook. After several histrionic threats to leave the kingdom (which Christian pointedly ignored), Brahe departed Hven in 1597—with his printing press, furniture, library, paintings, and family members and attendants, and even Jeppe, his dwarf, in tow. After briefly stopping at Rostock and Hamburg, he settled in Bohemia, where he became imperial mathematician to the Emperor Rudolph II. Rudolph, who was a great patron of learning, could scarcely believe his luck and placed the castle of Banetek, northeast of Prague, at his disposal. There Kepler joined him on February 4, 1600, as his assistant, in an apprenticeship that lasted eighteen months.

For all his titanic bluster, Brahe could also be too obliging for his own good. One evening in the fall of 1601, he remained too long at a state dinner and politely allowed his own bladder to burst rather than excuse himself early so as not to insult his host. He lingered in a delirium for several days—exclaiming over and over again, "Let me not seem to have lived in vain!"—before dying on October 24, 1601. Kepler was by his side and noted, as Brahe drew his last breath, that the Moon was in opposition to Saturn and in square to Mars. In a fitting tribute from his adoptive land, Brahe received a state funeral in Prague and was buried in a suit of armor in Prague's Teyn Church. His once-splendid, now untended, observatory on Hven eventually fell into ruin, and in 1623 its building materials were sold for scrap.

Kepler succeeded Brahe as imperial mathematician, and by a wondrous happenstance of history, the multitudinous calculations Brahe had compiled were entrusted to the one person capable of construing them aright. Brahe had lacked the ability to extract a general theory from them. Yet all the data to support one was there. On their basis, Kepler discovered the three great laws of planetary motion governing the elliptical orbits of the planets, which in turn led to Newton's law of universal gravitation. No one today would contest Kepler's right to be called the father of modern astronomy; but such eminence would hardly have been predicted from the life.

∗   ∗   ∗

BORN IN THE SMALL TOWN of Weil der Stadt, Germany, on December 27, 1571, Kepler was nothing if not odd—bow-legged, myopic, afflicted by boils and other ailments. By his own description he resembled a small lap dog, with a fawning eagerness to please. His father had been a mercenary soldier and a drunk; his mother was a psychic who, according to Timothy Ferris, "had been raised by an aunt who was burned alive as a witch." For most of his twenties, Kepler struggled. He worked as a dishwasher, taught mathematics in a local high school in Graz, Austria, and tried to make a living casting horoscopes for pitifully small fees. "Yet this was the man," writes Ferris, "who would discern the architecture of the solar system and discover the phenomenological laws that govern the motions of the planets." Immanuel Kant would call him "the most acute thinker ever born."

Endowed with none of Brahe's advantages of wealth and rank, Kepler had been raised in the crowded, turbulent home of his paternal grandfather, along with a host of near relations who came and went with the erratic flow of transients in a boardinghouse. A number of them were bizarre, if not unstable, while his constantly wrangling parents were often at each other's throats. In 1576 the family moved to nearby Leonberg, where his father, briefly home from the wars, opened a tavern. When it predictably failed, he went off to fight in the Netherlands and was never seen again.

Despite the chaos of these years, Kepler received a solid education. At the age of eight he entered the German and Latin Schreibschule at

Leonberg, advanced to the lower seminary or monastic school, and in 1586 to the higher school at Maulbronn, where the historical Doctor Faustus had studied a half-century before. After a fumbling start—in grammar school most of the children got through Latin twice as fast as he did—his grades improved, and by the time he entered Maulbronn his capacities had begun to make themselves known. In 1588 he earned the academy's highest honors, and in 1589 won a coveted scholarship to the Protestant University of Tübingen, where he studied Latin, Greek, Hebrew, geometry, theology, philosophy, mathematics, astronomy, and music. His astronomy teacher was Michael Maestlin, a convinced Copernican and prominent at the time, who recognized Kepler as his ideal pupil and took him under his wing. In fairly short order, Kepler earned his B.A. and M.A. degrees, and excelled in all subjects—including Latin, which he would come to write superbly, both in prose and verse.

As a student, there wasn't much he couldn't do. He had a gift for poetry and a passion for riddles and acrostics. He wrote, as he put it, many compositions on "unusual subjects," such as "the sight of Atlantis through the clouds." He loved mathematics, philosophy, and music, and was obsessed with working out elaborate allegories "to the minutest detail." He seems to have enjoyed anything of an intellectual nature, "argued with men of every profession for the profit of his mind," and in exploring mathematics did so, he recalled, as if it were his own invention, and in the process rediscovered its first principle and inner workings for himself. Astronomy, too, held Kepler in thrall. A little book, called *Somnium* or *Dream,* published posthumously but drafted in part at this time, was the first modern science fiction story, about a space trip to the Moon. For all the diversity of his interests, there was no doubt where his passions lay. But he was not yet free to make his own career. By the terms of his scholastic stipend, he was obliged to prepare for the Church, and so in 1593 he gave himself up to theology, which he had also begun to teach. In 1594, however, destiny played its hand. He reversed course, abandoned all his prospects, and took a poorly paid position as a high school teacher of mathematics at the Lutheran seminary in Graz. There, as district mathematician and calendar maker, he supplemented his paltry salary by publishing astrological almanacs.

As a classroom teacher he did poorly at first, and interspersed his lectures with mumbling digressions punctuated by sudden revelations that only he could understand. For the most part they had a mystical cast. Inspired by the Pythagorean doctrine of the music of the spheres, he believed that the movement of the planets along their orbits produced an ethereal harmony, and that the intervals between these orbits had a mathematical relationship that corresponded to the frequencies of the tones of the musical scale. He developed this concept, and ultimately became convinced that each planet had its own specific melodic line. Shakespeare expressed the same general idea in the *Merchant of Venice:*

> Look how the floor of heaven
> Is thick inlaid with patines of bright gold.
> There's not the smallest orb which thou behold'st
> But in his motion like an angel sings,
> Still quiring to the young-eyed cherubims:
> Such harmony is in immortal souls.

On one occasion, Kepler interrupted his own lecture to announce that there was a correspondence between astronomical and musical intervals, based on number. A second epiphany was inspired by his contemplation of a diagram showing the geometrical pattern made by recurrent Jupiter-Saturn conjunctions, which made a "nesting" configuration of angular shapes. That prompted Kepler to exclaim suddenly that "the five Platonic solids, nesting one inside the other, specified the relative distances of the six planets in their orbits around" the Sun. "It happened on July 19th, 1595," he later recalled,

> as I was showing in my class how the great conjunctions of Saturn and Jupiter gradually pass from one trine [triplicity] to another, that I inscribed within a circle many triangles, or quasi-triangles, such that the end of one was the beginning of the next. In this manner a similar circle was outlined by the points where the lines of the triangle crossed . . . And then it struck me . . . The earth's orbit is the measure of all things: circumscribe around it a dodecahedron and the circle containing this will be Mars; circumscribe around Mars a tetrahedron, and the circle containing this will be Jupiter; circumscribe around Jupiter a cube, and the circle containing this will be Saturn.

Now inscribe within the Earth [i.e., the earth's orbit] an icosahedron, and the circle contained within it will be Venus; inscribe within Venus an octahedron, and the circle contained within it will be Mercury.

What had begun as an astrological thought became an astronomical one, though still mystical in form. His students, perhaps not privy to "the five Platonic solids," listened in baffled silence—understandably enough, for Kepler had given birth to a wholly new idea.

In the end, he was wrong. But he was also right, for he had started down the path that would eventually lead him to his goal. Two years later, in 1597, Kepler published his first important work, *The Mysterious Cosmos,* in which he explained his hypothesis and expounded some other novel notions. But it would not be until 1619, in a subsequent work entitled *The Harmony of the World,* that he set forth the idea that each planet corresponded to a note in the musical scale: the Sun to C, Saturn to D, Mercury to E, the Moon to F, Mars to G, Venus to A, and Jupiter to B. In a choral sense, he even described Saturn and Jupiter as the bass, Mars the tenor, Venus the contralto, and Mercury the soprano of the "celestial choir."

✳   ✳   ✳

KEPLER'S EARLY WORK as a theoretical astronomer was well received, and did much to win acceptance for the Copernican system, which he enthusiastically embraced. He had also published an astrological almanac for some years running, and achieved instant fame by his accurate prediction of two major events: the record-breaking cold of the winter of 1595–96, and the subsequent invasion of Austria by the Turks. As he wrote that year to his former astronomy teacher, Michael Maestlin: "So far the almanac's predictions are proving correct. There is an unheard-of cold in our land . . . As for the Turks, on January the first they devastated the whole country from Vienna to Neustadt, setting everything on fire." Over the next three decades, Kepler would publish eighteen other such almanacs, and was in the process of gathering them all into a single volume when he died.

✳   ✳   ✳

IN HIS PERSON, Kepler was a rather small man, with coal-black eyes, black hair, and a delicate but graceful build. He had been sickly as a child, and as an adult was often beset by fevers, stomach problems, sundry skin diseases, and "suffered from multiple vision in one eye." His diet was spare, even spartan, with a self-abnegating penchant for dry bread and pungent fare; his idea of exercise rusticly plain—a good hike over rough mountain paths or a bit of a scramble up a thicket-strewn slope. By his own account, he was somewhat abrasive, yet slavishly obliging, out of a frantic fear of poverty and debt.

His family preoccupied him, and when he was twenty-six he wrote a sketch of some of its members based on their astrological charts. He described his grandparents as insufferable; his mother, as "small, thin, swarthy, carping, and morose"; and his father, as vicious, rigid, ill-tempered, and "doomed to a bad end." "Saturn in the seventh house brought him to gunnery," Kepler wrote, "while the condition of his Mars magnified his malice. Jupiter combust [conjunct the Sun] on the cusp of the Descendant made him a pauper but gave him income from his wife." An aunt named Kunigund was poisoned ("her Moon could not have been worse"), and an uncle who was a Jesuit "wandered in extreme poverty through France and Italy," presented himself as a psychic, and led an "impure" life. If Kepler's judgments seem harsh, he was just as hard on himself, and unblinkingly acknowledged that he was cut from the same celestial cloth. He considered himself incredibly reckless ("due to Mars in square to Mercury trine Moon"); wasteful of his time, yet ever regretful of that waste; frugal to a fault; anxious to expiate any wrong he had done by some abject public confession; yet a reprobate at heart. His first sexual encounter (at age twenty-one) was a disaster. "On New Year's Eve I was offered union with a virgin," he tells us. "I achieved this with the greatest possible difficulty, experiencing acute bladder pains."

Kepler also had a younger brother named Heinrich, whose life history stood as a warning to him as to how bad things could go. At one time or another, when still a child, Heinrich had been badly burned, nearly drowned, and mauled by a wild dog. At the age of twelve, "he had run away from home when his father threatened to sell him," and in subsequent years "was a camp follower with the Hungarian army in the Turkish wars, a street singer, baker, nobleman's valet, beggar, regimental drummer, and halberdier. Throughout all this, he remained the hapless victim of one misadventure after another—always ill, sacked from every job, robbed by thieves, beaten up by highwaymen—

until he finally gave up, begged his way home to his mother, and hung on to her apron strings until he died at forty-two."

In Kepler's own childhood, two moments of familial warmth and illumination stood out. Both pointed him toward his career. In 1577, his mother had led him up to the top of a hill to point out a comet in the sky; three years later his father took him out one night to witness a lunar eclipse.

★   ★   ★

AT GRAZ, Kepler did his best to establish a normal life. He agreed to teach the poetry of Vergil in order to hold on to his job (after no one showed up for his astronomy class), and married and started a family. But nothing seemed to hold. The marriage was difficult, and several children died almost as soon as they were born. Then in 1600 the Counter-Reformation swept through his district and drove him from Graz. Tycho Brahe, who had read Kepler's work, and disagreed with his theories but admired the brilliance they displayed, took sensible advantage of his plight and invited him to Prague as his assistant, though on somewhat penurious terms. Kepler bargained a bit over salary, working conditions, and so forth, but in full appreciation of Brahe's stature and achievements inwardly leapt at the chance.

Under Brahe's direction, Kepler at first worked on the orbit of Mars, and in correlating the observational data with his own elliptical hypothesis filled nearly one thousand folio sheets with arithmetic. He also collaborated with Brahe on a voluminous new compilation of stellar and planetary tables, ultimately called the Rudolphine Tables (in honor of the Emperor Rudolph), which, upon their belated publication in 1628, became the accurate standard for the next hundred years. As Brahe's successor after 1601, Kepler also published a landmark work *(On the New Star in the Foot of the Serpent Bearer)* about an exploding star, brighter than Jupiter, that appeared in 1604—today called "Kepler's Star," and the last nova to be observed in the Milky Way; gave theoretical coherence at last to the data bequeathed to him, and in 1609 set forth in *The New Astronomy* his first two laws of planetary motion: (1) that a planet moves in an elliptical orbit with the Sun at one focus; and (2) that a planet sweeps out equal areas in equal times. He also published a work in support of Galileo on Jupiter's moons; began his treatise on the chronology of Christian history, in which he correctly pointed out that the Christian calendar was off by five years; and wrote the first treatise to explain how a telescope works.

But industry and genius gave no guarantees. In 1612, upon the abdication of Rudolph II (by reason of insanity) and soon after the death of his wife and a seven-year-old son, Kepler was obliged to leave Prague for Linz, Austria, where for the next sixteen years he once more served as a "district mathematician" while teaching in a local school. Rudolph's successors, Matthias and Ferdinand II, contributed a token stipend, but all in all he was just scraping by. In the following year he remarried, and—such was his inventive mind—based on a casual observation about the capacity of some beer kegs at his wedding devised a method of finding the volumes of solids of revolution that set a precedent for infinitesimal calculus. In subsequent years, he offered the first mathematical treatment of the close packing of equal spheres (leading to an explanation of the shape of the cells of a honeycomb); gave the first proof—based on Book 5 of Euclid's *Elements*—of how logarithms work; wrote a textbook on Copernican astronomy (while he was also trying to get his mother released from prison as a witch); and in *The Harmony of the World* set forth his third great law of planetary motion—that for different planets, the square of the period of their orbit is proportional to the cube of their mean distance from the Sun.

The principle of gravity itself was also plainly within his grasp. In fact, he seems to have grasped it—though for some reason he failed to apply it to his planetary laws. In his introduction to *The New Astronomy,* we read: "If two stones were placed anywhere in space near to each other, and outside the reach of force of a third cognate body, then they would come together, after the manner of magnetic bodies, at an intermediate point, each approaching the other in proportion to the other's mass." For example, "if the Earth and the Moon were not restrained, each in its own orbit, the Earth would move up toward the Moon and the Moon would come down toward the Earth and they would join."

During all this time, Kepler's astrological speculations had continued apace. In a letter written in 1599 (when he was twenty-eight) to Johann Herwart, his friend and the chancellor of Bohemia, Kepler ruminated about the influence of planetary patterns upon a man's character at birth. "It influences a human being," he wrote, "as long as he lives . . . the way the slings a peasant wraps around a pumpkin do, for these do not make the pumpkin grow, but they determine its shape."

In his own case, he thought his Sun-Saturn trine had not only determined his disposition, inclinations, and taste, but even his physical form. He fully subscribed to the great-conjunction theory of historical epochs (in particular, those marked by a triplicity shift); thought a Mars-Saturn conjunction conduced to civil strife; and that planetary aspects had an effect on weather conditions of all kinds: for example, that Venus conjunct Mars brought turbulence, or that Jupiter and Mars conjunct in July brought clear skies. Just as the Moon affected the tides of the oceans, so it pulsed the bodily humors, and any physician worth his salt, he thought, kept a weather eye on the Moon. "The whole business of crises [in illness]," he wrote, "depends upon the return of the Moon and its configurations with the planets, and it is vain to seek explanation for it anywhere else."

It is sometimes said of Kepler that his interest in astrology faded with age. But the opposite is true. In a letter written in 1601, the prime of his professional life, he declared that "the belief in the effect of the constellations derives in the first place from experience, which is so convincing that it can be denied only by those who have not examined it." And he held to that conviction to the end. In astronomy books, this side of Kepler tends to be slighted; but for the historian, at least, that is a fool's game. For it also falsifies the history of ideas. For example, Kepler believed that the soul of each individual contains a geometric blueprint of the zodiac at birth. The "intermediary light reflected at right angles off the moving planets strike the triangles, squares, and hexagons inscribed on the soul, thus forming a direct and intimate connection between the planetary movements and man." Just as the astrological musings of Hipparchus led to the development of trigonometry—and those of John Napier, Baron of Merchiston, to logarithms, which he first described in 1614—so Kepler's theory of refracted planetary light, in its physical sense, led to modern optics, for he was the first to postulate the ray theory of light to explain how vision works. He also gave the first correct explanation of the structure of the human eye.

His devotion to astrology throughout his life is not in doubt. In 1601, for example, when he was thirty, Kepler wrote a book, *On the Fundamentals of Astrology,* by which he hoped to reform some of the tenets of the art. He deemphasized the division of the horoscope into twelve equal parts, or houses; raised questions about the traditional rulership of signs; but extolled the planetary aspects as having fundamental force. "Regard this as certain," he wrote to a friend in 1602: "Mars never crosses my path without involving me in disputes and putting me in a quarrelsome mood." He also introduced three new planetary aspects—the quintile (72 degrees), the biquintile (144 degrees), and the sesquiquadrate (135 degrees)—based on the division of the circle by five.

Two years later, in his *Report on the Fiery Triplicity,* which concerned great-conjunction theory, he wrote, "Philosophy, and therefore genuine astrology, is a testimony of God's works, and is therefore holy. It is by no means a frivolous thing. And I, for my part, do not wish to dishonor it." Three years later, in his treatise *On the New Star in the Foot of the Serpent Bearer* (1606), he wrote: "Although the sky is constantly making all kinds of possible changes—nonetheless the character of that configuration is preserved, namely the configuration that was in the heavens when the life of the human being was ignited at birth and was, so to speak, poured into the mold . . . Somehow the images of celestial things are stamped upon the interior of the human being, by some hidden method of absorption . . . The character of the sky flowed into us at birth." Again, in 1610, in *The Third Man Intervening* (a work in which he also scoffs at unbending skeptics), he explained how a progressed horoscope can predict future developments:

A human being's nature, at the beginning of life, receives not only an instantaneous image of the sky, but also its motion, as it appears down here on earth, for several successive days; and derives from this motion the way in which it will discharge this or that humor; and the time at which this nature will, very accurately, time these is determined by the directions based on the first few days of life. This is a truly marvelous thing, and is like an image or outflowing of the natural proportions of a day to a year.

He then described, in a technical way, how this was done:

The sun should be directed on the ecliptic according to its diurnal motion; the Medium Coeli [Midheaven, or tenth-house cusp] by right ascension; the ascendant by oblique ascension; and always, referring to the birth hour, to the right ascension of the directed place of the sun;

thus the horoscope is always to be recast anew. The moon must also be directed on the ecliptic, consistent with the diurnal motion of the sun, but not the Part of Fortune because it is not a celestial body nor a part of the sky. The remaining planets are not to be directed either, because they do not have, of themselves, anything in common with the motion of the earth.

In explaining how the images of celestial things are stamped upon an infant, he wrote: "This character [of the heavens] is not received in the body, which is much too ungainly for that, but rather in the nature of the soul itself, which is like a point." Ten years later, in *The Harmony of the World* (the same work in which he set forth his first two laws of planetary motion), published when he was forty-eight, he continued this very same thought: "The natural soul of the human being is no greater than a single point, and on this point the form and the character of the whole sky is potentially impressed." How so? "Inasmuch as the soul bears within itself the idea of the zodiac, or rather of its center, it also feels which planet stands at which time under which degree of the zodiac, and measures the angles of the rays that meet on the earth." In a progressed horoscope, therefore, "the new-born babe is marked for life by the pattern of the stars at the moment it comes into the world, and unconsciously remembers it, and remains sensitive to the return of configurations [that is, the repetition of certain patterns] of a similar kind."

In the same work, Kepler also firmly expressed his belief in astral heredity.

I believe that at the birth of children, and particularly of the first-born, the planets as well as the Ascendant and Midheaven are usually in the same zodiacal degrees, or in square or opposition to those areas, as at the birth of the father, or (especially) of the mother. I also believe that the same or similar aspects occur . . . I was born when the Moon was short of an opposition to the Sun by forty degrees. With my first-born the Moon was short by the same amount of a conjunction with the Sun. With the second child the Moon had passed the opposition to the Sun by the same number of degrees. With the fourth child the Moon was 38 degrees from the opposition to the Sun. With the third child it was not much different, for the Moon was at a separation of forty degrees from the Sun plus one day of lunar motion: the birth was expected the previous day. I pass over further examples that agree with these.

Such convictions only gathered force with time. In a 1598 letter to a skeptical friend, he once remarked:

> Look at the relationship between births. You were born under a conjunction of the Sun and Mercury and so was your son. Mercury was behind the Sun at the moment of both your births. You were born at the trine of Saturn with the Moon; your son was almost born at the Moon-Trine sextile. There was virtually a Saturn-Sun trine at both your births. You were born under Saturn, while your son was born under the Sun and Mercury. Venus was in opposition at the time of your birth and at the time of your son's birth.

Since, as we know, genetics provides an analogous pattern or system of inheritable traits, one might begin to wonder if the universe of man is not, after all, a hall of mirrors, in which one thing reflects another as well as itself.

For Kepler, astrology—however debased it might be in common practice—was a divine form of natural philosophy. He chastised its incompetent adherents for allowing it to become "the wanton little daughter of astronomy, selling herself to any and every client willing to pay," and thought its mish-mash of half-baked doctrine, as popularly expressed, had made such a mess of its teaching that it was almost impossible "to separate the gems from the slag." In his own attempt to do so, he compared himself to "a hard-working hen scratching for a grain of corn." Yet "no one should regard it as impossible that from the follies and blasphemies of astrologers, there should emerge a sound and useful body of knowledge, or that from mud and dirt there should come oysters, mussels, eels, good food." Scientists would greatly err, Kepler warned, if "while justly rejecting the stargazers' superstitions, they threw out the baby with the bath." By "superstition" he did not mean its power to describe and predict, but the silly generalizations that blighted it, akin to the Sun-sign readings of today. On the contrary, he considered it worthy of considerable study—as his own writings attest—for, as he wrote, a person "can devote as much industry to

astrology as a botanist to herbal medicine . . . Indeed, that much industry is required. And every day there is some new proof." No one
knows how many horoscopes Kepler cast, but at least eight hundred
are still extant from among the thousands he may have prepared. In
1999 another one turned up in a collection of astronomical papers in
the archives of the University of California at Santa Cruz.

Scientists today tend to be embarrassed by all this—by Kepler's love
of Neoplatonic and Pythagorean theory, or his belief in the Music of
the Spheres. They would, if they could, create him anew after their
own image. Yet if one looks at the whole of what he had to say, it is
clear that despite some scorn for how astrology might be presented, he
would not discard it. And despite all his railing against charlatans and
inept practitioners of the art, he ultimately tried to reform or fine-tune
the traditions, not cast them out. This was so even to his dying day,
when he thought he read his own demise clearly in the stars.

Kepler lived in a time of tremendous political and religious strife.
The whole map of Europe was changing, as Protestant and Catholic
powers clashed, new states were emerging, and displaced persons were
constantly on the move. In 1618, the decay of the Holy Roman Empire
had tempted various states and parties to advantage and engulfed the
whole continent in the Thirty Years' War. Kepler paid as little attention
as he could to the turmoil around him, believing, as he put it, that
"when the storm rages and the shipwreck of the state threatens, we can
do nothing more worthy than to sink the anchor of our peaceful studies into the ground of eternity." Yet he could hardly escape the whirlwind of events. He had been dislodged from his home on a number of
occasions, and the welfare of his family, the stability of his livelihood,
and even the continuity of his work had seemed at times to hang by a
thread. Not surprisingly, his talents as an astrologer were also in demand. All sorts of people wanted to know what the future held for
them, including Rudolph II, who asked him to predict the outcome of
the war between the Republic of Venice and Pope Paul V.

Kepler could flatter, of course. Rudolph had made Prague a cultural
oasis by his patronage of learning, and Kepler congratulated the German emperor on his horoscope, because it so nicely "contained" Mars.
Mars, he explained, ruled the sign on his Midheaven (Scorpio), was
exalted in the sign of his ascendant (Capricorn), at home in his Sun-
sign (Leo)—because Leo and Aries, which Mars also ruled, were both

fire signs—and because Aries ruled Germany, which (lo and behold!) Rudolph also ruled. As a bit of flattery, this was neatly done. In 1608, he had also been contacted by the agent of "an unnamed lord" who commissioned him to cast his horoscope. Kepler agreed to do so, gave him a reading, and also, by progression, predicted the circumstances of his life.

As a blind reading, it deserves our utmost respect. "I might truthfully describe this man," wrote Kepler, "as one who is restless, industrious, quick, alert, impatient with all convention, scornful of human law and custom, treacherous, brave, utterly war-like, with a brutal heart." In the course of his career, he might accomplish "great deeds," wrote Kepler, but in his "thirst for glory and striving for temporal honor and power," he would "make a number of great enemies," and with "Mercury so exactly in opposition to Jupiter, it almost looks as though he will yield to wild schemes." Kepler was candid, and his client might have taken some offense. This would surely have put Kepler at risk because the "unnamed lord" was Count Albrecht von Wallenstein, a soon-to-be-mighty figure, though at the time a mere twenty-five-year-old captain in the imperial guard. He kept Kepler's horoscope reading close by him for over a decade—as he rose through the ranks to become the greatest, and most ruthless, German general of the Thirty Years' War.

Born at Hermanice, Bohemia, on September 24, 1583, Wallenstein belonged to Czech aristocratic stock. From his youth he had demonstrated a strong affinity for military life, had joined up and earned his share of honors, and in 1621, three years after the Thirty Years' War began, received an independent command. His service against the prince of Transylvania endeared him to the Emperor Ferdinand II, and after his decisive victory at the battle of White Mountain he took possession of northern Bohemia—thereafter called Friedland—which he ruled with an iron hand. As the champion of the Habsburgs and the Catholic Church within the empire, he raised an auxiliary army of twenty thousand men, and in gratitude the emperor appointed him "captain of all imperial forces" and made him a duke. His star continued to rise. In the latter part of 1625, he occupied Magdeburg and Halberstadt, two rich and strategically important districts, and in April 1626 won a notable victory over Protestant forces at the bridge of Dessau over the Elbe. After an unexpected, but brief, reversal in

Silesia, he vastly increased the size of his army, took Silesia by storm, united with another force for a campaign against Holstein, and advancing as far as Jutland, occupied Mecklenburg. In January 1628, the emperor granted him the Duchy of Mecklenburg in fief for life.

Other Catholic princes mortally feared Wallenstein's growing power, even as he began to imagine himself dictator of all the lands he held. But as Kepler had predicted, he was destined to spoil his achievements by succumbing to "wild schemes." These included forging an unholy alliance between Catholic and Protestant states; cutting a canal to the Baltic; raising a fleet to match the Swedes; and taking Constantinople from the Turks, which he declared he could do in three years. In the fog of his ambition, his luck began to turn. In the summer of 1628, his siege of Stralsund was checked; his attempt to take Gluckstadt turned aside. The Swedes advanced. When he proved unable to stop them, he was dismissed from his command. Over the next three years, he nursed his resentment like a festering wound until the twists and turns of the war brought him back into the field. In a remarkable engagement, he fought the Swedes to a draw at Lützen, but then entered into negotiations that led to treason against the imperial throne.

Well before then, however, when the Count's fortunes were still at their height, Kepler, to his astonishment in December 1624, sixteen years after his first blind reading, had received the horoscope back with marginal notations in Wallenstein's hand. Wallenstein, now the Duke of Friedland, asked Kepler to extend his forecast and appended a list of questions that he hoped Kepler would address. Kepler obliged as far as he dared, but declined to extend his calculations beyond the end of February 1634, predicting for about that time a "horrible event." That was certainly prescient. For Wallenstein was murdered on February 25. What did Kepler see? One may surmise. Wallenstein, born at 4:36 P.M. on September 24, 1583, had Mars at the time in Aries in his eighth house, and his Midheaven, progressed by right ascension, square to the place Mars occupied at his birth. According to Ptolemy's *Tetrabiblos,* that combination would strongly indicate sudden death by stabbing. And that is how Wallenstein died.

By then Kepler himself was dead. Some thirty years before, in his almanac or calendar for 1601, he had written: "Almost every motion of the body or soul in its transition to a new state occurs at a moment when the figure of the heavens corresponds to its figure at birth." That could mean a number of things, depending. One of them was death. In November 1630, as Kepler set out from Sagan (in Prussia, now in Poland), for Linz to collect some back pay and interest on some bonds, he seems to have known the journey would be his last. Indeed, he made a conscious effort to put all his affairs in order, and assembled "all the documents pertaining to his financial affairs." He was in such a gloomy state, recalled his son-in-law, that "his wife and children would have sooner expected the Day of Judgment than his return." Biographers like to say that Kepler had a "foreboding." In fact, astrologically speaking, he thought he knew. At the beginning of each year, Kepler had calculated the directions and "revolutions" of his chart, and had done so, as a matter of course, for his fifty-ninth year. That figure is preserved, with a note: that the planets were about to occupy the same positions they had at his birth.

He got as far as Regensburg, crossed the Stone Bridge, and collapsed. He was bled by some doctors to no purpose, became delirious with fever, and at noon, on November 15, he died. His last known conscious gesture was to point with his index finger first to his head, then to the sky. His mortal remains were laid to rest in the Protestant cemetery of St. Peter on November 18—"a day fittingly followed," wrote one admirer, "by a lunar eclipse." Poor though he was, fame attended him, and many, both high and low, came to pay their respects. The inscription on his tombstone rightly commended him as "the prince of astronomy"; below it was an epitaph he had recently composed for himself: "Mensus eram coelos, nunc terrae metior umbras / Mens coelestis erat, corporis umbra jacet." ("My soul being from heaven, I measured the star-filled skies; / Now I measure a hollow plot of earth where a shadow of me lies.")

Neither the grave nor the tombstone survive. A year and a half after Kepler's death, the army of Gustavus Adolphus swept southward through Bavaria and laid the churchyard waste. No one can even say in what plot of ground his bones are laid.

★　★　★

IN A SENSE, given the eclectic breadth and depth of his understanding, none of Kepler's successors was his peer. But in the pantheon of history, there is also an equality of largeness to those who loom larger than life. In fame, if not in our story, then, of kindred stature was

Galileo—"the lynx-eyed astrologer," as one Jesuit called him—whose work in terrestrial dynamics (by legend, conducted from the top of the leaning tower of Pisa), experiments with the pendulum, and astronomical observations made him a giant of his age. Galileo was responsible for the final public acceptance of the Copernican theory of the universe, designed the first practical telescopes of any real power, and discovered the moon's craters, Jupiter's moons, the rings of Saturn, sunspots, and the fact that the Milky Way is made up of innumerable stars.

He also cast horoscopes. Galileo had first come to astronomy because of the excitement attending a Jupiter-Mars conjunction; drew up charts for his daughters as well as wealthy clients; and attempted to identify his own exact birth time by "rectifying" his natal chart. In the dedication of his landmark book, *A Message from the Stars,* about Jupiter's moons, he could find no better way of exalting his patron (Cosimo de'Medici, or Cosimo the Great) than by praising his horoscope. In it, Jupiter—"the most benign star, after God the source of all good"—held pride of place. "It was Jupiter, I say, who at Your Highness's birth, having already passed through the murky vapors of the horizon, and occupying the midheavens and illuminating the eastern angle from his royal house, looked down upon Your most fortunate birth from that sublime throne and poured out all his splendor and grandeur into the most pure air, so that with its first breath Your tender little body and Your soul, already decorated by God with noble ornaments, could drink in this universal authority and power." After Galileo's death, a horoscope was found on the back of one of his drawings of the mountains of the Moon. Today, however, he is commonly remembered for an attempt to reconcile the Copernican system with Christian theology that brought him into conflict with the Church. The Inquisition summoned him to Rome, and after a sensational trial, he recanted. He died, greatly discouraged, in 1642, the year of Isaac Newton's birth.

Chapter 8

THROUGHOUT EUROPE, almost to the end of the 17th century, science and astrology remained closely allied. Valentine Naibod, who is chiefly remembered today for calculating the mean annual motion of the Sun, did so to assist astrologers in the art of progressing their charts; Cyprian Leowitz, mathematician to the elector Palatine, prepared a history of the world based on conjunction theory from the beginning of Rome to his own time; and John Napier, Baron of Merchiston near Edinburgh, invented logarithms, originally called "Napier's bones" or "rods," to simplify the calculations required in making horoscopes. (Napier also designed prototypes of the motorized battleship and tank.)

In the brave new world that marked the dawn of modern science, astrology also extended into other fields. Robert Burton, author of the great *Anatomy of Melancholy,* published in 1621, was a priest of the Anglican Church, and took up astrology to prove to his parishioners that it was false. Instead his studies convinced him it was true. He applied his newfound art to psychological analysis, of which he was a master, as well as to predicting the events of his own life. His principal contribution was a collection of observations on mental illness. He noted that Saturn and Jupiter conjunct in Libra tended to produce a mild melancholia, while Saturn and the Moon (rulers of opposite signs) conjunct in Scorpio made it severe. By contrast, a conjunction of the Moon and Mars in opposition to Saturn and Mercury induced a manic condition; Venus in Leo in aspect to the Moon fostered lust; and an uncontrollable depravity resulted when Venus and Mercury were conjunct the ascendant, or Mercury conjunct the Moon.

Astrology provoked sharp debate. As the 17th century began, Sir Christopher Heydon, knighted for bravery at the siege of Cadiz when Essex sacked it, wrote a *Defence of Astrology* in response to a *Treatise Against Judicial Astrology* by the Anglican canon John Chamber of Windsor. Chamber was bested (or thought he was) and killed himself in despair. A more stalwart skeptic was William Fulke, a Puritan divine who was remarkably "modern" in denying the promise of alchemy, the magical virtue of stones, the medical value of astrology, and the influence of comets on civil or political affairs. "Sickness and health depend upon diverse causes," he wrote, "but nothing at all upon the course of the stars, for what way soever the stars run their race, if there be in the body abundance or defect, or from outward by corruption of the air infection it must needs be sick: and if none of these be, though all the stars in heaven with all their oppositions and evil tokens should meet in the house of sickness, yet the body should be whole, and in good health." Similarly, he derided any reliance upon the stars for guidance in farming or commerce: "Good days to buy and sell, be market days, and all other whensoever a man can get a good bargain. I think but few merchants will lose their mart in waiting for heavenly help from the stars." Even so, Fulke kept one foot in the opposite camp. In his *Book of Meteors* he expounded the view that rings around the Sun and Moon—like rainbows or mirages—had a moral purpose and "do not occur by chance, but are sent by God as 'wonderful signs, to declare his power, and move to the amendment of life.'" In the end, he allied himself with a moralized notion of astronomy as expressed by the Tudor mathematician Robert Recorde: "There was never any great change in the world, neither translations of empires, neither scarce any fall of famous princes, no dearth and penury, no death and mortality, but God by the signs of heaven did premonish men thereof, to repent and beware betimes."

Astrology still mingled with every sphere. John Perrin, chaplain of St. John's College, Oxford, regius professor of Greek, and (a few years later) part of the committee that created the King James Bible, asked the young astrologer Robert Fludd to help him recover some money stolen from his rooms. Fludd drew up a horary chart for the question, identified the thief, and recovered the gold coins. A few years later, Fludd traveled to Avignon, where his penchant for astrology outraged some Jesuit priests. They hauled him before the papal vice legate there as a possible heretic, but to their surprise the legate dismissed the charges, exclaiming: "Is there a single Cardinal in Italy who does not have his horoscope cast?" After that, the Jesuits couldn't get enough of his learning and urged him to stay on. However, he soon returned to England, where he earned his doctorate in medicine at Christ's Church, Oxford, became a fellow of the Royal College of Physicians, and was consulted by Sir William Paddy, physician to King James I. The king's other chief physician, also close to Fludd, was the great William Harvey, who discovered the circulation of the blood. Harvey's hypothesis about blood flow, in fact, was apparently inspired by an astrological idea—that just as the planets make their circular rounds above, so man, as an image of the heavens, might find a parallel motion, or circulation, of the blood through his arteries and veins. In the dedication of his great treatise to the king, he left little doubt as to the worldview out of which his discovery was born. "Most Illustrious Prince!" it began, "The heart of animals is the foundation of their life, the sovereign of everything within them, the Sun of their microcosm, that upon which all growth depends, from which all power proceeds . . ."

Among the nobility the science of astrology was, as ever, "ravened, embraced, and devoured." Under Elizabeth, the Earl of Leicester; William Cecil, Lord Burghley; the Earl of Essex; Sir Christopher Hatton; and the Earl of Oxford (to name a few) had been deeply absorbed by the subject; so were many of their successors in the court of King James. Under the latter, a horoscope was often cast at the beginning of Parliament as a means of telling how it would fare. That practice continued under Charles I. In 1628, the king's proto-Catholic archbishop, William Laud, alluded to the fact in the sermon he delivered for the opening of Parliament on March 17. "Join then and keep the Unity of the Spirit, and I'll feel no danger though Mars were Lord of the Ascendant, in the very instant of this Session of Parliament, and in the Second House, or joined, or in aspect with the Lord of the Second, which yet Ptolemy thought brought much hurt to Commonwealths." Yet it was a danger he ought to have felt, for the king was about to embark on a course that would make the Parliament his mortal foe.

One source of popular ferment was the astrological almanac, which assumed a number of forms. Some belonged, like today's tabloids, to the lowest strata of their kind, with a "garbled astrological nostrum lifted from more serious texts"; others were more worthily consulted by merchants as well as housewives, kings as well as thieves. At its best, the almanac, like the herbal reader of its day, was designed to satisfy the demand of the general public for enlightenment about technical subjects, even as it "laid the foundations for the vast literature of popular science" that has captivated (and instructed) the public ever since. At a time when most people still "lived close to the land, with its cyclic round of the seasons, cycles of planting and harvest, in the constant company of the Sun, Moon, and stars," almanacs noted the astronomical events of the coming year, such as eclipses, conjunctions, and movable feasts; church festivals; holidays; fairs; the tides, and so on, all combined with religious and political commentary, weather prediction (often tied to the phases of the Moon), and a forecast of important events. The first such English almanac had appeared in 1545; by 1600, over six hundred had been published; and by the century's end, four million copies of two thousand or more would flood the realm. During this period, almanacs outsold even the Bible and had a remarkable hold on the public mind. "Who is there," one contemporary wondered, "that maketh not great account of his almanac to observe both days, times, and seasons, to follow his affairs for his best profit and use?" Or, as the Prince in Shakespeare's *Henry IV, Part 2* exclaims: "Saturn and Venus this year in conjunction! What says th'almanac to that?" Most wanted to know. A good deal of the advice was often simplistic—as in this stellar instruction for begetting a male child: "If thou want'st an heir, or man-child to inherit thy land, observe a time when the masculine planets and signs ascend, in their full power and force, then take thy female, and cast in thy seed."

Another garbled astrological idea was that certain days in the month were automatically auspicious while others brought ill luck. These were sometimes presented in tabular form for reference, as in a table ascribed to Tycho Brahe, which he had supposedly left in a monastery wall on Hven. It contained thirty-two days in the year when "nothing important should be undertaken or brought to term."

These were
January 1, 2, 4, 6, 11, 12, and 20
February 11, 17, and 18
March 1, 4, 14, and 24
April 3, 17, and 18
May 7 and 8
June 17
July 17 and 21
August 20 and 21
September 10 and 18
October 6
November 6 and 8
December 6, 11, and 18

A birthday on any of those days also condemned the hapless person to a short impoverished life.

The Protestant Reformation was arguably responsible in part for some of this. Catholicism had always been more hospitable to astrology (and magic) because the rituals of the Church, including transubstantiation—by which the bread and wine of Communion are miraculously transformed by invocation into the body and blood of Christ—are "white magic" acts. Wherever the Protestants held sway, such rituals were ridiculed as "hocus pocus" (a slurred mockery of the sacramental words *hoc est corpus,* "this is my body"), and any kind of "magic" was regarded as in some sense diabolical, just as "popery" and superstition were said to be the same. As a result, the common people, who had relied on "white" Church magic, guardian angels or spirits, and the saints to protect them from evil, now turned instead to back-street astrologers and other "cunning folk," as they were called, who appeared to possess the hidden knowledge that would help and protect them in their lives.

At the same time, when the sometimes abstruse doctrines of astrology were made available to the public in popular form, it raised—as the vernacular Bible raised—the general level of discussion and debate. The idea of the art of astrology as a noble branch of learning found some resonance among the folk, even if the technicalities of it were beyond the common reach. One has only to look at the confidence and ease with which contemporary poets, playwrights, and others freely allude to and play with its tenets in their work. There are over two hundred references to astrology in Shakespeare's plays alone, for example, quite a few of them arcane. Astrology is often garbled—for example in *Twelfth Night,* when Sir Andrew Aguecheek and Sir Toby Belch, two clownish knights, argue over whether Taurus rules the sides and heart of the body, or the legs and thighs. Both are wrong. It rules the neck and throat. Most of the audience had to know this for the Bard to get his laughs. A few lines later, Sir Andrew alludes to some pseudo-astrological babble Sir Toby had uttered in a drunken state the night before: "In sooth, thou wast in very gracious fooling last night, when thou spok'st of Pigrogromitus, of the Vapians passing the Equinoctial of Quebus." It might be thought that Shakespeare is making fun of astrology here; more likely, he is making fun of those who make a mish-mash of it, since that is what is being mocked.

His characters, as the Shakespeare scholar Johnstone Parr reminds us, speak of stars, planets, comets, meteors, eclipses, planetary aspects, predominance, conjunction, opposition, retrogradation, and so on. They know

> that the Dragon's Tail exerts an evil influence, that Mercury governs lying and thievery, that Luna [the Moon] rules vagabonds and idle fellows, that Saturn is malignant and Jupiter benevolent, that planets influence cities and nations, that each trigon or triplicity pertains to one of the four elements, that stars rule immediately as well as at birth. They also frequently express the idea that various celestial phenomena—comets, meteors, eclipses—are harbingers of disastrous events to come.

In *Richard III,* part of the king's army hastily disbands when the "meteors fright the fixed stars from heaven," because it is believed they portend Richard's demise; in *Henry IV, Part 1,* Worcester is likened to a meteor and "a portent of broached mischief" when he deserts his king; and in *Henry VI, Part 1,* comets are invoked as celestial agents of revenge: "Hung be the heavens with black, yield day to night / Comets, importing change of time and states / Brandish your crystal tresses in the sky / And with them scourge the bad revolting stars / That have consented unto Henry's death!" In one way or another, it

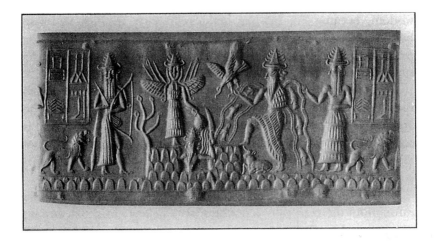

A greenstone cylinder seal from Mesopotamia, ca. 2300 B.C.
Venus (known to the Babylonians as Ishtar) is the second figure
from the left. *(Courtesy of the British Musem)*

A cuneiform planisphere (ca. 800 B.C.) showing the
planets and stars. *(Courtesy of the British Musem)*

A Roman coin minted by Caesar Augustus (63 B.C.–A.D. 14), with his portrait. On the obverse, the Goat-Fish of Capricorn, the Emperor's Moon sign, which he adopted as his emblem. The creature is holding a globe to signify that Augustus is master of the world.
(*Courtesy of the British Musem*)

Claudius Ptolemy (ca. A.D. 100–178), one of the foremost astronomers and astrologers of ancient times. His *Tetrabiblos,* written in Greek in Alexandria, Egypt, under Roman domination, is one of the touchstones of classical astrology.

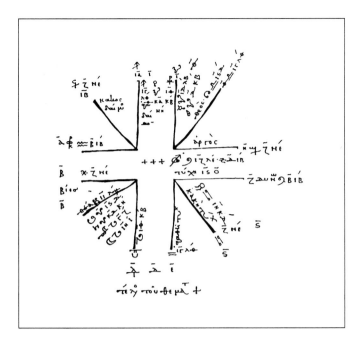

This detailed horoscope—one of the earliest to survive in manuscript—
was drawn by Rhetorius the Egyptian in A.D. 497.
(Courtesy of Tamsyn Barton, Ancient Astronomy, *Routledge, 1994)*

Muhammed ibn Ahmad al-Biruni (973–ca. 1048), one of the greatest
of all Arab astrologers, whose competence in every field of learning
exemplified the broad knowledge the discipline required.

Two signs of the zodiac as depicted in a 16th-century Turkish treatise on astrology. Aries *(left)* and Taurus *(right)*. The three decans or faces of each sign are also shown.

Sagittarius as depicted in a Persian *Book of Stars and Constellations*, 1630.

Geoffrey Chaucer (1342–1400), the reigning English poet of the Middle Ages, whose expert knowledge of astrology found reflection in his work.

An astrolabe, ca. 1400. Astrolabes were used throughout ancient and medieval times to observe the positions of the heavenly bodies and determine their altitude.

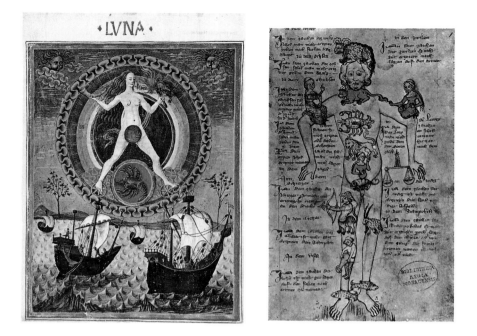

An illumination (ca. 1250) *(left)* from the textbook *De Sphaera* by John Sacrobosco, or John of Holywood. Here the Moon is depicted as ruling the sign of Cancer and the watery world.

An image of man as a microcosm (ca. 1300) *(right)*, showing the parts of the body each sign of the zodiac rules. *(Monacensia Library, Munich)*

A page from a so-called *Magic Book* of divination ascribed to Michael Scot (ca. 1175–ca. 1235), astrologer to the Emperor Frederick II. The characters, though of an Arabic cast, are of no known language.

Marsilio Ficino (1433–1499), the great 15th-century scholar, translator, philosopher, and astrologer whom some regard as a central figure of the Italian Renaissance.

In this woodcut (ca. 1500) a client consults an astrologer, who draws a horoscope. Note that the chart itself is drawn in a square, which in turn is subdivided into twelve equal triangular sections, each representing a house. This was the usual form until modern times. The horoscope may be a birthchart, but since the astrologer is pointing to the planets and stars at that moment in the sky, it is probably a horary or election chart in response to a question the client has asked.

Girolamo Cardano (1501–1576) of Milan. A superb mathematician
(he was the first to present the whole doctrine of cubic equations
to the world), Cardano was also one of the premier astrologers
and physicians of his age.

A woodcut from a pamphlet on the conjunction of all seven planets in
Pisces in 1524. That conjunction had seemed to threaten a torrential
flood and was afterwards blamed for the great Peasant Revolt
which shook Central Europe that year.
(By permission of the British Library, 8610.e.4)

The English mathematician, alchemist, and astrologer John Dee
(1527–1608) *(left)*, who determined the coronation date for Queen
Elizabeth I, served her in a number of capacities, and helped to foster
the full range of scientific studies in the English Renaissance.
(Ashmolean Museum, Oxford)
A woodcut *(right)* (ca. 1525) depicting Mercury, ruler of Gemini and
Virgo, as shown by the figures adorning the wheels of his chariot.
Also shown are activities, such as art, music, mathematics, and
astrology, over which he is said to preside.

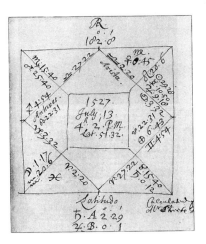

Dee's birthchart, drawn by himself, showing the fixed star Antares
conjunct his ascendant. As in almost all horoscopes, east is at the left,
west at the right, south above, north below—a form ultimately
derived from ancient Egyptian maps.
(The Bodleian Library, University of Oxford, MS. Ashmole 1788, fol. 137r)

Tycho Brahe (1546–1601) *(left)*, the greatest observational astronomer since Hipparchus and court astrologer to Frederick II of Denmark. Johannes Kepler (1571–1630) *(right)*, perhaps the greatest mathematical astronomer who ever lived and a convinced astrologer. As an astronomer, he discovered the laws governing the elliptical orbits of the planets and essentially grasped the law of gravity, which Isaac Newton later defined. As an astrologer, he cast hundreds of horoscopes, tried to reform some aspects of the art, and accurately forecast the death of his patron, Count Albrecht von Wallenstein, as well as the year of his own demise.

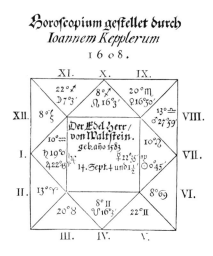

The famous horoscope which Kepler cast for von Wallenstein in 1608. From it, Kepler divined the course of the count's spectacular career and dreadful death.

William Shakespeare (1564–1616), whose work is drenched in the astrological lore of comets, eclipses, nativities, the signs and dignities of the planets, and the planetary hours. There are over 200 allusions to astrology in his plays and not one of their astrological predictions goes unfulfilled.

Simon Forman (1553–1611), physician and astrologer to London's population, high and low. The Royal College of Physicians and Surgeons made him a scapegoat, in part, for some of their own failings, but he outfoxed them in the end.

William Lilly (1602–1681) *(left)*, perhaps the finest astrologer England produced and the leading figure in the art's Silver Age. His skill at horary was unsurpassed, but his ability at elections and nativities (birthcharts) almost as grand. Among his notable public achievements, he predicted the outcome of several battles during the English Civil War, the fate of the Rump Parliament, the death by beheading of Charles I, and the Plague and Great Fire that afflicted London in 1665 and 1666.

A horoscope cast by Lilly *(right)* at the time of the king's death to forecast the future course of state affairs.

A woodcut from one of Lilly's pamphlets of 1648, which showed the Twins of Gemini (the sign of the zodiac on London's ascendant) falling headfirst into the flames. This was seen to confirm his prediction of a Great Fire which would afflict the city "in the year 1665 . . . or near that year . . . of sundry fires and a consuming plague."

Elias Ashmole (1617–1692) *(left)*, Lilly's friend, compatriot, and fellow adept, who served as unofficial astrologer at the court of Charles II. Ashmole's assiduous antiquarian studies and acquisitions are enshrined in the famed Ashmolean Museum of Oxford.

Jean-Baptiste Morin (1583–1656) *(right)*, Regius Professor of Mathematics at the College de France and the most accomplished of all French astrologers, before or since. He was present at the birth of Louis XIV, predicted the spectacular scale of his kingship, and was a power at his court during his reign.

Sibly's famous horoscope for the American Revolution, published ca. 1788. He predicted that the fledgling nation would eventually become an international power and the beacon of freedom to the world.

(By permission of the British Library C.71.h.14)

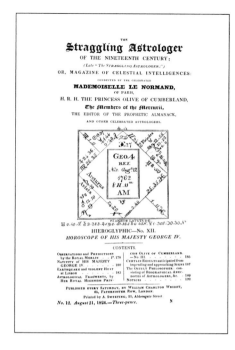

The cover of the first weekly astrological magazine to be
published in any language—in London in 1824.

Alan Leo (1860–1917) *(left)*, a follower of the Theosophical teachings of
Madame Blavatsky and the father of modern-day Sun-sign astrology.
Walter Gorn Old *(right),* also known as "Sepharial," Alan Leo's friend
and mentor, who later abandoned Theosophy and attempted
to bring astrology "back down to earth."

Evangeline Adams (1868–1932) *(left)*. Said to be a direct descendant of President John Quincy Adams, she was the most popular and success-ful astrologer in America during the first third of the 20th century. This photo was taken a few years before her death.
Carl Gustav Jung (1875–1961) *(right)*, the famed psychologist, who sometimes cast the horoscopes of his patients to assist his analytical work.

Louis de Wohl *(left)*, the unofficial astrologer of Winston Churchill's Political Warfare Department during World War II.
A copy of the bogus German astrological magazine *Der Zenit (right)*, which de Wohl helped to produce as part of a disinformation campaign during the war.

William D. Gann (1878–1955), perhaps the single most successful trader in Wall Street history, who used astrological calculations to help him predict commodity prices and general market swings.

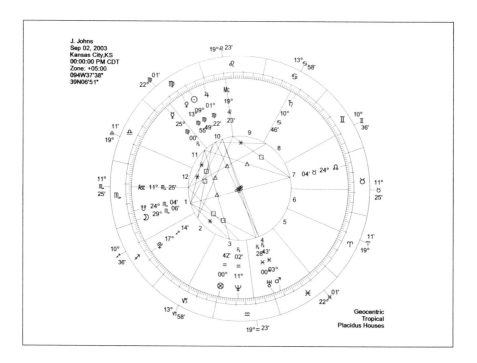

A typical horoscope (in this case using Win★Star Express software) produced by a computer today. The signs of the zodiac are shown in the round, with the planets in their houses and the degree and minute of each sign ruling a house marked on the house cusp. Aspects between the planets are indicated in the center, where connecting lines crisscross like the strands of a ball of yarn.

was axiomatic that comets signaled the fall of the mighty or some rip in the social fabric or peril to the state. "When beggars die," wrote Shakespeare, "there are no comets seen." Or as Cyril Tourneur, his colleague, put it in *The Revenger's Tragedy:* "When stars wear locks, they threaten great men's heads."

The dire events of *King Lear* follow a series of eclipses; in *Julius Caesar,* Antony believes an eclipse of the Moon foreshadows his fall. Othello insists that an eclipse should accompany his crime; and in *Hamlet* a Moon "sick almost to doomsday with eclipse" heralds the events that unfold. Planets in ill aspect were also to be feared, as Ulysses in *Troilus and Cressida* explains:

> . . . when the planets,
> In evil mixture, to disorder wander,
> What plagues and what portents! what mutiny!
> What raging of the seas, shaking of earth!
> Commotion in the winds! Frights, changes, horrors,
> Divert and crack, rend and deracinate
> The unity and married calm of states
> Quite from their fixture!

Thus Pericles attributes the loss of all his fortune to the "ire" of "angry stars"; Romeo and Juliet are hopelessly "star-cross'd" lovers; Hermione in *The Winter's Tale* can only assume that "some ill planet reigns" to explain her plight.

Did Shakespeare himself believe this? Perhaps. Those who would cast him as an opponent of astrology are obliged to credit the voice of his vilest villain—Edmund in *King Lear.* In *Lear,* the long-suffering Gloucester blames his condition, and other discords in the state, on "these late eclipses in the sun and moon." Edmund overhears this remark and scoffs at

> the excellent foppery of the world, that when we are sick in fortune, often the surfeits of our own behavior, we make guilty of our disasters the sun, the moon, and stars; as if we were villains on necessity; fools by heavenly compulsion; knaves, thieves, and treachers by spherical predominance; drunkards, liars, and adulterers by an enforced obedi-

ence of planetary influence; and all that we are evil in, by a divine thrusting on. An admirable evasion of whoremaster man to lay his goatish disposition on the charge of a star.

In proof, he offers himself as an example: "My father compounded with my mother under the Dragon's Tail, and my nativity was under Ursa Major, so that it follows I am rough and lecherous. Fut! I should have been that I am, had the maidenliest star in the firmament twinkled on my bastardizing . . . Fa, sol, la, mi!"

Shakespeare is expert in astrology here, and letting the knowledgeable reader know that Edmund was born under just those aspects that in fact describe who he is—a perverse and degenerate soul. Unlike Sir Toby and Sir Andrew, Shakespeare had his astrology right. Edmund's position is that of the modern "realist": a person born with any other horoscope would be the same. That sounds perfectly sensible to us, because Edmund's point of view is the respectable one today. But to Lear the disparate characters of his three daughters can only be explained by the astrological distinctions at their birth. "It is the stars / The stars above us," he cries, "govern our conditions; / Else one self mate and make could not beget / Such different issues." If Shakespeare had wanted to debunk astrology he would have given Edmund a wholly different chart. In sifting all this, it might be useful to remember Juliet's famous lines: "What's in a name? that which we call a rose / By any other name would smell as sweet." Readers today tend to take that as common sense. But there, too, Shakespeare is playing another game. In the play, it turns out that there is a great deal in a name: in Verona, the names of "Montague" or "Capulet" could get you killed. As often as not, he is teaching the audience to rethink some common assumption they are prone to make.

In the end, as Johnstone Parr remarks, "All of the astrological predictions in Shakespeare's plays are fulfilled."

Similar sophistication was shown by George Chapman in *Byron's Conspiracy,* an Elizabethan tragedy based on the downfall of the Duke of Byron, the favorite of King Henry of Navarre (Henry IV of France). At one point, the Duke decides to consult an astrologer in connection with his plot against the throne. He arrives incognito with his horoscope in hand, but the astrologer (La Brosse) quickly sees through his disguise and divines his fate:

BYRON:

I would entreat you, for some crowns I bring,
To give your judgment of this figure cast,
To know, by his nativity there seen,
What sort of end the person shall endure
Who sent me to you and whose birth it is . . .

LA BROSSE:

My son, I see that he, whose end is cast
In this set figure, is of noble parts,
And by his military valor rais'd
To princely honors, and may be a king;
But that I see a *Caput Algol* here
That hinders it, I fear.

BYRON:

    A *Caput Algol?*
What's that, I pray?

LA BROSSE:

    Forbear to ask me, son;
You bid me speak what fear bids me conceal . . .
You'd rather wish you had been ignorant.

The Duke, however, threatens to kill him if he doesn't make himself
clear. The astrologer yields: "The man hath lately done / An action that
will make him lose his head." In a rage, the Duke beats him senseless,
then defiantly exclaims:

Spite of the stars, and all astrology,
I will not lose my head; or if I do
A hundred thousand heads shall off before.
I am a nobler substance than the stars,
And shall the baser overrule the better? . . .
I have a will, and faculties of choice,
To do, or not to do: and reason why
I do, or not do this; the stars have none.
They know not why they shine more than this taper,

Nor how they work, nor what; I'll to my course.
I'll piece-meal pull the frame of all my thoughts,
And cast my will into another mould:
And where are all your *Caput Algols* then?
Your planets all, being underneath the earth
At my nativity: what can they do?
Malignant in aspects? In bloody houses?
Wild fire consume them!

As in Edmund's aside in *King Lear,* we learn from this soliloquy that the Duke has a nasty horoscope ("malignant in aspects," "bloody houses") though his lines about free will—"I am a nobler substance than the stars"—sound, to our romantic ears, reminiscent of Hamlet's rhapsodic comparison of man to an angel or a god. But despite his will, the Duke follows his star. That star is "Caput Algol"—a white binary and variable fixed star in the constellation Perseus marking the Gorgon head of Medusa. To the East Indians it was known as "al Ghul," the Demon's Head; to the Chinese as "Tsieh she," or "piled-up corpses"; and to the ancient Hebrews as Lilith—the nocturnal vampire who was said to be Adam's first wife. It is one of the two or three most malevolent of all fixed stars, associated with murder, mob violence, and beheading, and would soon be linked in the chart of Charles I to his own shocking fate. In the Duke's chart, it occupied a ruling (if unspecified) position—by implication, conjunct his Sun or Mars. Ptolemy had explained this long ago: "If Mars be in imperfect signs, or near the Gorgon *(Caput Medusae)* of Perseus, it will produce death by decapitation, or by mutilation of limbs. If the Sun be found with the Gorgon's head (Caput Medusae), and not aspected by any benefic star, and if there be no benefic present in the eighth house, and the lord of the conditionary luminary [in this case, the Sun] be opposed to Mars, or in quartile [square] to him, the native will be beheaded." And that is what happens to the Duke.

Shakespeare's own view overall was not quite fatalistic. Rather it was classical, in the astrological sense. In *Julius Caesar,* Cassius remarks: "Men at some time are masters of their fates. / The fault, dear Brutus, is not in the stars, / But in ourselves, that we are underlings." Shakespeare is quoting Manilius here, almost word for word: "Wonder not at the grievous disasters which betide man and man's affairs, for the fault oft lies within us: we have not sense to trust to heaven's message." What did Shakespeare mean? He likely meant, in context, what Manilius meant, which is the opposite of what the modern reader tends to think: not that it is all in our power, but that if we don't pay attention to what the heavens are telling us, we have only ourselves to blame for how things go. So Prospero in *The Tempest* notes that the planets at the time of the wreck of his enemy's ship are favorably placed for the accomplishment of his revenge: "I find my zenith doth depend upon / A most auspicious star, whose influence / If now I court not, but omit, my fortunes / Will ever after droop." In other words, he needs to take advantage of the opportunities the stars provide. Perhaps the most famous descant upon this theme may be found in *Julius Caesar:* "There is a tide in the affairs of men / Which taken at the flood leads on to fortune; / Omitted, all the voyage of their life / Is bound in shallows and in miseries."

Most of the literary figures of the English Renaissance came down on astrology's side. At times they seem to argue for some tenet of its doctrine; at others they simply incorporate it as a given into the fabric of their work. In the twenty-sixth sonnet of his *Astrophel and Stella,* Sir Philip Sidney tells us, in the context of a witty conceit, that only fools can doubt astrology—that "bodies high reign on the low." In Spenser's *The Faerie Queene,* as Thomas Allen points out in his indispensable study, *The Star-Crossed Renaissance,* one misbegotten character was born when Saturn sat "in the house of agonies" (that is, the twelfth); and in his *History of the World,* Sir Walter Ralegh upheld scholastic doctrine when he wrote that the stars "have complete power over all the reasonless things in the inferior world; they have a definite influence on the disposition of men and signify at their births the nature of their mortal career." By "reasonless things," Ralegh meant animal, vegetable, and mineral—and those human souls without a conscious direction to their lives. They were, by that token, "dull." That is perhaps what Shakespeare's Helena means in *All's Well That Ends Well* when she exclaims: "Our remedies oft in ourselves do lie, / Which we ascribe to heaven: the fated sky / Gives us free scope, only doth backward pull / Our slow designs when we ourselves are dull."

The lingo of astrologers came in for some mocking, of course. In John Marston's *The Malcontent,* a whore parlays the following advice: "Court any woman in the right sign, you shall not miss: but you must take her in the right vein then: as when the sign is in Pisces, a Fish-monger's wife is very sociable; in Cancer, a Precisian's wife is very flexible; in Capricorn, a Merchant's wife hardly holds out; in Libra, a lawyer's wife is very tractable, especially, if her husband be at the term [in court]: only in Scorpio 'tis very dangerous meddling" (because of the association of Scorpio with venereal disease). Such wordplay is everywhere. In John Fletcher's *The Sea Voyage,* one character tells another: "Y'are much inclined to melancholy: and that tells me, / The sullen Saturn had predominance / At your nativity, a malignant Planet, / And if not qualified by a sweet conjunction [i.e., copulation] / Of a soft and ruddy wench, born under Venus, / It may prove fatal."

Most allusions, however, were schooled in the traditions of the standard texts. In John Webster's *The Duchess of Malfi,* an astrologer glancing at the horoscope of a newborn child (who happens to be the duchess's son) remarks flatly, "The Lord of the first house, being com-bust in the Ascendant, signifies short life: and Mars being in a human sign, joined to the tail of the Dragon, in the eighth house, doth threaten a violent death." That was succinct. In another scene in the same play, a horary chart is cast. "Ah: how falls your question?" some-one asks. "Do you find it radical?" This was more arcane. It meant that such a chart had to accord with the "radix"—"root" or birth chart—to be true.

★   ★   ★

AS ASTROLOGY ENTERED ITS SILVER AGE, some conspicuous fig-ures emerged. One was Simon Forman, "the notorious astrological physician of London," as some have called him, though others held him in esteem. If John Dee's clientele was primarily found at court, that of his compatriot encompassed every class. At one time or an-other, Forman was consulted by bishops, knights, merchants, actors, prostitutes, and admirals. He drew up charts for members of Shake-speare's troupe, helped a captain of Elizabeth's navy contest the Span-ish on the high seas, and was implicated in the most sensational murder of the age.

Born on December 30, 1532, in the village of Quidhampton, Wilt-shire, to free tenant farmers of "good yeoman stock," Forman was an avid learner, did well in his early schooling, and as he entered his teens began to yearn for higher things. The family income, however, could not support his advancement, and after his father's sudden death he was apprenticed to a local grocer who compounded medicines and drugs. That opened his eyes to the world of "physick," in which he received a decent grounding, but after six years as the grocer's assistant, decided he had learned all the man could teach him and struck out on his own. After five months as a schoolmaster at the priory of St. Giles, Forman set off for Oxford with his hopes now set on a gentleman's career. He enrolled in Magdalen College, but in order to make ends meet had to hire himself out as a servant to two aristocratic wastrels who, he tells us, seldom cracked a book. While he exhausted himself attending to their needs, they spent all their time hunting, feasting, and chasing girls. One of the two, he later noted wryly, became a bishop of the Church.

After a year of this, Forman was fed up. He left Oxford and thereafter held a number of poorly paid teaching positions in country schools. At some point he seems to have gone abroad to further his education and apparently sought to obtain a medical degree in Holland. Meanwhile, he had also been studying medicine and astrology on his own. In time he began to establish himself as competent in both, and opened an office on the outskirts of London, where he acquired a devoted clientele. His reputation spread and from 1580 on (much to the chagrin of the Royal College of Physicians, affiliated with Oxford) his unauthorized practice thrived. As a result, they hounded him for years, tried successfully at times to bar him from practice, stigmatized him as a "bold and impudent imposter," and on at least two occasions had him arrested and jailed. But in 1592, when the plague swept through London, Forman earned the gratitude of the public when he declined to flee the city for safer precincts, as most of the licensed physicians did. Forman himself was among the thousands infected, but he effectively treated himself and helped many others besides.

When the physicians returned, they demanded "by what authority [he] meddled in their precinct," and accused him of having "learned his skill under a hedge." But Forman would not be cowed. He scorned their "pisspot physic," as he called it, and sensibly questioned the use of bloodletting and enemas for chest complaints. In some things, to be sure, "he had the right of it," but in 1594, after he had been practicing for sixteen years, the Royal College once more hauled him up for review. He was examined in medicine and astrology, judged woefully inadequate in both, fined, and branded a quack. But he was determined to outwit his detractors, and after many efforts, at length received his license (on June 27, 1603) from Cambridge, infuriating the Oxford dons. After years of ridicule, he burst into joyful rhyme:

> Throughout the world I had their seal
> To practice where I list
> And back to Lambeth I returned
> And O how they were pissed!

Meanwhile, as a result of his expanding London practice, he had bought a house, invested in real estate, and (as gentlemen did in those days) had his portrait drawn. It showed, in one description, "a small face beneath a bushy beard and beetling brows; a small sensitive mouth; a large exploratory nose; broad forehead, and a magnetic, obsessive look in his large eyes." Once down-and-out, now stylishly groomed, he got himself up in a tailor-made purple gown, with velvet breeches and a taffeta cloak, let his hair and beard grow (according to fashion), bought a sword with an embroidered scabbard, complete with silver trim, and had money enough to "lend out sums on the security of plate or jewels."

His new comforts were not unearned. In addition to the plague, he treated syphilis, measles, gynecological disorders, colds, gonorrhea, sciatica, jaundice, dropsy, strangury (slow and painful urination), and other ills. In fact, he seems to have been as good a doctor as most, and better than many with advanced degrees. He emphasized, as all doctors then did, the four humors and the need to right imbalances between them for the preservation of health. And he was perfectly able to do the basic things a doctor did: take a pulse, tie off a vein, lance a boil, make a sling, dress wounds, let blood, and peer at urine with a knowing expression on his face. As a young man, he had absorbed on faith most of what the standard textbooks taught. But with experience he had grown more skeptical, and eventually developed his own treatments for a number of ills. He was highly eclectic in his approach, and willing to borrow something from any tradition if it worked. But in his diagnoses, he also relied heavily on medical astrology or decumbiture charts. In itself this was not at odds with official medical practice, as espoused, for example, by the Royal College of Physicians, most of whom, if asked where to look for a disease, would have promptly answered "the sixth house."

Forman's own general advice on how best to proceed is given in a little poem he wrote in 1594:

> If thou wilt a good physician be
> And of a sickness judge aright
> Mark well what I do write to thee
> For urine is of little might,
> But study well Astronomy,
> The course of heavens aright to know
> For they will tell thee certainly
> That which the urine will not show.

Even so, he evidently kept up with the medical literature, such as it was, and also with new experiments and theories. It seems unlikely he was incompetent, whatever his methods, since so many prominent people—who might have had any licensed doctor they chose—came knocking at his door. His notebooks, indeed, provide very good descriptions of a number of conditions, mental as well as physical, including madness, which he calls "frenzy," and scurvy, a vitamin deficiency disease. His account of "frenzy," as one scholar notes, might well guide a Shakespearean actress to perfection in playing Ophelia's mad scene: "They will sing and rhyme and never leave talking and playing with their hands and whirring them round about one another, & will laugh much, and will out of the bed naked and about the house and climb up to the top of the house if they get loose . . . & sometimes they are fearful and will cry and weep suddenly & call & keep much ado, & say come kiss me to the standersby . . . and tell 20 bawdy tales. And they eat little and sleep not at all. I have seen many both men and women in this disease."

His description of scurvy likewise hit the mark:

Those that have the scurvy have much pain in their head, jaws, gums, and teeth, and their gums do swell and the flesh groweth over the teeth and is very sore and full of waterish blood, so that they can hardly eat or join their teeth together. Their breath doth stink, they have much pain in their stomach, and pricking in their head like needles, and sores all over their flesh, and swelling in their joints and legs. And sometimes they have knobs in their flesh, and sometimes all their body from the crown of the head to the soul of the foot doth swell with pain and aching. And most commonly they that have or are like to have this disease are spotted in the body and members with black spots as though they had been pinched.

As a remedy, Forman prescribed eight spoonfuls of lemon juice a day—this, a full century before lemon, lime, and other antiscorbutics were acknowledged as the cure.

Forman treated high and low, and in charging his patients, developed a sliding scale. Sometimes he refused to accept payment if his remedies failed. He was principled in that, and in much else besides, being diligent and indefatigable and devoted to his work. But he had no compunction about sleeping with his patients, or clients, and by dint of a certain swarthy allure seems to have drawn an unending stream of women to his bed. Forman's lovelife, indeed, was hectic. His sexual awakening had come late, but he made up for it with a randy appetite after the age of thirty that gave him "the sexual appetite of a goat."

A sizeable portion of his notebooks is taken up with his adventures, especially his tormented love affair with Avis Allen, a recusant Catholic and a cheesemonger's wife. Their four-year affair was wracked by irrepressible longings, all-night sweats, and jealous fits, with an intensity of passion matched only by distrust. She railed at him "when she found another woman's apron in his house, or sex-stained sheets on his bed, or met some woman she was suspicious of hanging around his garden gate." He fretted that she might be falling for one of his servants, deceiving him with another, or trying to conceal where she had been. On occasion their arguments erupted into violence, only to end in an embrace.

After her death, which he had predicted almost to the day two years before, he made this obituary note:

She was begotten about 4 in the afternoon against a birch [tree] the Friday after St. George's Day. She stayed in her mother's womb 270 days, 1 hour & a half . . . was married to a merchant. She was a Catholic & would not go to church. She died of a catarrhus [apoplexy],

1597 the 13 of June about 7 of the clock in the morning and was buried out of Christian burial. Her mother did never love her well. She was much troubled about her religion. She had 2 children, a maid and a boy, and 11 or 12 scapes [miscarriages] besides. Her children died before her. She was a woman somewhat tall, [with] a good motherly fair face and of a good nature and disposition.

She had had a hard life.

If Forman ever worried about scandal, he never let on; but then the sexual mores of the age had a latitude equal to our own. No one, in any case, shied away from his door. The high-born among his clients, or patients, included knights of the realm, government officials, leading merchants, at least two ecclesiastical deans, one bishop, a prominent Puritan theologian, and Sir William Monson, the greatest naval commander of Queen Elizabeth's waning years. On April 13, 1595, Monson asked Forman whether he was likely to return safely from Cadiz, as he prepared to set sail with a powerful fleet to finish off Spain. Forman cast for an answer and concluded that Monson would have a close scrape, but return safely. In the expedition, Monson served as Essex's flag captain in the *Rainbow,* landed at Cadiz, and was nearly killed as he fought his way through the town. Yet, as predicted, he made it safely home.

Another regular client was the Dean of Rochester, who "could hardly make a move" without having a chart for his actions cast. The Dean had preached before King James himself (James having now succeeded), held a number of lucrative livings, including three rectories, and was chaplain to the queen. But he wanted more than anything to be a bishop, with all its public prestige. From the start, however, his prospects for advancement had been subverted by the antics of his wife. She was young and pretty with a wide mouth and reddish hair, but "proud and inconstant," squandered her husband's wealth in compulsive shopping, and "given to lust and a diversity of men." She was always in love with someone, and even had an affair with a fellow dean. Forman himself slept with her twice, in June and July of 1593.

At this time he was also visited by Martha Webb, the mistress of Sir Thomas Walsingham; she wanted to know if Sir Thomas would be true. Forman found her attractive—she was "very fair, of good stature, plump face, little mouth, kind and loving; desired to go gay"—and despite the risks of dabbling with a nobleman's mistress, he possessed her, as he tells us, on March 15 at "10 past 2 p.m., plene et volenter [fully and freely]," noting the exact time for astrological reasons, in case she conceived. Yet that coupling seems to have been a mere diversion for them both, for Martha continued to pine for Sir Thomas, and in order to help her secure his love, Forman made a sigil for her out of tin (the metal of Jupiter) as Jupiter rose one morning in the sky.

Hugh Broughton, a prominent biblical scholar and Puritan divine, also hoped for preferment, and might have had it if his tactless tongue and cantankerous ways had not tended to spoil each chance that came his way. Broughton wanted to become bishop of Chester, and asked Forman to determine when to approach Lord Burghley with his letter of appeal. Forman looked at the stars and saw that "he shall not be bishop" and had "best go home." Broughton petitioned Burghley anyway, and was rebuffed. (A few years later, when the translation committee for the King James Bible was selected, Broughton, who had been working for some years on his own translation, failed to make the cut.)

Curiously enough, Elizabeth's archbishop of Canterbury, John Whitgift, with whom Broughton often clashed, also secretly consulted Forman about his ills.

Then there was the daughter of Viscount Howard of Bindon. She first consulted Forman when she was nineteen, and was told that she would marry three times. In succession, it turned out, she wed a well-to-do vintner named Henry Pranell; the Earl of Hertford; and finally the Duke of Richmond. In between, she pursued the Earl of Southampton (Shakespeare's patron), and once asked Forman whether she might gain his hand. He told her no, and again he was right, for at the end of the year the Earl married someone else.

As others came for answers to their questions, he constantly drew charts to find answers to his own. He cast to determine the best times for his commercial transactions, to ascertain the loyalty of a servant, the whereabouts of a mistress, which of two applicants to hire as an assistant, to find things lost or stolen, as well as to diagnose and cure disease. In connection with his prognostications, he sometimes crafted costly sigils to harness and direct the power of the stars. Once, when he found himself attracted to some girl he met on the street, he cast to know what kind of wife she might make. The answer was disconcerting: "She will prove a whore." On another occasion, some scissors were missing. The figure showed they had been taken by the maid. And sure enough, they turned up in her room.

Larger questions of public import also engaged him. In 1599, for example, he managed to divine with remarkable clarity the tragic fate of Robert Devereux, Earl of Essex, when he cast to know how Essex's foray into Ireland would turn out. In answer to his question, Forman got a complicated answer that proved just about right: "There seems to be in the end of his voyage negligence, treason, hunger, sickness, and death . . . The end will be evil to himself, for he shall be imprisoned or have great trouble. He shall find many enemies in his return and have great loss of goods and honor; much villainy and treason shall be wrought to the hazard of his life." After an unsuccessful campaign against the Irish rebels, Essex concluded an unfavorable truce. Upon his return to England he tried to vindicate himself before the queen, but she stripped him of his honors and placed him under house arrest. He conspired to raise the population of London against her, but failed, and was beheaded for treason on February 25, 1601.

Forman's chief following was probably among merchants, "particularly those trading overseas, with their liability to losses", along with the mothers, wives, and girlfriends of seamen concerned about their safety on the seas. One such merchant was the ironmonger Nicholas Leate, who had pioneered the direct trade route with India via the Cape of Good Hope, and whose ships had to run a gauntlet of pirates through the straits to the Levant. He often consulted Forman about their fortunes, since their safe return was always in doubt. Forman seems to have read their fate correctly in just about every case.

There was scarcely a question he would not take. One girl asked Forman whether the man offering her marriage did so in good faith; another came to him to find a lost ring. One fellow wanted to know if he was going to meet some woman again whom he had encountered at Sturbridge Fair. The answer was mixed: he would, but he might be sorry he did! Indeed, clients did not always get the answers they hoped for. One woman wanted to know when her husband would return from a trip. Forman cast and found that he had already been back for several days.

Forman also rubbed elbows with some of London's famed theater

folk. For a time, his rooms had been on the northern edge of town—
"beyond Cripplegate out along Red Cross Street"—not far from
the house on Silver Street where Shakespeare lived. He was consulted
by Richard Field, who printed Shakespeare's *Venus and Adonis* and
Spenser's *The Faerie Queene;* by Philip Henslowe, the most important
theater manager of the age; and Richard Burbage, the leading actor of
Shakespeare's troupe. On November 22, 1597, Shakespeare's own
landlady came to see him about some valuables lost from her purse.
Still another client may have been the "Dark Lady" of Shakespeare's
Sonnets (said to be one Emilia Bassano) if the scholar A. L. Rowse may
be believed. The grandson of Sir Thomas More also paid a call, as did
Jane Sondes, the wife of Sir Thomas Flud, Treasurer at War in the
Netherlands, and the mistress of Sir Thomas Gates. Gates later be-
came governor of Virginia and it was the shipwreck of his flagship on
Bermuda in a hurricane that gave Shakespeare his idea for *The Tempest*
with its astrological theme.

Forman often went to plays himself and in 1611 saw, among others,
four by Shakespeare—*Cymbeline, Richard II, The Winter's Tale,* and *Macbeth.* Not surprisingly, Forman crops up in some contemporary plays
himself. Ben Jonson in particular did not like him, and mocks him in
Epicoene, or the Silent Woman; The Alchemist; and *The Devil Is an Ass.* In
the last, he derisively linked Forman with those who "have their crystals, I do know, and rings / And virgin parchment and their dead men's
skulls / Their ravens' wings, their lights, their pentacles / With characters: I have seen all these." In *The Alchemist* (perhaps inspired more by
John Dee), the hero, or villain, is a doctor of impoverished skill whose
deficiencies are obscured by the occult trappings of his trade: "this
Doctor . . . he is the Faustus / That casteth figures and can conjure,
cures / Plague, piles and pox by the Ephemerides / And holds intelligence with all the bawds / And midwives of three shires."

These lines, however, were written a few years after Forman's
death, when he had been posthumously (and unfairly) implicated in a
famous political murder and his reputation trashed. That murder was
the death by poison of Sir Thomas Overbury, as arranged by Lady
Frances Howard, Countess of Essex, at the court of King James. The
context was this: at the age of thirteen Lady Frances had been unhap-
pily wed to the juvenile son of the Earl of Essex. The king himself had
sponsored the union, but to the scandal of the court, she despised her

husband and had steadfastly refused to share his bed. This went on for some time, despite the pleadings of many parties; but she was unable to overcome her feelings and then became infatuated with the king's favorite, Robert Carr. In her frantic state, she made contact with Forman through one Anne Turner, the wife of an alchemist, in quest of a magic sigil "to compel Carr's love." It is doubtful, however, that he ever supplied it, for soon thereafter—on Thursday, September 8, 1611—Forman died.

Meanwhile, the beleaguered Countess had managed to get her marriage annulled—after "a jury of matrons examining her under a cloud of veils" determined she was still a virgin—and Carr had begun to requite her love. The two were therefore free to wed. Plans for the event went forward, but met stiff opposition from Sir Thomas Overbury, Carr's intimate companion and friend. As it happened, Overbury had recently been committed to the Tower for declining to become England's new ambassador to Russia (then justly regarded as a hardship post) and the Countess took advantage of his confinement to plot his demise. Over the course of four months, rosalger (a compound of arsenic), sublimate of mercury, and white arsenic were surreptitiously mixed into his food. When he finally succumbed, his death was ascribed at first to syphilis because of the numerous blisters and abrasions that the poison produced on his skin. Some two years passed before the crime was discovered and the culprits arraigned. A state trial ensued. In the course of it, Forman's name "was dragged up" and "his memory held up to execration by the Lord Chief Justice Edmund Coke." Various love objects, pictures, and talismans were exhibited, as well as a supposed list of adulterous ladies at court. Coke opened it, but refused to read it aloud when supposedly he saw the name of his own wife at the top of the list. As A. L. Rowse remarked, "The fact that some of these people had been clients of Forman's in love matters had nothing to do with their misdeeds in regard to Overbury two years after Forman's death." Even so, the posthumous notoriety he gained from the trial fixed his reputation as a kind of Faust. Two hundred years later, Hawthorne would exploit it in *The Scarlet Letter,* in which Forman is adventitiously linked to Roger Chillingworth, the dark villain of the tale.

All this would have been a disappointment to his hopes. Throughout his life Forman had sought the social status his humble origins had denied him. He had married a woman of station (the niece of a knight, with a coat of arms), reinvented his family history to give it a faintly aristocratic shine, and in his autobiography even introduced himself in biblical fashion, with all the genealogical pretensions of a patriarch or saint:

This is the book of the life and generation of Simon, the son of William, the son of Richard, the son of Sir Thomas Leeds, the son of Sir Thomas Forman of Furnival's and of Anne his wife, daughter of Sir Anthony Smith, etc. Born in the year from the nativity of our Lord Jesus Christ, 1552, the 30th of December, being Saturday and New Year's Eve, at 45 minutes after 9 of the clock at night, of the natural body of Mary, wife of the said William Forman aforesaid, and daughter of John Foster, esquire, by Marian Hallam, his wife. In a village called Quidhampton in the county of Wilts, situate in the valley on the north side of the river between Wilton and Salisbury. Whose parents were well descended and of good reputation and fame.

The Sunday before he died, Forman had calmly informed his wife, "I shall die ere Thursday night." According to one contemporary: "Monday came—all was well. Tuesday came—he was not sick. Wednesday, still in health: with which his impertinent wife did much twit him in the teeth." On Thursday, after dinner, "he went down to the waterside, and took a pair of oars to go to some buildings he had in Puddle-dock. Being in the middle of the Thames, he presently fell down, only saying [to the boatman], 'An impost, an impost,' and so he expired."

Forman's manuscripts were acquired by his friend and student Richard Napier, the rector of Great Linford in Buckinghamshire and the nephew of the baron who had invented logarithms. Napier in turn passed them on to Elias Ashmole, the revered founder of the Ashmolean Museum at Oxford. A renowned bibliophile and scientist, Ashmole was also a committed astrologer, one given "to magical practices to make spirits appear" and a man, Rowse writes, who "spent much of his time in casting horoscopes, engraving sigils and, when he gathered peony roots and dittany for his purposes, he did so at the full moon with the proper rites."

✳   ✳   ✳

WELL BEFORE ROBERT CARR HAD FALLEN OUT of favor with King James, he had been replaced in the king's affections by George Villiers, afterward the first Duke of Buckingham. James nicknamed him "Steenie," because, like St. Stephen, he was said to have "the face of an angel," and adorned him with a number of high posts and titles, including Master of the Horse and Lord Admiral, culminating in that of Duke in 1623. Villiers had strong Catholic ties; married Katherine Manners, the Catholic daughter of the Earl of Rutland; tried to concert an alliance with Spain through the marriage of James's son, Charles I, to the Infanta Maria; and subsequently negotiated the marriage of Charles to Henrietta Maria, sister of Louis XIII of France. From 1625 to 1628 he was the power behind the throne, enriched his family through his control of royal patronage, and incited the distrust and enmity of Parliament by all his acts. After the miserable failure of an expedition against Cadiz, Parliament tried to impeach him on charges of corruption, but the king intervened and dissolved the body before it could vote. Enter the astrologer John Lambe. Just as the king's blunders were blamed on the Duke, so the Duke's influence was traced to Lambe, who was blamed in turn for every decision Buckingham made.

For all his reputed power, Lambe remains the shadow of a name, without a single known horoscope of his casting to examine, or any means to assess the quality of his counsel or judgment beyond the aspersions cast upon him by his foes. All that is known about his early years, for example, is that he had studied some "physick" and tutored the children of a number of gentlemen-nobles before Villiers made him his confidential aide. Nevertheless, according to an anonymous pamphlet printed in Amsterdam in 1628 and entitled *A Briefe Description of the Notorious Life of John Lambe,* he was charged with sorcery and witchcraft (how else to account for Villiers' meteoric rise?) and other "evil Diabolical and execrable arts." In particular, he was said to have bewitched Thomas, Lord Windsor "to disable, make infirme, and consume his body and strength"; to have foretold the death by drowning of the Earl of Mulgrave's three sons; and made various men impotent, according to his whim. As a result, we are told, he was sure to be hanged had not "the High Sheriffe, the Forman of the Jury, and divers other of the Justices, Gentlemen there present, and [members] of the Jury, to the number of forty dyed." Then in 1627, however, Lambe was apparently incarcerated in London's King's Bench prison for allegedly raping an eleven-year-old girl. The evidence was doubtful, but the trial was politically charged, and he was convicted anyway and sentenced to death. The king intervened to free him, but Lambe was now doomed by sullied renown. The following June, in an instigated attack of mob violence, a crowd of apprentices—calling him the "duke's devil"—beat him to death in a London street.

Lambe's orchestrated death was a warning to the Duke. After Lambe was killed, it was said:

> Let Charles and George do what they can,
> The Duke shall die like Doctor Lambe.

In August 1628, a young naval officer named John Felton, moved by popular sentiment, took it upon himself to do the people's will and stabbed the Duke to death at Portsmouth, where he was preparing to relieve the Huguenots of La Rochelle.

Chapter 9

LAMBE REMAINS AN ENIGMA. Forman, despite all his notoriety and faults, had clearly done much to bring affordable health care to London's poor. That, for many, was enough to expiate his sins. Another in that noble line was the English botanist, physician, and astrologer Nicholas Culpeper, whom all modern advocates of herbal medicine revere. Born in Sussex in 1616, the son and grandson of Puritan divines, Culpeper had been a child prodigy of sorts, studiously well-read far beyond his years, and fluent by the age of twelve in Latin and Greek. By then he had already become aware of astrology, through Sir Christopher Heydon's great treatise in defense of the art, and was captivated by William Turner's *New Herbal,* published in 1551, with its striking illustrations of herbs and plants. Meanwhile, he had begun to explore the botanical wonders of his environment, and to learn anatomy with the help of a graphic text prepared by Thomas Vicary, the barber-surgeon to Henry VIII. At sixteen, he went up to Cambridge to study classics and, according to his family's expectations, prepare for a career in the Church. But that was not his heart's desire. After a couple of years of youthful dereliction, he tried to elope to Holland with an heiress but she was killed en route to their rendezvous when lightning struck her coach. Disowned by his family, Culpeper left Cambridge and settled in London, where he worked as an apothecary's assistant before opening his own shop in Red Lion Street, one of London's slums. In 1642, he was briefly charged with witchcraft for practicing medicine without a license, but like Forman he was actually more of an empirical physician than many of those the Royal College turned out. "They have learned little since Hippocrates," Culpeper noted, and "use blood-letting for ailments above the midriff and purging for those below." That, he thought, left a lot to be desired. Meanwhile, in addition to his exploration of herbal cures, Culpeper had begun to study astrology, "an art which teachest by the book of creatures," he wrote, "what the universal Providence mind and the meaning of God toward man is." Hippocrates, the great Greek physician, had insisted that his students master the art, and Culpeper likewise came to regard it as indispensable, once remarking that "physic without astrology is like a lamp without oil."

After serving as a field surgeon and infantry captain on the parliamentary side in the English Civil War, he established a clinic in Spitalfields, just outside London, where "he treated all comers, even those who could not afford to pay." The classical anatomy of Galen and the doctrine of the four humors were joined in his practice to the medieval doctrine of a sympathy or correspondence between all created things. Astrology formed the keystone. "If you do but consider the whole universe," he wrote, "as one united body, and man as an epitome of this body, it will seem strange to none but madmen and fools that the stars should have influence upon the body of man, considering he, be[ing] an epitome of the Creation, must needs have a celestial world written in himself . . . Every inferior world is governed by its superior, and receives influence from it."

Culpeper would prove even more unpopular with the medical establishment than Forman had been. He earned its undying hatred when he translated the standard medical reference work of the time— *A Physical Directory, or the London Pharmacopoeia*—from Latin into English. Like the vernacular translation of the Bible, this rebellious act made once-arcane knowledge suddenly available to all. In a rage, the Royal College denounced him as an atheist (he was nothing of the sort) and tried to discourage the sale of his version by impugning its skill. In fact, it was learned and exact, though he had taken the liberty of appending his own informed commentary about the uses and virtues of each drug. Two years later, he published an important guide, *The Astrological Judgment of Diseases,* as well as *A Directory for Midwives,* with advice for women and mothers at every stage of their childbearing years. His magnum opus, however, appeared in 1653. This was *The English Physician, or an Astrologo-physical Discourse on the Vulgar*

[*Common*] *Herbs of This Nation,* a compendious astrological primer for herbal cures. Known today as *Culpeper's Herbal,* it is the only book in English other than the King James Bible from the 1600s that has been continually in print since it appeared.

"First consider what Planet causeth the Disease," he wrote. "Secondly, consider what part of the Body is afflicted by the Disease, and whether it lie in the Flesh, or Blood, or Venticles . . . Thirdly, consider by what Planet the afflicted part of the Body is governed. Fourthly, you have in this book the Herbs for Cure appropriate to the Several Diseases, and the Diseases for your ease set down in the Margin, whereby you may strengthen the part of the Body by its like, as the Brain by Herbs of Mercury, the Breast and Liver by Herbs of Juniper, the Heart and Vitals by Herbs of the Sun, &c." This was all according to a kind of unified field theory of interconnectedness and correspondence, as nicely expressed by Sir Walter Ralegh in *The History of the World:* "Why should we rob the beautiful stars of their working powers?" he asked. "For, seeing they are many in number and of eminent beauty and magnitude, we may not think that in the treasury of his wisdom who is infinite there can be wanting, even for every star, a peculiar virtue and operation; as every herb, plant, fruit, flower, adorning the face of the earth hath the like." After all, the movement of the heavens produced the diurnal cycle of day and night, as well as the course of the seasons; the moon controlled the tides; flowers opened and turned to greet the rising Sun.

In medical astrology, the natal chart was used to identify the patient by temperament, which helped the doctor select the right treatment; while a decumbiture chart—from the Latin word *decumbo,* meaning "to fall or lie down"—was cast for the onset of the illness (or for the time the patient decided to consult a doctor) and served to help diagnose the ailment, indicate its course or progress, and suggest its cure. Different parts of the body were held to be ruled by different signs— the head by Aries; the cerebellum and neck by Taurus; the arms and lungs by Gemini; the stomach by Cancer; the heart and spinal cord by Leo; the intestines by Virgo; the kidneys by Libra; the sex organs and rectum by Scorpio; the hips and thighs by Sagittarius; the knees by Capricorn; the ankles by Aquarius; and the feet by Pisces. Different herbs were held to be ruled by different signs according to a related scheme. According to the doctrine of signatures, appearance often de-

noted function. For example, citron apple, spikenard, mint, and pars-
ley, which bear in their leaves a heart formation, were thought to be
congenial to the heart. In medical astrology, they were therefore ruled
by Leo and the Sun. Other herbs, ruled by Gemini, that simulated the
shape of the lungs, such as sage and lungwort, were thought to be good
for pulmonary complaints. Yet it would be a mistake to think of these
as *accidentally* linked. They were (or were believed to be) essentially
connected, the appearance bespeaking the Idea.

Culpeper's holistic approach, as John Frawley writes, at least "had
the virtue . . . of treating the patient rather than the illness—of regard-
ing, that is, the patient as an individual organism with individual qualities, not as one of a race of identical machines which had developed a
mechanical fault."

After an examination of the patient and his horoscope, Culpeper applied the appropriate herbs. Nettles proved useful for sore throats;
rhubarb acted as a sunburn cool; caraway assisted digestion; myrtle relieved diarrhea; the kidneys were cleaned by broom and furze; knap-
wort (a marsh weed that acts as an astringent) helped with diseases of
the skin. Culpeper had wandered all over England, collecting and clas-
sifying the plants and herbs he found, and "in each case meticulously
recorded its character, the conditions in which it was found, its time of
flowering, its zodiac ruler, and the diseases it could cure." The best
days for bloodletting and purging and for administering certain po-
tions or drugs were also astrologically determined, especially by the
phases of the Moon. It was axiomatic in medical astrology, for example, not to perform surgery (if you could help it) when the Moon was
waxing (lest the blood flow increase) or to operate on some particular
part of the body when the Moon was in the sign by which that body
part was ruled.

Despite his demonstrable success in treating others, Culpeper lost
the battle with his own failing health. He had been wounded at the
battle of Reading during the English Civil War and while in the hospital had contracted tuberculosis, which he was unable to cure. When he
died on January 10, 1654, he was just thirty-eight years old.

Culpeper's methods are still used today to treat everything from
heart failure to bee stings. "In each decumbiture chart," one modern
practitioner explains, "look at the Ascendant and the planet which
rules it and see how the symbols describe the patient, their health, and

vitality. Note also the element linked to the zodiacal sign, and how the nature of the planet—hot, cold, moist, or dry—interacts with the elemental nature of the sign it is in." The condition of the ruling planet specifically describes the patient's vitality—if dignified or exalted, strong; if in its detriment or fall, weak. It will be weaker still if retrograde or aspected by Saturn or Mars. The Sun and Moon (if different from the ruling planet) must be assessed in the same way. All this reveals what the general strength of the patient is. To identify the disease, one must look to the sign on the sixth house cusp, the planet that rules it, and the sign that planet is in. Again, the condition (by dignity) of that planet indicates how powerful the illness is. Different planets may signify its nature—Saturn (chronic), Mars (acute)—and the quadruplicities its likely course: cardinal signs, a disease that goes rapidly one way or the other; fixed, one that is entrenched; mutable, an illness prone to wax and wane. Whether the patient (unaided by therapeutic care) can prevail or not will normally be indicated by which of two planets—the lord of the Ascendant or the sixth house cusp—is overall in better shape.

✳ ✳ ✳

ENDURING AS CULPEPER'S REPUTATION IS, still greater renown must be accorded to William Lilly, "the English Merlin," universally considered one of the foremost astrologers who ever lived. It was Lilly who had given Culpeper his first tutorial in astrology when Culpeper had come to London at the age of nineteen, and had lent him an ephemeris and a book of "aphorisms for physicians" to guide his steps; and it was Lilly whose work would be looked to as the standard by most astrologers of talent for the next hundred years.

Born in the village of Diseworth in Leicestershire, on May 1, 1602, Lilly's beginnings were not unlike Forman's. He showed promise as the gifted son of a yeoman farmer, longed for higher education, seemed likely to get it, then saw his early prospects dashed. When he was a youngster, his father was imprisoned for debt, and though some land was sold to defray it, the family still had to scrape. Even so, Lilly received a good grammar school education—"my mother always intending I should be a scholar"—learned Latin, Greek, and a little Hebrew; read Cato, Cicero, Ovid, Horace, Vergil, and Homer; and "could make extempore verses" in any meter on any theme. He was also, it seems, the star of his school debating team. "If any scholars from remote schools came to dispute," he tells us, "I was ringleader to dispute with them; and if any minister came to examine us, I was brought forth against him, nor would I argue with him unless in the Latin tongue, which I found few could well speak without breaking Priscian's head."

After his mother died, Lilly taught school for a year, which agreed with his bookish bent; but that upset his father, who wanted his help on the farm. Lilly demurred, in part out of a pronounced aversion to "country labor," as he put it, but also because he could see "no hope by plain husbandry to recruit a decayed estate." Not surprisingly, his father denounced him as useless and threw him out of the house. Lilly accepted his fate, bought a rough new suit of traveling clothes, and (like Dick Whittington with his cat) set off for London with twenty shillings in his pocket to make his way in the world. He walked the whole way, through wind and storm, and arrived with seven shillings and sixpence left. Yet his prospects were not as bad as they seemed. Through a family friend he obtained a letter of reference to Gilbert Wright, a prominent merchant in the salt trade who was also "an upper servant to Sir John Puckering, the Lord Keeper of the Great Seal, sometime Speaker of the House of Commons, and a great favorite of the Queen." Lilly became his general assistant, which, to his chagrin, entailed all sorts of menial chores in addition to clerical tasks. He had to clean Wright's shoes, sweep the street before his house, fetch water in a tub every morning from the Thames, weed his garden, empty his slops, and "ceremoniously walk before him to Church on Sundays." During the seven-year term of this indentured servitude, as he saw it, his master went through three wives. The second of the three, who died of breast cancer, had apparently consulted Simon Forman at some point, for one of his sigils was found among her belongings at her death.

Wright's third wife, Ellen Whitehaire, however, outlived him, and it fell to Lilly to settle Wright's affairs and pay his debts. Lilly did so in a competent fashion, which kept the estate intact, and in grateful admiration Wright's widow decided to make him her own. "She had many suitors," Lilly tells us, "old men, whom she declined; some gentlemen of decayed fortunes, whom she liked not, for she was covetous and sparing," but his own combination of acumen and youth had the right appeal. On September 8, 1627, they were married, and though Lilly was scarcely more than half her age, they lived quite happily together for the next six years. She enjoyed his youthful vigor, and he "reveled in the opportunity to live comfortably" without having to work. Who could blame him? After drudging for Wright, he now went fishing in the Thames, passed his time at games such as primero (a form of poker) and basset, and frequented the theaters and fairs. Sometimes he went out to Lincoln's Inn Fields, to watch the long-bow archers bully up their marks, or took leisurely walks with his own servant now in tow. Yet he was not an idle soul. His passion for learning was prodigious, and in addition to his private medical studies (precipitously begun when Wright's second wife was dying), he kept up with the classics and his knowledge of Latin and Greek.

One Sunday in 1632, Lilly met a law clerk at church who afterward introduced him to an astrologer. That astrologer was John Evans, a Welsh divine who had been forced to flee his Staffordshire parish for some offense. He was an odd character and not the best of the breed. When Lilly first met him, he was "suffering from a monumental hangover" and living in a flophouse in Gunpowder Alley, off Shoe Lane. In his person, Evans was somewhat deformed—"a squat little man, dark and beetle-browed, with splay feet and a flattened nose"—shrewd but irascible, and a bit of a rogue who often got into street fights and tavern brawls.

Still, Lilly was intrigued, engaged Evans to instruct him, and after seven or eight weeks of intensive study "could set a figure tolerably well." That was something, for in London at the time, according to Lilly, an in-depth academic knowledge of the art was fairly rare. Evans himself, in fact, had almost no library to speak of and relied mainly on *The Judgment of the Stars* by Haly Abenragel, the court astrologer to an 11th-century Tunisian prince. Lilly made the rounds in search of other adepts, but didn't have much luck. For example, there was one Alexander Hart, who lived in Houndsditch, "a comely old man of good aspect. He professed Questionary [Horary] Astrology and a little Physick. His greatest skill was to elect young gentlemen fit times to play at dice, that they might win or get money. I went unto him for resolutions for three questions at several times, and he erred in every one. To speak soberly of him, he was but a Cheat." Hart was later indicted for fraud but rescued by the intercession of John Taylor, "the

Water Poet," who talked the Lord Chief Justice into granting him bail. Once released, Hart fled to Holland, where he died. Another pretender was a certain "Captain Bubb," who lived in Lambeth Marsh; there was also a William Poole, "a nibbler at Astrology, sometimes a gardener, a drawer of linen, a plasterer and a bricklayer. He would brag many times he had been of 17 professions; was very good company mostly for drolling," and had a penchant for scatological verse.

A couple of others were more worthy. There was Jeffrey Neve, the former mayor of Yarmouth, who had "a little smattering in Astrology, could resolve a question of theft, or love-question, and something of sickness," which wasn't bad. Better still was William Bredon, the vicar of Thornton, who went strictly by Ptolemy, "which he well understood," and was "absolutely the most polite person for Nativities in that age." Bredon had collaborated with Sir Christopher Heydon on his *Defense of Astrology,* and was a reverent soul, but so addicted to tobacco that for lack of it he once cut the bell ropes of his church to smoke in his pipe.

Not long after Lilly met Evans, his wife, Ellen, died and left him with a comfortable income, which he used to lease his house and buy a part share in thirteen houses in the Strand. That allowed him the freedom to turn his full attention to the stars. He bought up everything on the subject he could find, and eventually his library included almost every relevant book in print. The heart of it was acquired from the Reverend Arthur Bedwell, former chaplain to Sir Henry Wotton, whose collection contained many choice works, including Valentine Naibod's *Commentary upon Alchabitious,* which Lilly especially prized. His study of astrology now preoccupied him wholly, by his own account, up to eighteen hours a day, even as he began to experiment with various forms of occultism, including astral magic and crystal-gazing, though he later gave these up.

From Evans he had learned the basics: how to draw up and interpret a birth chart, cast for horary questions and elections, read a solar return, ingress charts, and so on, as well as the importance of shrewd judgment in getting a reading right. Evans himself, according to Lilly, "had the most piercing judgment naturally upon a figure of theft and many other questions that I ever met withal," which was high praise indeed, considering what a master Lilly became. However, the two eventually had a falling out when it became apparent to Lilly that

Evans was willing to doctor his readings for a fee. One day, it seems, a woman came to see Evans and "I standing by all the while and observing the figure," Lilly tells us, "asked him, why he gave the judgment he did, since the significators showed the clean contrary, and gave him my reasons; which when he had pondered, he called me 'Boy,' and must he be contradicted by such a novice? But when his heat was over he said had he not so judged to please the woman, she would have given him nothing, and he had a wife and family to support."

Lilly may have had his opportunistic side, but in astrology he was a purist, and his developing standards made him impeccable in the practice of his craft. Years later, in laying out a code of ethics for aspiring astrologers, he offered this sound yet lofty advice: "Be humane, courteous, familiar to all, easy of access: afflict not the miserable with terror of a harsh judgment [prediction]; direct such to call on God, to divert his judgments impending over them. Be civil, sober, covet not an estate. Give freely to the poor, let no worldly worth procure an erroneous judgment from thee, or such as may dishonor the art."

To Lilly's credit, he heeded his own counsel, and devoted many years to the practice of astrology before venturing into print. By then he had a solid name. His clientele, like Forman's, ran the gamut, from nobles and high state officials to chambermaids. In the course of his career, he drew tens of thousands of charts (four thousand in one two-year period alone) entailing every kind of question, though an increasing number had to do with matters pertaining to the English Civil War. In the 1640s, he attracted the interest and friendship of a number of men of power, including Bulstrode Whitelocke, later keeper of the Great Seal; Sir Philip Stapleton (a prominent parliamentary leader); Robert Reynolds; Sir Robert Pye; Anthony Ashley Cooper, later first Earl of Shaftesbury; Philip, fourth Earl of Pembroke; and Sir Christopher Wray. All of these men had come to him at one time or another for astrological or medical advice. An ailing Whitelocke, for example, had brought him a sample of his urine, and by setting a figure for the moment he received it Lilly determined (correctly, it seems) a proper treatment for the disease. Indeed, by 1644 Lilly's practice was already thriving when he published his first almanac, modestly entitled *Merlinus Anglicus Junior,* followed by *A Prophecy of the White King and Dreadful Deadmen Explained.* From then on until his death he issued a new almanac each year.

Of all the many predictive manuals that crowded the bookstalls in those years, his were by far the most popular and widely esteemed. He had some trouble though over the way the first one came out. It so happened that the government licenser of all such texts at the time was John Booker, also an astrologer, "who had earned a considerable reputation by correctly predicting the deaths of Gustavus Adolphus and the Elector Palatine." Booker published an annual almanac of his own, and was initially wary of Lilly as a potential rival. So before he agreed to license Lilly's work, he "made many impertinent obliterations," as Lilly put it, which tempered and abridged the text. He also ran it past some religious censors who grilled Lilly about his personal beliefs. Even so, it was a great hit, read avidly by members of Parliament as well as the general public, and Booker was soon forced to reissue it uncut.

Meanwhile, in 1642 England had erupted into Civil War. Astrologers took sides, of course, and their forecasts were often used as propaganda—each tending to emphasize those planetary trends that seemed to favor his own cause. Lilly cast his lot with Parliament, and proved so potent in the influence he wielded by his predictions that the king declared he would trade half a dozen regiments to get him on his side. One day when Cromwell's army was in Scotland, about to do battle, a soldier stood with Lilly's almanac in his hand, crying out as the troops passed by, "Lo, hear what Lilly saith; you are in this Month promised Victory; fight it out, brave Boys!" which gave them a confidence only Cromwell himself could have inspired. Among important battles, he successfully predicted the crushing Royalist defeat at Naseby in June 1645, and the capitulation of the Royalist stronghold of Colchester in 1648. George Wharton, a Royalist astrologer, attempted to challenge Lilly at every turn, but in the competition of their predictions was almost always wrong. Lilly, in truth, didn't think much of Wharton's skill. Once, when Wharton ventured to criticize Lilly's remarks on a Mars-Saturn conjunction in Taurus, Lilly wrote to him scornfully: "You are yet in your ABC's; when you have learned your Accidence you may be able to prate, but not correct." That only fueled Wharton's ire. In 1647, they clashed again. Lilly had publicly announced, "God is on our side; the Constellations of Heaven after a while will totally appear for the Parliament, and cast terror, horror, amazement, and frights on all those Dammee-Blades [Cavaliers] now in Arms against us." Wharton rashly rushed to proclaim: "The Stars are now so propitiously inclined to our Party, and so averse and destructive to theirs . . . that we cannot fail of Victory." Yet Parliament prevailed.

Lilly rebuffed a number of Royalist attempts to win him over, and on one such occasion declared that "he would rather eat bread and water with the Parliament than roast meat with the Cavaliers." Yet throughout the conflict he never lost some personal affection for the king. Indeed, even while he was being consulted by Cromwell's General Thomas Fairfax (who professed to "know nothing of astrology" but hoped it "was lawful and agreeable to God's word"), he could not turn away when asked to help save the king's life. It so happened that after Charles I had been captured, he sent an emissary to Lilly to ascertain, by astrological means, "in what quarter of this nation he might be most safe" to go should he escape. Lilly drew up a chart in response to the question; it indicated that he would be wise to seek refuge in Essex. Instead, the king made for Southampton, where he hoped to take ship for France. No ship was available, so he crossed to the Isle of Wight, where he was caught. Lilly later remarked: "I told him to go east and he went west."

Lilly was an adept in every branch of astrology—horary, natal, elective, and mundane—as well as on the impact of comets, eclipses, and parhelia on public affairs. To a remarkable degree, he got things right. In his *Merlinus Anglicus Junior* of 1644, for example, he had analyzed the relations between the king and Parliament according to a figure for the Sun's entrance into Aries for that year. The planetary pattern showed a mutual desire for peace (since the planets that signified the two parties formed a benevolent aspect); on the other hand, those planets representing the articles of peace themselves, then under negotiation, were skewed in favor of the king. (The telltale planets, it seems, were in trine aspect to the tenth house, "representing his Majesty" but not to the Ascendant, by which Parliament was shown.) In time, the articles were rejected precisely because the king sought an unfair advantage by their terms. In another pamphlet published the same year—*England's Prophetical Merlin*—Lilly linked the Civil War to a recent Jupiter-Saturn conjunction, and thought the solar eclipse that had taken place in

Gemini in 1639 portended the king's demise. This seemed confirmed by two mock Suns, or parhelia, which had been observed flanking the Sun itself on the king's own birthday, November 19, 1644.

Mock suns had been of interest to astrologers and natural philosophers for a long time. Aristotle thought they were produced by the refraction of sunlight by ice crystals in the "upper air"; others, by some effusion from the Earth. Almost all thought they had a meaning for events. Pliny reported that parhelia had hung over Rome when Augustus was enthroned; and a vision of three suns in 1434 was supposed to have predicted the fall of Constantinople in 1453. On February 2, 1461, at the battle of Mortimer's Cross during the War of the Roses, Edward IV had also observed three suns, which Shakespeare recounted in *Henry VI, Part 3*. Edward says: "Dazzle mine eyes, or do I see three Suns?" Richard, the Duke of York, replies:

> Three glorious suns, each one a perfect sun;
> Not separated with the racking clouds,
> But sever'd in a pale, clear-shining sky.
> See, see! they join, embrace, and seem to kiss,
> As if they vow'd some league inviolable:
> Now are they but one lamp, one light, one sun.
> In this the heaven figures some event.

No one doubted that the Sun was the natural symbol of the king, so it seemed natural enough that his fate would be reflected in solar events. Mock suns could appear for good or ill. At Mortimer's Cross they had merged in surpassing brilliance to spook the foes of Edward IV. But, Lilly wrote in *An Astrological Prediction* (1648), "when mock Suns like Yeoman of the Guard, so encompass the true Sun," as they had on the birthday of King Charles, it meant his "captivity," which indeed he suffered, as events played out. Lilly also made the following judgment on a horoscope that he cast in 1647 for the king's fate: "Luna is with Antares, a violent fixed star which is said to denote violent death, and Mars, approaching Caput Algol, which is said to denote beheading, might intimate that." At the time, few thought it would come to such a pass. Two years later the king's head fell on the block.

In 1647, in the midst of the Civil War, Lilly published *Christian Astrology,* a massive work of 841 pages that offered the first comprehensive guide in English to the theory and practice of the art. The whole tradition found reflection in its pages, which in three parts summarized the fundamental precepts, the art of horary astrology, and nativities or natal charts. In his account of the planets and signs, he is often pithy and succinct. As a master of horary astrology he was perhaps beyond compare. His book is full of example charts, many of them remarkable for their kind, with a focused grasp always of the salient points. If an astrologer is asked, "Whether one absent will return or not, and when?" he tells us, "Consider by what house the absent party is signified, and [by] what Planet" (in this instance, the Lord of the ascendant); "and if the journey is short, look to the third house; if of some length, to the fifth; if long, to the ninth; if very long, the twelfth . . . But if he be in a Cadent house [the third, sixth, ninth, or twelfth], and not behold his own ascendant, he neither cares for his return, or hath any thoughts thereof, nor could he return even if he wished." Again, to know "Whether a Damsel be Virtuous or not, Behold the lord of the seventh, the cusp of the seventh, and the Sun; and if they be in fixed signs and well aspected, you may judge that she is correct."

His description of the planets was both pungent and entertaining. For example, Mercury, "being well-dignified, represents a man of subtle and politic brain, intellect, and cogitation; an excellent disputant or Logician, arguing with learning and discretion, and using much eloquence in his speech, a searcher into all kinds of Mysteries and Learning, sharp and witty, learning almost anything without a Teacher." Ill-dignified or poorly placed, on the other hand, it signified "a troublesome wit, his tongue and Pen against every man, wholly bent to fool his estate and time in prating and trying his conclusions to no purpose; a great liar, boaster, prattler, busybody, false." Again, Mars well placed, or dignified, "in feats of War and Courage is invincible, scorning any should exceed him, subject to no Reason, Bold, Confident, immovable, Contentious, challenging all honor, Valiant, lovers of War." Ill-dignified or poorly placed, "a prattler without modesty or honesty, a lover of Slaughter and Quarrels, Murder, Thiefery, a promoter of Sedition, Frays and Commotions, as wavering as the Wind, a Traytor of turbulent Spirit, Perjured, Obscene, Rash, Inhumane."

Lost objects were a staple of horary work, and according to the rules laid down by Lilly, to ascertain the thief and the object's location, the ascendant, in most cases, signified the place from which it was stolen; the planet ruling the ascendant, the person robbed; the Moon and the planet ruling the second house, the object; the fourth house and its ruling planet, where the object was concealed; and the seventh house and its ruling planet, the thief. Prospects for recovering the lost object were good if the Moon was moving toward the ruler of the ascendant, or the planet ruling the twelfth house, or the planet ruling the house in which the Moon itself was placed. Otherwise, not—unless the Moon was in the ascendant or in the second house.

Here is a horoscope Lilly judged concerning the theft of some money. "Question: Money lost. Who stole it? Is it recoverable? 24th May, 1647, 5 P.M." Result:

Here Scorpio ascends, and partly describes the questioner's person; Mars, lord of the ascendant, show his mind and disposition . . . Finding Mercury in an angle, having no essential dignities, and in partile conjunction of Saturn, and square of Mars, I took him to signify the thief . . . The Moon was in a masculine sign, applying to a masculine planet (Mars), and Mercury was in conjunction with Saturn, and square to Mars, both masculine planets; I judged therefore that the sex was male. As Mercury ever signified youth . . . I said he was a youth, of some 15 or 16. I described him of reasonable stature, thin visaged, hanging eyebrows, with some scar or blemish in his face, because Mars cast his square to Mercury; bad eyesight, as Mercury is with evil fixed stars (the Pleiades) of the nature of Mars and Luna [the Moon]; dark hair, because of his closeness to Saturn; a scurvy countenance, and one formerly accused of knavery and theft. The youth significator being in conjunction with Saturn, lord of the third and fourth houses, I judged him the child of some neighbor; and as Luna is in Gemini, and Mercury in Taurus in the seventh house, I said he dwelt either opposite the questioner, or a little south-west.

Lilly turned out to be right on all these counts. Moreover, "as Luna applied to sextile with Mars, lord of the ascendant, and was within about four degrees of the aspect, I judged [the victim of the theft] should . . . have his money again within four days." And within that time, it was returned.

There was a general admiration for his work, though from time to time religious zealots attacked him—rather to his surprise, since he firmly believed astrology was consistent with Christian belief. When one Reverend Thomas Gataker denounced him for a perfectly ordinary observation, Lilly joked to a friend: "I only wrote 'senes bis pueri' [old men are twice children] and he wrote 42 pages against astrology and myself."

But it was his astounding skill at political prediction that amazed the public mind. There "is nothing appertaining to the life of man in this world," Lilly wrote in *Christian Astrology,* "which in one way or other hath not relation to one of the twelve houses of heaven." For example, in 1646 Lilly was asked by a member of the House of Commons, "Whether Presbytery shall stand?" that is, whether the Calvinist Presbyterian faction of the government would succeed in taking power. The Royalists were then in retreat, but those on the other side were split: between the Independents, on the one hand, who believed in freedom of worship for all Protestants, and the Presbyterians, who did not. The Independents looked to Cromwell and the army for their strength; the Presbyterians had the House of Commons on their side. Lilly drew his chart. In it, all the angles were in mutable signs and only Saturn among the planets was "fixed." This indicated that the situation was extremely volatile and could go either way. He then looked to the ninth house (of religion), which gave him three planets to assess: Jupiter, the natural ruler of the ninth and "the general significator of religion"; Saturn, which occupied it; and Venus, the ruler of Taurus, the sign on the ninth-house cusp. Jupiter, though exalted in Cancer, was impaired because it was about to enter Leo, where it would aspect malefic fixed stars and be in its detriment by term and face; Venus was in her detriment in Aries, in the twelfth house from her own (when counting counterclockwise from the ninth as the first), and would have to pass through a square to Jupiter and Mars before reaching her own sign; finally, Saturn (representing the Presbyterians) was also in poor shape: it was "peregrine, occidental, with no essential dignity"; failed to make any aspect with the Part of Fortune; and, of paramount importance, failed to aspect the Moon (which stood for the people) in the chart. From all this, Lilly concluded that the Presbyterians would fail in their bid for public support.

As events transpired, he was right. He also correctly predicted that "within three years" the king (then still living) would be gone and "a more amenable government rule." Three years later, on January 30, 1649, Charles I was beheaded and a republic was formed. From still another chart, prepared for the date of the king's execution—which he took to be the birth chart of the commonwealth itself—it appeared that the commonwealth would be replaced in turn by a new monarch within twelve years. Just so: it ended with the coronation of Charles II on April 23, 1661.

Of course, he could also be wrong, though that was rare. He did uneven service for Charles Gustavus, the king of Sweden, who gave him a medal and a gold chain (while the service was still good); and was certainly off the mark when he predicted in 1659 that Cromwell's son, Richard, would be able to fill his father's shoes. On at least one occasion, the range of Lilly's acumen also skirted the law. The use of astrology in medicine was generally accepted, and in horary castings few topics were out of bounds. But in 1655, Lilly was indicted for having "unlawfully given judgment in return for payment respecting the recovery of stolen goods." The recorder entered a plea in Lilly's favor and the case was dismissed.

Throughout these years, Lilly kept his practice going at his house on the Strand, where he based his fees on a sliding scale. He charged half a crown or more from those who could afford it; a shilling from those of lesser means. On occasion, he gave medical advice to the poor for nothing, and urged his colleagues to do the same. Meanwhile, in 1648 he had received a handsome pension from the government (later revoked by the Rump Parliament, when he predicted its fall); remarried (upon the death of his second wife); and weathered the transition back to monarchical rule. Lilly accepted the shift without resentment, and fervently shared in the hope of the Lord Chancellor, Lord Clarendon, who in his speech to Parliament on the Act of Indemnity, prayed that the nation's horoscope would now assume a brighter shape. "The astrologers have made us a fair excuse," said Clarendon, "and truly I hope a true one: all the motions of these last twenty years have been unnatural, and have proceeded from the influence of a malignant star; and let us not too much despise the influence of the stars. And the same astrologers assure us, that the malignancy of the star is expired; the good genius of this kingdom is become superior, and hath mastered that malignancy, and our own good old stars govern us again."

Yet the height of Lilly's fame still lay ahead. The most celebrated of his forecasts were those made in 1648, when he foresaw the plague that would sweep through London in 1665, and the Great Fire that nearly consumed it in 1666. "In the year 1665," he wrote, "or near that year . . . more or less of that time, there will appear in this kingdom so strange a revolution of fate, so grand a catastrophe . . . as never yet appeared . . . it will be ominous to London . . . to all sorts of people . . . by reason of sundry fires and a consuming plague." Hieroglyphics of this twofold horror subsequently appeared in his pamphlet entitled *Monarchy, or No Monarchy,* published in 1651: images "representing a great sickness and mortality; wherein you may see the representation of people in their winding-sheets, persons digging graves and sepultures, coffins, etc." On the next page, "after the coffins, and pickaxes, there is a representation of a great city all in flames of fire." On the cover was also a striking woodcut of twins (representing Gemini, the sign on London's ascendant) falling headlong into a great conflagration and men struggling to put out the blaze.

Thirteen years later, ominous signs of the plague itself began to appear in the blistering hot London summer of 1664, when flies arrived in swarms. They lay in matted clusters on the ceilings, and clung in clumpy nests to windows, posts, and doors. The first reported deaths, with their horrible, black lymphatic swellings, occurred in St. Giles and Drury Lane, but the contagion soon spread from one parish to another, until by the spring of 1665 the mounting toll was 400 a week. The plague then spread eastward, encompassing the heart of the city, before crossing the Thames to Rotherhithe and Deptford. In a vain effort to contain it, the taverns and theaters were closed, and all sporting and other crowd-drawing events were banned. A nine o'clock curfew was also imposed, but the deaths exponentially increased each month until, by summer's end, four to five thousand were perishing each week. All who could flee the city did—the king and his court to Oxford; the nobility to their country estates; clergy, doctors, and well-to-do merchants to other retreats. By September, there were not enough able-bodied men left to bury the dead. Dwellings where infection had been diagnosed were marked with a red cross and "Lord have mercy upon us" scrawled upon the door. Every night, carts were driven through the empty streets to collect the corpses, the drivers shouting "Bring out your dead," which were then thrown into huge burial pits covered with lime. Although flea-laden rats were of course the real emissaries of contagion, in one counterproductive and misguided measure—inspired by the idea that pets helped spread the disease—the mayor ordered all dogs and cats in the city killed. But that just gave the rats, freed of their natural predators, more room.

For some obscure reason, the scent of nosegays was believed to ward off the germs, and many who succumbed were found with the gentle flower in their hand. Hence that riddling rhyme:

> Ring a ring o'roses
> A pocketful of posies.
> Atishoo! Atishoo!
> We all fall down.

The ring of roses was the circular red pattern of the telltale rash; the posies were the nosegays; "atishoo" was the violent sneezing fit that plague victims often suffered toward the end; and "we all fall down" told the common fate. Indeed, before the plague subsided completely in the winter of 1666, some hundred thousand Londoners had died.

One year later, on the night of September 2, in the house and shop of Thomas Farynor, baker to the king, in Pudding Lane, the Great Fire of London began. Farynor had neglected to douse the coals in his oven the previous evening, and embers scattering from it ignited some kindling nearby. At one o'clock in the morning, a servant woke to find the house ablaze. Sparks from the house fell on hay and straw in the yard of an adjacent inn; from there the flames flowed like lava across the thatched roofs of London's close-packed hovels to engulf warehouses on Thames Street filled with a combustible mix of hemp, spirits, and oil. The Lord Mayor, Sir Thomas Bloodworth, was awakened, but long since exhausted by his battle with the plague, thought nothing of it and said merely, "A woman might piss it out." Within five hours, however, the blaze had become an inferno, fanned by a strong east wind. It roared through the mostly wooden houses sealed with tar and pitch, and wrapped itself around St. Paul's Cathedral, whose bricks burst like grenades. On the second day, wrote the diarist John Evelyn, "we beheld that dismal spectacle, the whole city in dreadful flames near the waterside; all the houses from the bridge, all Thames street and upwards towards Cheapside, down to the Three Cranes . . . The people were so astonished that from the beginning, I know not by what despondency or fate, they hardly stirred to quench it, so that there was nothing heard or seen but crying out and lamentation, running about like distracted creatures without at all attempting to save even their goods." Belated attempts to create firebreaks by blowing up houses in the path of the flames failed to halt them and by the fourth and final night, "all the sky was afire, like the top of a burning oven," and London within a radius of ten miles as bright as day. Most of the surviving inhabitants fled to outlying fields as far as Highgate, where they put up makeshift shelters as best they could. In all, fully 80 percent of the city, including 13,000 houses and 89 churches, was reduced to charred rubble. Almost the whole of Shakespeare's London was gone.

After both catastrophes came to pass, Lilly was summoned to appear before a committee of the House of Commons and closely questioned about his knowledge of the conflagration and its cause. (They could hardly suspect him of having conspired to cause the plague.) But he succeeded in vindicating himself completely, for as he afterward wrote: "The Committee seemed well-pleased with what I spoke." With respect to his prescient pamphlet, he told the committee that he had found it "most convenient to signify my intentions and conceptions thereof in Forms, Shapes, Types, Hieroglyphics, etc., without any commentary, so that my judgment might be concealed from the vulgar, and made manifest only unto the wise . . . Having found, Sir, that the City of London should be sadly afflicted with a Great Plague and not long after which an exorbitant Fire, I framed these two hieroglyphics as represented in the book, which in effect have proved very true." Lilly added that he had subsequently attempted to discover by astrological means whether the fire had been deliberately set but that seemed not to be the case. He concluded, therefore, that it was "the finger of God."

How did Lilly foresee the fire? Here a distinction between natal and mundane astrology must be drawn. In natal astrology, a conjunction of the Sun, Moon, or one of the four angles of the chart with a powerful fixed star was considered of great moment—for example, when Charles I was born, his Sun stood at 8 degrees 3 minutes Sagittarius and was therefore conjunct Antares, a malevolent fixed star that indicated "a rash, ravenous, and head-strong person, destructive to himself by his own obstinacy," which, few would contest, Charles proved to be. In mundane astrology, a judgment from a direction of the ascendant to a fixed star had more weight. Why? As Guido Bonatti explained: "The Fixed Stars are slow in motion, and consequently in mutation; whence it comes to pass that their impressions require subjects and patients of the same nature, that is to say, such as are the more lasting [like cities, states, or nations], and carry a conformity with them to perfect or accomplish their effects." In foreseeing the Great Fire of 1666, Lilly had noted the close proximity of the Bull's North Horn—a star of the second magnitude, "of the nature of Mars"—to London's ascendant (at 17 degrees 54 minutes Gemini), and calculated the time of its conjunction for 1666. (In a similar way, the transit of Saturn through London's ascendant had also boded ill. Rebellion and pestilence had accompanied such transits in 1408, 1439, and 1554; they had coincided with the English Civil War; and in modern times, with the London Blitz.)

After his acquittal, Lilly obtained his medical license and retired to Hersham, Surrey, where in his last years he pursued his practice on a more modest scale and carried out his duties as the local warden of St. Mary's Parish Church. Except for a brief, bitter dispute with John Gadbury (an astrologer of rank whose career he had helped to foster), his retirement was a quiet one in which he was greatly assisted by Henry Coley, an able young astrologer who became his adopted son and scribe. Beginning in 1676, Coley came every summer to Lilly's home in Hersham to work with him on his almanac, and, when Lilly's sight began to fail, on English editions of excerpts from some of the classic texts. These were later published in 1676 as *Anima Astrologiae: Or, a Guide for Astrologers, being the Considerations of Guido Bonatus [Bonatti] and the Choicest Aphorisms of Cardan [Girolamo Cardano]*. Meanwhile, Lilly's health rapidly declined and on June 9, 1681, he died of a stroke. Mourned widely, he was exalted as the "English Atlas" of his arcane craft, and ceremoniously interred beneath a black marble gravestone (which his friend and colleague Elias Ashmole had provided) near the chancel or altar of his church. George Smallridge, later bishop of Bristol, supplied a Latin epitaph.

Chapter 10

T HOUGH ASTROLOGERS MIGHT WRANGLE with one
another over issues ranging from politics to the technicalities of
their art, all belonged to the same confraternity of sorts, and occasionally suspended hostilities in a civic get-together, organized by the
Society of Astrologers, called Astrological Feasts. These were apolitical,
nonpartisan events in which about forty astrologers, on average, often
otherwise at odds, met convivially to enjoy a good meal, "some technical gossip," and "drown their differences" in wine. The mayor of London himself sometimes presided at these events, which were usually
hosted at notable clubs or taverns—such as the White Hart in the Old
Bailey, Painters' Hall, or the Three Cranes in Chancery Lane. Lilly
himself (despite his disputes with Gadbury, Booker, and Wharton) had,
for the most part, an excellent rapport with most of his colleagues;
though he had sided with Parliament throughout the Civil War, he
established friendships across party lines. Indeed, after the war, when
George Wharton was imprisoned by Cromwell and might have been
hanged, Lilly's great friend and colleague Elias Ashmole had appealed
to him for help, and he graciously forgot Wharton's former insults and
successfully appealed to Bulstrode Whitelocke, now president-elect
of the State Council, to intervene. It was with the staunchly Royalist
Ashmole, in fact, that Lilly would form the closest bond.

Born at Lichfield on May 23, 1617, Ashmole, the son of a saddler,
had studied mathematics at Oxford, law under the guidance of Sir
James Pagit, Baron of the Exchequer, and had fought with the Cavaliers. During the Civil War he had met Wharton, who first introduced

him to astrology, and a year later (in 1646) met Lilly through Jonas Moore, mathematical tutor to the Duke of York. It was Moore who would later persuade Charles II to build the Royal Greenwich Observatory for John Flamsteed, the first astronomer royal of the realm.

According to the contemporary diarist John Evelyn, Ashmole was "addicted to astrology." There is no doubt that once he got a taste of it he could scarcely get enough. Over time, he became learned in all its aspects, and adept enough that (like Lilly) in fifteen minutes on average he could calculate a chart. Alchemy and magic also cast their spell. Ashmole's mature life was utterly shaped by these interests and he was constantly on the lookout for related books and tracts. He eventually acquired the libraries of John Dee, John Napier, Simon Forman, and Lilly himself among others, which formed part of the nucleus of the great collection that later helped to make his Ashmolean Museum at Oxford a unique center for scientific and antiquarian research.

Ashmole was especially fond of elections, and in his *Theatrum Chemicum,* a work mainly devoted to alchemy, he wrote in a hopeful vein: "By Elections we may Governe Order and Produce things as we please." That of course depended on one's skill. At times, he seemed to have the gift. In 1647, when Ashmole undertook to prepare an index for Lilly's *Christian Astrology,* he elected a time he thought would make it go well: "About 10 after noon I began to make an index of this book," he noted in his diary, "Aquarius ascending Moon applying to [a] sextile with Mercury who was Lord of the [planetary] hour / she [the Moon] being in the 7th house / Mercury descending into the 5th / Moon separating from [a] square [with] Saturn, Lord of the Ascendant, and appling to Mercury. I had been long in determining to do this but did not go about it till now. But now Moon entering Virgo and being slow in motion, and reception between Mercury and her, and he very slow, I believe I shall go through it speedily"—which (all indexers take note!) he did. In short, Mercury, as the planet of writing, was linked, at this elected time, to Saturn, by the Moon, which entered Virgo "a sign associated with attention to detail, where Mercury (writing) also rules."

His choice for a wedding day was not so well done. But that was not for lack of trying. Ashmole's first marriage had been a love match. For his second bride, he chose a wealthy widow twenty years his senior so he could live "in that condition I had always desired, which was that I might be enabled to live to myself and studies without being forced to take pains for a livelihood in the world." However, several things had to fall into place for this to work out. He considered three possible times as best or most promising within a three-day period (November 14–16, 1649) for tying the knot. In the end, he opted for 8 A.M. on the 16th—admittedly an "uncomfortably early hour," as he put it, for gaieties and wedding cake. But the Moon was just then separating from a trine with Mercury, and applying to a sextile with Venus, Saturn, and Jupiter—thus giving, as he thought, "good aspects between the Moon and the benefics, Venus and Jupiter, as well as Saturn." However, he had failed to take into account the directions, or progressions, of his own birth chart for that year, as Lilly later pointed out. The marriage was a disaster, and his wife so incorrigibly morose that he came to use the glyph (♄) of Saturn to signify her in his notes.

Astral magic was an enthusiasm of many learned men in those days. Thomas Hyde, reader of Hebrew at Queen's College, Oxford, and later curator of the Bodleian Library, provided Ashmole in 1661 with information about fashioning sigils based on Islamic texts, and Lilly sent him an assortment, though he soon learned to craft them himself. One group, fashioned under a Saturn-Mars conjunction, was designed to rid his home of vermin, such as fleas and mice. Strategically placed throughout his house—in the kitchen, cellar, storeroom, and elsewhere—they apparently performed admirably, for he developed a "high reputation for disposing of vermin" and others came to him for help. In order to better his chances as a parliamentary candidate for Lichfield in 1678, he also cast magic sigils (in this case unsuccessfully) "for increase of honor and estimation with great men." In truth, they could be made for any purpose. Another astrologer in the 1690s developed sigils that sold for four shillings each "for use as contraceptives by servant-girls."

Under Charles II, Ashmole became a prominent court official, served as Windsor herald, comptroller and auditor of the excise for London, and as secretary and clerk of the Courts of Surinam for life. He was also a founding member of the Royal Society, established in 1662, and wrote *The Institution, Laws and Ceremonies of the Order of the Garter,* "the fruit of prolonged and industrious research." As unofficial court astrologer, he gave coveted advice, was consulted by a number of cabinet officials, including the lord high treasurer, and even occasionally by the king himself—for example, on the timing of some of the speeches he gave to Parliament.

His own professional standards were high. Like Lilly, he regarded the office of astrologer as a priestly one by nature, and in his code of ethics emphasized a chaste and sober devotion to his art. Guido Bonatti, whom Lilly and Ashmole both revered, had insisted that clients consult him only about matters of importance; Lilly, in turn, noted that "those who take this course shall find the truth in what they enquire after; but whosoever do otherwise deceive both themselves and the artist [astrologer]; for a foolish Querent may cause a wise Respondent to err." Yet even under the best circumstances, astrology could hardly be expected to get everything right. "The Planets and Stars are ministers not masters," Lilly reminded his readers on one occasion. "Expect not that all accidents shall precisely happen to a day or a week." Yet he worried that the ignorant in their zealous demand for infallibility would nevertheless deride its mistakes. No appeal to its venerable traditions would help, since most would hardly know what they were. As George Wharton—in agreement with Lilly on this—remarked, "Ptolemy may be something to eat for aught they know." Ashmole in turn had a horror of astrological quacks—the "multitude of Pretenders that pester'd the age"—and was sure their blundering, unschooled pronouncements would ultimately subject the art to "scorne and contempt." In *Theatrum Chemicum,* he warned his readers, "Trust not to all Astrologers . . . There are in Astrologie (I confess) shallow Brooks, through which young Tyroes may wade; but withal there are deep Fords, over which the Giants themselves must swim."

Such warnings were inevitably lost on determined skeptics, of course, and anathema to satirists, who leapt at the chance to skewer the foibles of the craft. No one leapt with more zest than Samuel Butler, as the "scorn and contempt" that Ashmole feared came tumbling out in abundance from his pen. In *Hudibras,* a mock-heroic poem in octosyllabic couplets, he held up to promiscuous ridicule every aspect of contemporary life, including the sectarian squabbles of the Presbyterians and Independents, popular pastimes like bear-baiting, and the pretended skill of astrologers in matters of love and war. To Butler, all astrologers were thieves, liars, vagabonds, and rogues. As a Royalist and vicarious aristocrat (sprung of humble stock), he also seemed to single out Lilly in the character of Sidrophel for his sundry "seditious" predictions during the Civil War:

> Did not our great reformers use
> This Sidrophel to forebode news;
> To write of victories next year,
> And castles taken, yet i' the air?
> Of battles fought at sea, and ships
> Sunk, two years hence? the last eclipse?
> A total overthrow given to the king
> In Cornwall, horse and foot next spring . . . ?

In one of the poem's more memorable scenes, Sidrophel is consulted by the hero, Hudibras, about his prospects for wooing a widow, only to be thrashed when his pronouncements prove amiss. As summarized by the "Argument" for Part 2, Canto 3:

> The Knight, with various Doubts possest,
> To win the Lady goes in quest
> Of Sidrophel, the Rosy-Crucian,
> To know the Dest'nies' Resolution;
> With whom being met, they both chop Logick
> About the Science Astrologick,
> Till falling from Dispute to Fight,
> The Conj'rer's worsted by the Knight.

Butler, however, had to get around the fact that Lilly had often been right. He thought Lilly had simply been lucky in his public predictions, and had only succeeded with private clients by hiring private detectives to research their lives in advance. Thus Sidrophel has an assistant in the poem named Whachum, whose

> business was to pump and wheedle,
> And men with their own keys unriddle;
> And make them to themselves give answers,
> For which they pay the necromancers;
> To fetch and carry intelligence,
> Of whom, and what, and where, and whence . . .

> So Whachum beats his dirty brains
> T'advance his master's fame and gains.

A similarly wholesale rejection of the art may be found in John Wilson's *The Cheats,* a Restoration satire in which the mere list of an astrologer's claims was meant to elicit jeers. One rogue declares:

> I resolve these ensuing astrological questions: the sick whether they shall recover or not; the party absent whether living or dead; how many husbands or children a woman shall have; whether you shall marry the desired party or whom else, whether she has her maidenhead or no, or shall be honest to you after marriage, or her portion well paid; if a man be wise or a fool; whether it be good to put on new clothes, or turn courtier this year or the next; if dreams are for good or evil; whether a child be the reputed father's or not, or shall be fortunate or otherwise; ships at sea, whether safe or not; whether it be good to remove your dwelling or not; of law-suits which side shall have the better; and generally all astrological questions whatsoever.

That sort of satire was unremitting. The poet John Dryden had a defter touch, and made light of astrologers who sagely "predicted" past events. Again, Lilly by dint of prominence was made the butt of the joke. "Thus, Gallants, we like Lilly can foresee, / But if you ask us what our doom will be, / We by tomorrow will our Fortune cast, / As he tells all things when the Year is past." That wasn't Lilly's practice, of course, as Dryden well knew, but he saw a commercial chance to exploit Lilly's name. Doing so also gave Dryden cover, for he was an astrologer himself. He mined the subject for his plays, based his *Annus Mirabilis* in praise of the Restoration on an astrological idea, and had consulted Ashmole about his own chart. In the *Annus Mirabilis* ("Year of Wonders," 1666), Dryden alluded to the plague of 1665 and the Great Fire of 1666, both marked by comets, and gave London an astrological reprieve:

> The utmost malice of the stars is past,
>   And two dire comets which have scourged the town,
> In their own plague and fire have breathed their last,
>   Or dimly in their sinking sockets frown.

Now frequent trines the happier lights among,
    And high-raised Jove, from his dark prison freed,
Those weights took off that on his planet hung,
    Will gloriously the new-laid work succeed.

When his son Charles was born, Dryden strictly noted the moment of his birth, and after casting his horoscope "observed with grief that he had been born in an evil hour." As he explained to his wife, Jupiter, Venus, and the Sun were all "under the Earth," that is, below the horizon, and the lord of the ascendant was squared by Saturn and Mars. When Dryden progressed the horoscope, he found recurrent dangers of a violent death in his son's eighth, twenty-third, thirty-third, and thirty-fourth years. As it so proved. When Charles was eight, Dryden took him on holiday to the country seat of the Earl of Berkshire, who had arranged for a stag hunt to amuse his guests. To keep the child out of mischief, Dryden "set him a double lesson in Latin, with a strict injunction that he should not go out of the house." But it happened that while young Charles was poring over his declensions, the stag under chase fled straight toward the house. As the servants hastened out to see the sport, one of them took Charles by the hand. Just as they reached the gate, the stag and dogs together clambered over an old wall, which promptly collapsed and buried Dryden's son in debris. Charles, though badly injured, survived, only to be nearly killed again in his twenty-third year when, on a tour of Italy, he fell from a ruined tower on the Vatican grounds. Then in the thirty-third year of his life, he drowned while swimming the Thames.

✳   ✳   ✳

ASTROLOGY ENJOYED OFFICIAL STANDING in England longer than elsewhere, and the same King Charles II who encouraged and subsidized Butler's satirical work was loathe to release Ashmole and other court astrologers from their private obligations to the Crown. In those days, everyone seems to have consulted someone. The astrologer Richard Saunders (whose name Benjamin Franklin would later appropriate for his *Poor Richard's Almanac*) was consulted by Sir Walter Cope, master of the Court of Wards; John Booker by Lord Berkeley, Earl Rivers, Sir Edward Harington, Oliver Cromwell's son-in-law, and a number of titled ladies; Ashmole by Sir Robert Howard, Sir

John Hoskins (president of the Royal Society), the physician Thomas Wharton, and the linguist John Ogilby, who asked Ashmole for a propitious time to start learning Greek. The time chosen must have been a good one, for he became a celebrated translator of the Homeric tales.

The king himself was bound to the art from the start. At the time of his birth in 1630, both Venus and the Pleiades (called "Charles Waine" in the vernacular) were visible at midday, and "a noonday star" had appeared in the sky, which some took as a sign of heaven's blessing and proof that he was born by divine right to rule. For "as soon as Born, Heaven took notice of him, and ey'd him with a Star, appearing in defiance of the Sun at Noon-day, either to note, That his life shou'd shine like a Star; or else to prove, that if it be question'd, whether Sovereigns be given us by chance, or by the hand of the Almighty, it is here manifest, that this Prince came from Heaven." After Charles was crowned, the star was compared by Royalists to the Star of Bethlehem that had signified Christ's birth, since the Restoration itself was seen as a kind of Resurrection, with Charles cast in the role of savior of his flock. In part for political reasons, the king embraced all this, though even before his elevation, his horoscope, drawn by Sir Kenelm Digby when the two were in exile together in France, had forecast his recovery of the throne.

The fact that Charles took astrological advice was known to Louis XIV of France, who had his own court astrologer in Jean-Baptiste Morin. Morin had originally been employed by Louis's father, and at the dauphin's birth had been smuggled into the chamber at Versailles to record the child's first cry. From the chart he drew, Morin apparently predicted much that afterward transpired in the Sun King's reign.

Born on February 23, 1583, at Villefranche, Morin had studied medicine at Aix and Avignon, earned a name for himself as a superb mathematician as well as a doctor, and in 1630 was appointed royal professor of mathematics at the Collège de France. He was quite ingenious, and one of his early achievements was to develop a reliable method for calculating longitude at sea. The elevation of the Moon, as one scholar explains, "was measured from a star whose position was known exactly, and from this the right ascension and latitude as well as its longitude and declination were obtained. It was necessary then to calculate according to tables the time when the Moon had this same position in the sky in the place for which the tables were compiled and of which the longitude was known. The difference in the time when converted into degrees would give the position of the ship."

As an astrologer, Morin became an independent power at the French court. At one time or another he served the Bishop of Boulogne, the Duke of Luxembourg, the Duke of Effiat, and Cardinal Mazarin; developed a close, if uneasy, rapport with Cardinal Richelieu (who had his own court astrologer in Jacques Gaffarel, the royal chaplain); and elected propitious times for the Comte de Chavigny, the French secretary of state, to begin his official trips. According to the *Larousse Encyclopedia of Astrology,* Morin also correctly predicted the exact date of the death of King Louis XIII, and the demise of Cardinal Richelieu "to within ten hours." Few knew the science as well as he, as amply demonstrated toward the end of his life in a massive astrological treatise called *Astrologia Gallica,* or *French Astrology,* published in 1661 and sometimes compared to Lilly's *Christian Astrology* as a commanding work. Here is his analysis of some aspects of a solar return:

The more the figure of the revolution, whether solar or lunar, is similar to the radical position of the signs and planets, the more efficaciously it will bring forth the significations of the geniture, whether good or evil, and especially those that will be signified by a similar direction. For that similarity is not always favorable and a promise of some great good, but it only signifies the same things as the figure of the geniture, whether good or evil. Otherwise the planets would not act in accordance with their determinations . . . And in particular those revolutions should be watched for in which the same degree of the ecliptic is found in the Ascendant as was in the Ascendant of the radix; for then each planet rules the same houses in the revolution as it ruled in the radix, which does not usually happen without [producing] some noticeable effect signified by the nativity, since the force of signification of the signs will also be doubled, at least in the place of the nativity and thereabouts . . . [On the other hand] when the Ascendants of the radix and the revolution are opposed, it is evil and disturbing, and worse still when the degrees [themselves] are opposed, especially in the case of a solar revolution. For, since the revolution either brings forth or inhibits the effect of the nativity, and it can only bring it forth from a similarity of the figures, it is plain that this con-

trariety of position, both of the Sun and of the whole caelum, will in-
hibit the radical influx and prevent it from bursting forth into action,
but especially into good action, and will only bring forward ineffective
efforts in connection with the good things signified by the directions
in that year, with many contrarieties, damages, anxieties, sicknesses,
and dangers to life. And the reason is because the signs are then determined to significations contrary to the radix.

★   ★   ★

MORIN AND LILLY BELONGED TO A DYING BREED. After the
Restoration, despite the interest of the English king, astrology was
viewed by many Royalists as suspect, since, after all, it "had played an
important role in spurring the Roundheads on to victory over the Cavaliers." it was therefore "indelibly tainted with an aura of subversion,"
especially since Lilly and others had helped to "democratize" some of
the techniques of the trade. It was said, with horror, that every cobbler,
peddler, blacksmith, and roofer claimed the power to predict. Their
idle predictions—like any widely credited rumor—could also stir social unrest. As one writer points out, "to predict the weather was to
predict the harvest; to predict the harvest was to predict the discontent
which would follow a food shortage, and the rebellion which might
follow the discontent." The farmer in *Macbeth* who "hang'd himself
on expectation of plenty" (because it drove the price down) was an example of how some might be swayed; another was the aptly named
Sordido, in Ben Jonson's *Every Man out of his Humour,* who, based on a
prediction in his almanac, decided to hoard his corn. So the Earl of
Northampton was right enough to complain, "Pamphlets which prognosticated famine have been causes of the same; not by the malice of
the planets . . . but by the greediness of husbandmen, who, being put
in fear of such a storm . . . by forestallment, and . . . by the secret
hoarding up of grain, enhance the prices in respect of scarcity." Yet the
obverse was also true. The Royalist astrologer John Gadbury argued,
for example, that a better knowledge of astrology might have rescued
the nation from its fate. How? Well, for one thing, the king might have
seen the turmoil coming, and taken steps to prevent it. And once the
war began, the Royals might have avoided battles they were doomed to
lose.

Though the favor of astrology had begun to wane, few of those associated with the birth of modern science spurned it outright. The vanguard of the learned and scientific community on the whole remained convinced of its value, though some pressed for reform; but most thought there was something to it and viewed it as a discipline that had yet to be fully developed and explored. In their assessment or reassessment, their hope (in Kepler's phrase) was not to throw out the baby with the bath. Even among skeptics, that had long been the trend. In his essay *On the Increase of Knowledge,* for example, Sir Francis Bacon, the "father of modern science," had written: "As for Astrology . . . I would rather have it purified than altogether rejected . . . The last rule (which has always been held by the wiser astrologers) is that there is no fatal necessity in the stars; but that they rather incline than compel. We will add one thing more (wherein I shall certainly seem to take part with astrology, if it were reformed); that we are certain the celestial bodies have other influences besides light and heat. Let this astrology be used with greater confidence in prediction, but more cautiously in election and, in both cases, with due moderation." The great writer, antiquarian, and physician Sir Thomas Browne argued in his *Pseudodoxia Epidemica* for a "sober and regulated astrology," even though his work is studded with appreciations of the art.

For intellectuals generally astrology remained a topic of consuming interest, and engaged figures as diverse as Samuel Hartlib, the social reformer and advocate of universal education; Anthony a Wood, the antiquarian scholar who wrote a comprehensive history of Oxford; the social philosopher John Locke (ideological mentor to our own Founding Fathers), who, not incidentally, "believed in the astrological choice of times for picking medicinal herbs"; and Robert Hooke, the physicist who discovered the law of elasticity, explained the relationship between breathing and combustion (via oxygen), and built the first Gregorian reflecting telescope. William Gilbert, who pioneered magnetic theory, scoffed at the idea that metals were ruled by the planets, but did not doubt that children were influenced by the stars at birth. And as Patrick Curry points out in *Prophecy and Power,* Christopher Wren declared in his inaugural lecture as Gresham Professor of Astronomy in 1657 that there was "a true Astrology to be found by the inquiring Philosopher, which would be of admirable Use to Physick," if it could be correctly ascertained.

Many founding members of the Royal Society were similarly intrigued. Robert Boyle, the father of experimental chemistry, thought there must be something to astrology, otherwise, "we know planets only to know them," which seemed to him absurd. In an essay entitled "Of Celestial Influences or Effluviums in the Air," he expressed his stubborn conviction that "celestial bodies (according to the angles they make upon one another, but especially with the sun or with the earth in our meridian, or with such and such other points in the heavens) may have a power to cause such . . . changes, and alterations . . . that shall at length be felt in every one of us." When John Wilkins, a founding fellow of the Royal Society and afterward bishop of Chester, died on November 19, 1672, his colleague Hooke noted in his journal: "Dyed about 9 in the morning of a suppression of Urine . . . a conjunction of Saturn and Mars . . . Fatall Day." William Ramsey, physician to Charles II, considered astrology indispensable to his practice though he condemned the democratization of the art.

The learned diarist John Evelyn was likewise of two minds. On December 12, 1681, he wrote: "We have had of late several comets, which though I believe appear from natural causes, and of themselves operate not, yet I cannot despise them. They may be warnings from God, as they commonly are forerunners of his animadversions." And when a comet appeared in the winter sky of 1680, John Tillotson, later archbishop of Canterbury, wrote: "What it portends God knows: the Marquess of Dorchester and my Lord Coventry dyed soon after." Lord Brouncker, the first president of the Royal Society, cast the horoscope of Walter Charleton, sometime president of the Royal College of Physicians, at his request. The antiquarian John Aubrey was convinced that "we are governed by the planets, as the wheels and weights move the hands of a clock," but he fully appreciated the infinite vagaries of human personality, and the need for an astrological vocabulary to express them aright. Accordingly, when gathering the biographical data which became his *Brief Lives,* he was careful, as befitted a fellow of the Royal Society, to note the exact nativity of his subjects whenever it could be found. In the course of his research, he managed to get hold of the data for Sir Kenelm Digby, Robert Burton, John Dryden, John Evelyn, Titus Oates, William Penn, Anthony a Wood, Thomas Hobbes, Walter Charleton, Robert Hooke, Edmond Halley, and Christopher Wren, among others. Most of them "had more than a passing interest in their own nativities, and 'revolutions' based upon them," year to year. Aubrey ascribed his own struggles in life to "a crowd of ill directions" (progressed aspects), which he thought explained the many worldly disappointments he endured.

Jeremy Shakerley, an observational astronomer who correctly predicted the solar transit of Mercury in October 1651, embraced astrology, as did the academic astronomer Vincent Wing, who defended it as a "divine art" and produced an astrological almanac as well as horary charts. His colleague Thomas Streete, who also worked with both Hooke and Halley, devised a method of deducing longitude from the motions of the Moon, even as he toiled away on an astrological textbook, being "convinced," according to Curry, "that astrological aspects have great force." Well into the 18th century, the London Bills of Mortality would sometimes allude to those who had succumbed to some sudden, mysterious illness as having been "planet-struck."

Though the first astronomer royal, John Flamsteed, doubted astrology and hated almanacs—because "the Vulgar have esteemed them as . . . oracles" and their predictions had often been exploited to stir up unrest—he took the precaution of choosing an auspicious time (August 10, 1675, at 3:45 P.M.) for laying the foundation stone of the famous Greenwich observatory, and placed the chart for the occasion over the door. True, he may have meant the gesture as a jest; yet three years later on July 4, 1678, he also confided to a friend, "You know I put no confidence in Astrology, yet I dare not wholly deny the inferences of the stars since they are too sensibly impressed on [us]." On the whole, he thought it best to hedge his bets. Eighteen years later he still thought it best to hedge them, for on June 30, 1696, when he attended the laying of the foundation stone for the Royal Greenwich Hospital, he carefully "observ[ed] the punctual time by instruments," and noted the exact time the event took place.

In the records of the prosperous English merchant Samuel Jeake of Rye, who died in 1699, we get a glimpse of how a "split-screen" view of things might work out in a life. Jeake was the friend, or acquaintance, of Flamsteed, Halley, and others of scientific note; broadly shared their interests and followed the learned transactions of the Royal Society; and kept an astrological diary for much of his adult life. Though Jeake was skeptical of the heliocentric theories of Copernicus, supposing (as late as 1670) that "if the Earth moved round, it would be very reasonable to conclude, that a Man running towards the East should rid more ground than if he ran towards the West"—he was not dogmatic about it. He took a scientific interest in weather conditions as correlated to the phase and position of the Moon, and as a business-man, for example, was thoroughly up to date. He took care to diversify his investments, dealt in stocks, and ably navigated the price fluctua-tions of two of the principal commodities he traded (flax seed and hops)—all, however, with the help of horary and elective charts. In point of fact, Jeake was a man of the new age, quite in tune with the new aspirations of late Stuart England, yet every time he bumped his head, or tripped on a stair, he looked for an explanation to the stars.

"July 16, Thursday, About 11 p.m. Going into a Room by dark part . . . ; fell down into the Cellar almost up to the middle in Water. But through the good Providence of God, did neither fall into the Well, nor dash my head against the Cellar Walls, though near both: so that I had not the least hurt. Behold the position of heaven at that in-stant." At which he drew a chart: "Therein these Remarkables are to be observed. First that Saturn is just arising in Pisces a watry sign, & near the degree of the Cusp of the house of death in the Radix (as when my Leg [last January] was in danger of being broke)."

Again: "December 24 (1672), Tuesday, About 9:45 a.m. went with John Weeks junior & another in Company to Westfield. About 1:50 p.m. going over a hedge, I received a blow casually on my right Eye-brow by a Staff of one in the Company, which struck off the skin, & made the place swell & red, & afterward black for a fortnight. And it was a great Providence that the stroke did not light full in my Eye. The Sun neer the Cusp of the 8th opposite to the radical place of Mars & in square to Jupiter & Mars & in opposition of the Moon. All in Cardinal signs."

Once more: "May 14 (1677) Monday. About 8 p.m. Coming down out of a neighbor's Garret (where I had been to prevent a fire &c) at the bottom step, it being only a Stave; my foot slipped & I broke my left Shin, & it was a great mercy that I did not break my leg. Saturn & the Sun in conjunction in the beginning of Gemini in opposition to the Cusp of the 1st."

All this was an attempt to give objective coherence to his life. In that regard, he could be disarmingly candid in the way he described him-self: "My stature short . . . Complexion Melancholy, Face pale & lean, Forehead high; Eyes grey, Nose large, Teeth bad & distorted, Hair of a sad brown, & curling: . . . after 20 had a great quantity of it; but from thence it decayed & grew thin. My voice grew hoarse after I had the small pox. My body was always lean, my hands & feet small, I was partly left handed & partly Ambodexter [ambidextrous]. In my right hand was found [this is palmistry now] the perfect Triangle composed of the Vital [Life], Cephalic [Head], & Hapatick [Heart] Lines, all entire; but the Cephalick was broken in my left."

At various times, Jeake worked with transits, progressions, and solar returns; was heavily influenced by the French astrologer and mathematician Jean-Baptiste Morin, explored palmistry, the magic power of numbers, talismans, and even (briefly) poltergeists. In light of his varied experience and expertise, it is odd that, like Ashmole, he did not elect to marry at a more propitious time. His analysis of the astrology of it (after the fact) is poignant, if frank:

"March 1, Tuesday. About 9:35 a.m. I was married to Mrs. Elizabeth Hartshorn at Rye by Mr. Bruce . . . The Sun shone out just at tying the Nuptial Knot" but "the Positure of Heaven seems not very fortunate for . . . Jupiter is cadent detrified & opposed by the Moon, squared by Mars, Mercury detrified in opposition to Mars & in square of the Moon. Seeming to presignify divers troubles and discontents, as Saturn in the 2nd portends variance about . . . money. Mars . . . aspected in the 5th: Death, or mischief to Children." Alas!

★ ★ ★

AMONG LEADING ASTROLOGERS, the foremost champion of reform was John Gadbury, Lilly's sometime antagonist. Born in 1627, he had studied mathematics at Oxford, joined a "free love" sect in his youth, was pro-Parliament during (but a Royalist after) the Civil War, and in 1652 took up astrology under the tutelage of the mathematician Nicholas Fiske. In 1659, he published his *Doctrine of Nativities,* "a general treatise on natal astrology together with a set of astronomical tables that made it possible to erect and interpret a chart with no other guide." Gadbury also published a number of almanacs but is principally remembered today for his *Collectio Geniturarum* or *Notable Nativities,* in which 150 horoscopes (most of them of eminent persons, including himself) are reproduced. For each horoscope, "the principal accidents of the person's life is usually given along with comments on the natal chart and the primary directions associated with key events."

Like Aubrey, he endeavored to establish astrology as a legitimate science, according to empirical principles, in the hope that its "truth could be demonstrated through an objective study of cause and effect. 'One real experiment,' he declared, 'is of greater worth and more to be valued than one hundred pompous predictions.' "

At the same time, being wholly traditional in some things, Gadbury regarded comets as "beacons, whose use and office is to give warning to mankind of approaching dangers." He also followed a curious branch of received doctrine that held that horary astrologers could use a querent's physical characteristics to validate a chart. "For if the Moles &c. of the Person enquiring, correspond exactly with the Scheme erected, the Artist may safely proceed to judgement." The rules for this were elaborate and concerned the size, shape, color, and location of various marks:

1) Having erected your Figure, consider the sign ascending, and what part or Member in Man's body it Rules; for the Querent hath a Mole, Mark, or Scar in that part of his body. Example, if the Sign ascending be Virgo it is on the belly; if Libra the reins; if Scorpio the Secrets, &c. 2) Then consider in the next place, in what Sign of the Twelve the Lord of the Ascendant is posited, and say the Querent hath a Mole, &c. in that Member or part of his body represented thereby. 3) Observe the place of the Moon and tell the Querent that he hath another Mole, or Scar, &c. in that part of his body that is represented by the sign she possesses. 4) Consider the Sign of the 6th, and the Sign wherein the Lord of the 6th is located; for usually in those Members represented by these Signs, the Querent is also marked. 5) When Saturn shall signify the Mark, &c. it is generally an excrescence of a darkish obscure or black color. If Jupiter it is usually a purple or blueish Mole, &c. If Mars, 'tis commonly some scar, slash, or cut, chiefly in a fiery Sign; and sometimes a reddish Mole, or spots of Gun-powder. If Sun, generally of an olive or chestnut color. If Venus, of a honey color. If Mercury, it is sometimes whitish, and other times of a pale lead color. If the Moon, 'tis often white, yet many times participates of the color of that planet she is in aspect with. 6) If the Planet and Sign representing the Mark, Mole, or Scar be Masculine, the Mars &c. is then on the right side of the body; if feminine, judge the contrary.

And so on. Though Gadbury emphasized a semimodern statistical approach, others believed, on the contrary, that astrology had strayed too far from its classical roots. In John Partridge, Gadbury's rebellious student, they found their leading voice. Born in 1643 in Mortlake, not far from John Dee's ancestral home, Partridge had begun life as a cobbler, taught himself Latin, Hebrew, and Greek, became interested in astrology, studied under Gadbury, and earned his medical degree at Leyden, before he published his own astrological work in 1679. In time, he became completely enamored of Ptolemy's *Tetrabiblos* (the arbiter of all his judgments); "spurned horary work as an Arabic invention, tainted with divination"; extolled natal astrology as a "rational form of knowledge"; and therefore had no use for Gadbury's Baconian reforms.

In his *Opus Reformatum,* or *Treatise of Astrology in which the Common Errors of that Art are Modestly Exposed and Rejected* (1693), Partridge set out to correct "divers Errors in the Study and Practice of Astrology, especially in that Part of it that concerns Nativities," by a strict adherence to Ptolemaic methods, and, like Ashmole, was exasperated by the charlatans who had sullied the Art. He especially didn't like those who dabbled in magic, who "pretended to fetch people back that are absent or run away," or who crafted astrological sigils "to promote or prevent copulation, according to their clients desires, either out of Love or Malice to those they intend it, with abundance more of such stuff, as I could relate, that is practiced under the pretense of Astrology, by a Crew of Scandalous cheats." He set forth his own doctrine in a series of demonstration horoscopes. One was of Oliver Cromwell "with a Table of Directions from his birth to his Death; with each arc, its true measure," showing that he had to die when he did.

Giving Cromwell's birth as April 25, 1599, at 1:05 A.M. at Huntington, England, 52 degrees north latitude, we find: Aquarius rising, Saturn in Libra, Jupiter in Cancer, Mars in Aries, Venus in Taurus, Mercury in Taurus, Sun in Taurus, Moon in Virgo, the Dragon's Head in Aquarius, the Dragon's Tail in Leo. In charting a solar return for his last year of life, Partridge finds:

Saturn and Jupiter are both returned to their own Radical Places; and so is Mars and the Moon to the square of theirs; Mercury and Venus are in Taurus, where they were in the Radix, and not far from their own Radical Places. So that you see all the Planets are returned to their own places: except Mars and the Moon, and they are in square to them. Now, the use I shall make of the Revolution is this: the Moon, Mars, and Saturn are all of them Promittors [receivers of an aspect] by direction; Mars is in square to Saturn, Lord of the Radical Horoscope, who is returned to his Radical Place; and the Moon, though Hyleg [giver of life], yet she is here a Promittor also, and is going to the direct Opposition of the place of Direction, and to the Square of her own place; and besides this, Mars is going to the Mundane Parallel of the Sun. And to sum up all, we find both the Moon and Mars in violent Constellations, the Moon being with Aldebaran, of the Nature of Mars, and Mars with those Stars in the beginning of Cancer called Castor and Pollux, of the nature of Saturn. So that we may from the Sun's return, and its Configurations compared with the Directions, conclude, That according to second Causes, it could be no less than mortal.

To confirm his diagnosis, he also looked at the secondary directions and progressions of the chart "for if all concur, we may certainly judge that nothing but a Miracle can save." All, he found, concurred. "Under this Revolution, we find that the Ascendant by Secondary Motion was directed to the Opposition of Jupiter, the Sun under the Square of Saturn, . . . the Moon to the Opposition of Jupiter, the Moon to the Opposition of Mars, and that just toucheth about the time of his Sickness, all of which are ill, and show a bad year."

A number of other example charts—that of Nostradamus, of Cardano's unfortunate son, the renowned geographer Gemma Frisius, and so on—were likewise advanced to show the inexorable workings of fate. If Partridge's methods seem improbably complex, they were governed by time-honored rules. Kepler, for example, would likely have drawn the same conclusion—did so, in fact, when for his own last year of life he sadly remarked an analogous pattern in his chart.

★　★　★

THE IDEA, still common, that the enlightened scientific establishment had uniformly rejected astrology by Newton's day is not borne out. William Whiston, Boyle lecturer in 1707 and Newton's successor as Lucasian Professor of Mathematics at Cambridge, for example, thought it likely that a comet had signified the biblical Flood. He also thought "an omniscient God could synchronize the regular returns of comets to bring sinners to his knees," so that periodicity was not an argument against the meaning of the event. He therefore had no trouble believing that a comet would one day signify the destruction of the Earth by fire. In medicine, eminent doctors, such as Richard Mead, a vice president of the Royal Society who treated both Newton and Halley, linked epileptic fits, menstruation, and other conditions to "the quadratures of the Moon."

The decline of astrology is generally ascribed to the triumph of the heliocentric hypothesis of Copernicus, to Kepler's laws, Newton's mechanics, the discovery of new planets such as Uranus and Neptune, and, "last but not least to the healthy skepticism of modern man." Yet "with the exception of the discovery of the new planets," as John West and Jan Toonder point out, "these scientific milestones in no way affected either the principles or the practice of the art." "The astrologer himself careth not," Sir Christopher Heydon had long since declared, "whether (as Copernicus saith) the Sun be the center of the world." Indeed, it had made no difference in that sense to Copernicus either, or to Brahe, Dee, Galileo, Kepler, Lilly, or Morin. Again, it was said that after the appearance of "new stars" in a once-unchanging sky, or the discovery of Jupiter's moons, astrology was belied; but astrology was a self-contained system that operated on Neoplatonic and Pythagorean terms. The known periodicity of a comet's return, as established by Halley, was said to make nonsense of astrological prediction, since "it became harder to regard a comet in the sky as a heaven-sent warning of a particular disaster." But not to an astrologer, who, after all, had always dealt with cyclical, or recurrent, celestial events.

Newton himself had some acquaintance with astrology. He discussed its origins in his *Chronology of Ancient Kingdoms Amended* in 1728—in which he "assumed a relationship between the history of nations and the stars"—and thought some distant historical dates might be "rectified" or accurately determined by astrological means. According to a relative, John Conduitt, who married Newton's niece, Newton first began to study mathematics seriously after he had trouble with some of the math in an astrology book purchased at Sturbridge Fair. He was also a closet alchemist, wrote over a million words on alembics and the philosopher's stone, and "used astrological concepts and reasoning" in his alchemical work. For example, "he held that the best water for such work was drawn, 'by the power of our sulphur which lieth hid in Antimony. For Antimony was called Aries by the Ancients. Because Aries is the first Zodiacal Sign in which the Sun begins to be exalted and God is exalted most of all in Antimony.' " The idea of gravity was also inspired (at least in part) by an astrological idea—that of action by one body on another at a distance, which Newton came to regard as "a spiritual force."

Astrology declined not because it had "reason" to, in the Age of Reason, but because God went out of the world: that is, the Neoplatonic God of cosmic harmony and correspondence, compatible with the Christian God, in paradigm. Though Newton himself was clearly drawn to the occult, in the end the mechanical nature of the science he fostered left little room for the impalpable to breathe. "The wild dance of shadows thrown by the stars on the wall of Plato's cave," wrote Arthur Koestler, "was settling into a decorous and sedate Victorian waltz. All mysteries seemed to have been banished from the universe, and divinity reduced to the part of a constitutional monarch, who is kept in existence for reasons of decorum, but without real necessity and without influence on the course of affairs."

PART THREE

The gods have taken alien shapes upon them,
Wild peasants driving swine
In a strange country. Through the swarthy faces
The starry faces shine.

Under grey tattered skies they strain and reel there:
Yet cannot all disguise
The majesty and beauty of the fallen gods, the beauty,
The fire beneath their eyes.

They huddle at night within low, clay-built cabins;
And, to themselves unknown,
They carry with them diadem and scepter
And move from throne to throne.

   —A.E., "Exiles"

. . . These perturbations, this perpetual jar
Of earthly wants and aspirations high,
Come from the influence of an unseen star,
An undiscovered planet in our sky.

And as the moon from some dark gate of cloud
Throws o'er the sea a floating bridge of light,
Across whose trembling planks our fancies crowd
Into the realm of mystery and night,—

So from the world of spirits there descends
A bridge of light, connecting it with this,
O'er whose unsteady floor, that sways and bends,
Wander our thoughts above the dark abyss.

   —HENRY WADSWORTH LONGFELLOW,
    "Haunted Houses"

Chapter 11

I N 1680, one Oxford astrologer wrote to another in distress, "But do you hear the news from Alma Mater. All Astrology must be banished, and that also, as it shall not so much as find room in the imaginations of men!" By then, it had also met censure and resistance at the Sorbonne as well as the French Academy of Sciences, while in Rome a bull was about to be issued against it by the pope. Similar bans were soon to be promulgated throughout Europe—in Germany, Hungary, Bohemia, and other lands. In England, too, it was beginning to go underground. One of the last known high officials to place much stock in it was Sir John Trenchard, secretary of state to William III. Trenchard "confessed on his death-bed that everything his astrologer had predicted for him had come true." The last English sovereign known to be so counseled—by a Dutchman named Von Galgebroh—was Queen Anne, who assumed the throne as the 18th century began. At her insistence, he reluctantly revealed to her the likely day of her death, explaining that August 1, 1714, would be a hard day for her to get past. On that day she died. But on most fronts astrology was under siege. Its practitioners were classed under the vagrancy laws with rogues, fortune-tellers, and "Sowers of Discord or Sedition"; often equated with witches; or ridiculed as superstitious fools. Daniel Defoe (in his *Journal of the Plague Year*), Jonathan Swift, and Samuel Johnson added their influential voices to the scorn. In his great *Dictionary*, Johnson—who nonetheless believed in witchcraft and cherished Culpeper's *Herbal*—defined astrology as "the practice of foretelling things by the knowledge of the stars; an art now generally exploded as irrational and false." In a similar vein, the *Encyclopaedia Britannica* of 1771 dismissed astrology as "a conjectural science" that had long since "become a just subject of contempt and ridicule."

If Lilly had been posthumously skewered by Butler, John Partridge met a living death at the hands of Jonathan Swift. Early in 1708, Swift had brought out a mock astrological almanac under the pseudonym of Isaac Bickerstaff, in which he prominently predicted that "Partridge the Almanack-maker" would "infallibly" die "about eleven at night, of a raging fever" on March 29. The 29th came and went, and on the 30th, Swift duly published a full and detailed "Account of the death of Mr. Partridge, the Almanac maker, upon the 29th Instant," in the form of "a letter from a Person of Honor," replete with a supposed confession as part of the deathbed scene. The hoax was crowned with an "Elegy on the Death of Mr. Partridge" laced with malicious rhymes. It began: "Here five feet deep lies on his back / A cobbler, starmonger, and quack."

On April 1, Partridge was awakened by the local sexton who, mistaking him for some relation, rapped at his window wanting to know if the family had any orders for his funeral dirge. Later that morning, the undertaker's men arrived to measure the house for mourning drapes. Then came a little troupe of embalmers. And that afternoon, when Partridge went out for a walk, people occasionally stopped to tell him he looked exactly like someone they had known recently who died. The Stationers' Company, believing in turn that Partridge was dead, struck his name from their rolls, so that the following year he was unable to get his own work into print. And for several years thereafter he had to keep proving that he was still alive.

Partridge was eventually restored to his former eminence, and Swift died insane; but the latter's mischievous scorn for the common run of almanacs was not unjustified, for in general they had fallen—as esteem for the art had fallen—and were as vacuous as the Sun-sign columns of today. "Their observations and predictions are such as will equally suit any age or country in the world," complained one reader, "This month [it says] a certain great person will be threatened with death or sickness. And we find at the end of the year, that no month passes without the death of some person of note; and it would be hard if it should be otherwise, when there are at least two thousand persons of note in this kingdom, many of them old." Or, "This month an eminent clergyman will be preferred; of which there may be many hundreds, half of them with one foot in the grave." Meanwhile, as science began to concern itself primarily with the measurement of material phenomena, astronomy, the observation of the stars, became divorced from astrology, the interpretation of their significance to man. Forced into the shadows, astrology found a partial, covert refuge in such semisecret societies as the Freemasons and the Rosicrucians, where it was preserved in symbolic form. Masonic traditions in particular, with their roots in the great cathedral building of the Middle Ages, resonated with its teaching with its structural emphasis on the astronomy and astrology of time. The "Royal Arch Mason," for example, was so named to celebrate the arc of the ecliptic that extends from equinox to equinox—from 0 degrees Aries to 0 degrees Libra—which had its focal (or mid-) point at the summer solstice (0 degrees Cancer), the still point of the Sun's eternal round.

The practice of astrology itself did not entirely languish, though most of its representatives, for lack of knowledge or training, were plausibly suspect, and therefore—not implausibly—subject to the vagrancy laws. "It is not to be conceived," wrote Joseph Addison in the *Spectator,* "how many Wizards, Gypsies and Cunning Men are dispersed through all the Counties and Market Towns of Great Britain, not to mention the Fortune-Tellers and Astrologers." Their dubious activities were perhaps epitomized by those of the village witch. One typical item from the time tells us:

A few days ago, Robert Ashworth, residing at Marland, Castleton, went to a woman named Alice Platt in Rochdale, a fortune-teller and a seller of charms, spells, and small medicine, in the hope of being cured of epileptic fits. The following is the remedy prescribed:—At the full moon he was to procure a black tom cat, which he was to put in a reticule. He was then to draw out its tail, which he was to cut at the fourth joint. He was directed to catch thirty drops of blood, put them in a wine-glass of the best Hollands gin, and drink the mixture at midnight, on the same day the moon is full. Immediately after taking the medicine he was to go to bed, to place a sealed paper which she gave him under his head, and not rise until six in the morning. The woman, who is about fifty and wears a man's jacket, assured Ashworth that if he would follow her directions, and repeat the Lord's Prayer when he rose in the morning, he would never have fits again.

Others, however, did their best to keep the dignity of the art and its traditional practice alive. In that regard, perhaps the two most prominent and capable astrologers of the late 18th century were Ebenezer Sibly and John Worsdale, who (though they didn't much like each other) belonged to the dwindling confraternity of adepts. Born in 1766, Worsdale had been a disciple of Partridge. Like his mentor, he rejected the reformed, empirical approach of Gadbury; admired Lilly for his horary skill (though scorning some of Lilly's methods as contrary to classical precepts); and extolled Ptolemy as the lodestone of everything authentic in the art. "Immortal teacher of this Art divine, / Thy Rules on record, most refulgent shine." Worsdale's own strictly "rational" brand of astrology entailed, as one writer put it, "highly detailed mathematical calculations, particularly concerning directions, and arcane interpretive points such as the apheta and hyleg" concerned with the length of life. In the grim tradition of Balbillus—court astrologer to Claudius, Nero, and Vespasian, and whose only known astrological treatise, *Astrologumena,* seems (by its surviving synopsis) to have been almost entirely about how to determine when, and in what manner, a person will die—Worsdale used his expertise in *A Collection of Remarkable Nativities . . . Proving the Truth and Verity of Astrology* to demonstrate, with shocking efficiency, the ability of a horoscope to precisely predict the time and circumstances of a person's death.

He was not the least bit modest about his skills. In the Contents of his *Celestial Philosophy or Genethliacal Astronomy,* for example, he declares: "This work contains an exposition of the Errors of all Ancient and Modern Authors, impartially stated, proving that no Original Works on this Department of Astronomy have been published in this kingdom for several centuries past, and that all modern publications on Nativities, which have yet been given to the world, are pirated from ancient authors." He promised, for his part, to set the record straight and provide "the true method of delivering judgment in all cases . . . divested of every fallacious hypothesis," as exemplified in "thirty remarkable modern nativities, never before published, including" (in particular, for some unknown reason) "the geniture of the infant Duke de Bordeaux." In one chart reading after another, he then proceeded to predict the demise of each client or subject with unremitting success. For example, the fate of one Mary Dickinson, born February 1, 1798, at 3:48 A.M., latitude 53 degrees:

> When this unfortunate Native applied to me, as soon as I had viewed the alarming testimonies that were prevalent, and presented themselves in every part of the Heaven, I was then convinced that a violent Death was apparent, and that there was nothing that could possibly mitigate the fury of those inimical configurations that were conjoined with the Anaretical Directions, in their corresponding places, and Mortal stations. The Ascendant is afflicted by the Body of Mars (who is near the Heart of the Scorpion) and . . . that important Angle is likewise afflicted by the Moon . . . The two Lights of the World are also in opposition to each other, and Mars, just rising at the time of Birth, afflicts the Sun and Moon from obnoxious [malevolent] places and Terms, all of which portend a violent Death.

Jupiter, he noted, was also at the nadir of the chart, afflicted by Mars and Saturn, "Venus applying to the square of Saturn, . . . both benefics deprived of their natural and benign qualities, both by position and Direction," and so on. After continuing in this vein a while longer, he forecast death by drowning. And sure enough we learn that five months later the poor girl drowned.

Worsdale was not an elegant stylist, and his occult jargon could almost be taken for a parody penned by Swift. But beneath the stilted language was an iron logic that among his followers at least inspired both fear and awe.

Though death was his special dominion, he could also be right on other fronts. Eleven years before the defeat of Napoleon at Waterloo, Worsdale predicted that if Napoleon Bonaparte and the Duke of Wellington ever met in battle, Napoleon would lose. He thought so because Wellington's tenth-house Jupiter was exactly conjunct the place occupied by Saturn in Napoleon's chart. Wellington liked to say that the battle had been won long before on the fields of Eton; in an astrological sense, it had been won from the beginning of time. "Jupiter in the tenth house, the mansion of honor, authority, and preferment," we read, "sustains the dignity of the individual; Saturn in the tenth, through his dominant magnetism, brings elevation; but being of the earth, it leads carnal ambition to heights beyond its sustaining power. The very symbol of this planet points to the apparent mastery of Matter (+) over the sublimated essences of the Soul (crescent Moon); an inverted condition which may take on the semblance of security only for a time." In Wellington's chart, Jupiter and the Part of Fortune were also conjunct, while in his seventh house of open enemies or rivals Mars and Saturn (Napoleon, in effect) were in Cancer, the sign of their debility, which pointed to Wellington's success.

Ebenezer Sibly belonged to a different school. A Bristol-born master Freemason and doctor who had been influenced by the *Primum Mobile* of Placidus de Titus, a monastic astrologer and professor of mathematics at the University of Padua (and sometime adviser to Leopold William, the Austrian archduke and brother to the Holy Roman Emperor Ferdinand III), Sibly had been so enamored of astrology in his youth that he had locked himself in his attic for two years and had all his meals sent in through a hole in the door. He went on to become a prominent doctor, and in the course of his long career, published a number of books, including an expert edition of Culpeper's *Herbal;* an annotated *Collection of Thirty Notable Nativities by Placidus de Titus;* a general account of *Physick [Medicine] and the Occult Sciences;* and, in the mid-1780s his four-volume masterwork, *A New and Complete Illustration of the Celestial Science of Astrology,* which went through twelve editions by 1817.

Sibly's expertise in astrology covered all branches. While Worsdale based his own work on what he took to be the pure classical tradition, Sibly fully embraced its so-called magical strain of divination (as exemplified in horary practice and elections) and also worked with ingress charts and solar returns. In the second part of his *New and Complete Illustration,* he explains, in cosmic terms, why "The Art of Resolving Horary Questions" works: "A question of importance to our welfare cannot start from the mind but in a point of time when the planets and signs governing the person's birth, act upon the very subject that engages his thoughts and attention. Hence the birth of a question, like the nativity of a child, carries the story of the whole matter in hand upon its forehead. And from this follows also that skill in natural predictions by which the artist is enabled to demonstrate the particulars of the event required." Moreover, for those unable to ascertain the exact time of their birth, "the doctrine of Horary Questions" could fill in the gap. "So that from a question seriously propounded almost as much satisfaction may be given the querent upon many subjects of inquiry, as if his nativity were actually known." One can scarcely doubt that Bonatti and Lilly, among others, would have agreed. Sibly gives a full account in his own work of the rules for casting, and judging, horary charts, with deftly handled examples for such questions as: "an absent son, whether dead or alive"; "on the prospect of riches"; "on the fate of a ship at sea"; "shall the querent obtain the promotion desired?"; "shall the querent's husband be found guilty, or released?"; and so on—all quite standard for their kind. But the range of his work was complete. In the spirit of Lilly, it also included a guide to the planets in the signs; "the art of calculating nativities"; astrological disquisitions on topics such as "Children," "Friends and Enemies," and "Marriage"; a section on progressions or directions; and Tables of Right Ascension-North Latitude, Declination-North Latitude, Oblique Ascension, and so on at the back. For a taste of his manner or method: "Jupiter in Scorpio represents a person of middle stature, a well-compacted body, brown hair, a full fleshy face, a dull complexion; but in disposition, a lofty, proud, ambitious, person; one that desires and endeavors to bear rule over his equals, resolute, and ill-natured, covetous, and guilty of too much subtlety in all his actions; and therefore ought warily to be dealt with by those who shall be concerned."

Here is Sibly on the sexual proclivities of men:

> Their disposition, as to modest or vicious habits, very much depends on the positions and configurations of Mars, for, if he be separating from Saturn and Venus, and applying to Jupiter, men born at that time will be discreet and modest, decent in their intercourses with the other sex, and disposed only to the natural use. If Jupiter and Venus be configurated to Saturn and Mars, the native will be easily moved on, and have a secret desire to acts of venery; but will have an external show of chastity, and labor to avoid the shame. If Mars and Venus are alone configurated together, or if Jupiter bears testimony, the native will be openly lascivious, and indulge in the most luxuriant enjoyments of the opposite sex.

Sibly, in fact, was often consulted on matters of romance. On one occasion a woman asked him whether she should go ahead with her engagement. Sibly examined her horoscope, compared it with that of the prospective groom, and declared that he "could not find a single configuration in the one that bore the least harmony or similitude with the other . . . The benefic stars in the angles of one were opposed to the other's malefics. The masculine temperament was strongest in the female horoscope . . . while in the man's geniture, the effeminacy of the female influence was but too plain." He therefore advised her to break it off. But the man, it seems, was very rich and she was evidently dead set on the match for financial gain. They wed, and the marriage abruptly failed. Sex proved more important to her than she knew. "The newly-married pair were put to bed—where love and joy should take their fill: but [none being found], the bride rose up with the Sun, and finding no other allurement to supply that defect, she immediately deserted her husband, who never took the pains to retrieve her; and she has since had children by two other men."

Sibly also applied his skill to world affairs. Some of his mundane or political predictions, in fact, were remarkable. When the American Revolution began, his analysis of ingress charts for 1776 told him clearly how the struggle between the Crown and the colonies would go. Most expected the rebellion to be crushed; Sibly flatly predicted it would triumph and that "a total and eternal separation of the two" parties would be the result of the war. A decade later, in the third part of his *New and Complete Illustration,* published in 1787, he concluded (looking ahead) from an ingress chart cast for the Sun's entrance into Aries in 1789: "Here is every prospect, from the disposition of the significators in this scheme, that some very important event will happen in the politics of France, such as may dethrone, or very nearly touch the life of, the king, and make victim of many great and illustrious men in church and state, preparatory to a revolution or change in the affairs of that empire, which will at once astonish and surprise the surrounding nations." Two years later, the French Revolution occurred (on schedule, in 1789), and assumed the "astonishing" character predicted, exactly as described.

Sibly is even more famous for the chart he had cast earlier in 1784 to ascertain America's postrevolutionary fate—when the revolution, that is, had triumphed but the victorious colonies were still struggling to make themselves one. Included in plate 53 of the third volume of his great work, America's horoscope is depicted as borne aloft in the right hand of a trumpeting angel, who hovers in the clouds above the hoped-for federal union, shown as an infant below. On the basis of it, Sibly declared that America would one day "have an extensive and flourishing commerce; advantageous and universal traffic to every quarter of the globe, with great security and prosperity amongst its people," and, still more, eventually constitute "a new Empire that shall soon or late give laws to the whole world." For an unstable, experimental democracy—scarcely yet a nation—which America then was, that was a remarkably optimistic assessment, given that the Constitutional Convention, destined to meet three years later in Philadelphia, was scarcely yet a glimmer in Washington's eye. (Worsdale, not to be outdone, would later write that the "elevated and dignified" positions of the planets at the time of American independence "most clearly forebode, that the time will arrive, when THAT EMPIRE shall give laws to all Nations, and establish FREEDOM and LIBERTY in every part of the habitable Globe." Though written just after the War of 1812, that was still a rather bold pronouncement to make.)

Sibly's prediction, however, is more interesting because it is linked to a specific chart. And that is the foundation chart for the United States. When, exactly, was the United States born? That question has occasioned much debate. Charts have been drawn for the first colonial settlement (in Roanoke, Virginia); the Boston Tea Party; the First Continental Congress; the Declaration of Independence; the Articles of Confederation; the British surrender at Yorktown; the Treaty of Paris; the Constitutional Convention in Philadelphia; and the inauguration of George Washington as the first president of the United States. Most astrologers accept the Declaration of Independence as the seminal moment—as distinct from the "Proclamation of Independence," which was actually made at noon on July 8 when the Declaration was read out in Philadelphia's State House Square—but it turns out that that "moment" is fiendishly hard to fix. July 4 is not even the agreed-upon day. John Adams tells us in one of his letters that the vote for Independence—"That these United Colonies are, and of right ought to be, Free and Independent States"—was taken not on the 4th but the 2nd, which "will be celebrated," he had predicted, "by succeeding Generations, as the great anniversary Festival . . . with Bon-

fires and Illuminations . . . Pomp and Parade." But it was the 4th that
gained that honor in the end. As for the document itself, no one seems
to know when it was actually approved. On July 2, the date on which
the resolution for independence was passed, the document itself,
drafted by Jefferson and amended by Franklin, served merely as the
formal announcement of the seconding of the resolution. So in a sense
it was adopted; and in a sense it was not. Two days later, however, the
Journal of Congress records that "at a little past meridian, on the Fourth
of July, 1776, a unanimous vote of the thirteen colonies was given in
favor of declaring themselves free and independent states." That might
seem to clinch the matter. But a confirming copy of the document was
not placed in the *Journal* at the time. In fact, the lower half of the page
was left blank. Only later was a printed copy of the Declaration inserted, using three wafers to hold it in place. Even so, it was widely assumed that the Declaration had been signed on that day. Not so. In
1817, a letter written three years before to John Adams by Thomas
McKean, a member of the Second Continental Congress, turned up.
In it McKean stated that "no person" had then signed it—even though
Jefferson claimed his notes showed that almost every member had.

Jefferson was probably mistaken. Neither the rough nor the corrected congressional *Journal* mentions a general signing. But from a
legal standpoint, that is perhaps a technicality. A congressional resolution assumes legal status once it is signed by the president of Congress,
and John Hancock, in that capacity, did sign it on the 4th. Charles
Thomson, the secretary of Congress, also signed it, to authenticate
Hancock's signature at the time.

On behalf of July 4 we therefore have not only the mighty weight
of tradition, but the force of legal fact. We also have the document,
passed by Congress on June 20, 1782, that defined the design of the
Great Seal of the United States. The last paragraph reads: "This date
[July 4, 1776] underneath [the Pyramid, on the reverse side of the
Seal] is that of the Declaration of Independence and the words under
it [Novus Ordo Seclorum] signify the beginning of the new American
Era, which commences from that date."

That would seem to resolve the issue as to the day, but not the time.
An early morning hour has been proposed but is unlikely, since Congress had adjourned the evening before, and it is recorded in the *Journal* that nine o'clock was set as the time to reconvene. And there is no

reason to suppose that Congress did not reconvene at that hour. According to the Philadelphia Historical Society, "the Declaration was graced with its first signature [Hancock's] at approximately five o'clock in the afternoon after which the members all went off to dine." Such a time, it turns out, is supported by the horoscope drawn by Ebenezer Sibly, which was cast for the Declaration, and gives a time of 10:10 P.M. London, which corresponds to 5:10 P.M. in Philadelphia, when Hancock affixed his name.

How did Sibly get the time right? His Masonic connections probably hold the clue. There had been an explosion of Freemasonry in Western society in the second half of the 18th century and many (perhaps a majority) of America's Founding Fathers belonged to Masonic Lodges and rose to high rank. At least nine of the signers of the Declaration of Independence were Masons, including Benjamin Franklin and John Hancock; and George Washington was the Grand Mason of the United States. Benjamin Franklin's own lodge—"St. John's Lodge" in Philadelphia—had been founded on July 4, 1730, and three years later sought official recognition from the Grand Lodge of London. Masons in England and America had kept in touch with one another across the political divide, in part because most English Masons were sympathetic to the revolt. Sibly, a Master Mason, must have received the time from one of his American confreres.

In the end, the only known confirmation for the probable time of the nation's founding is an astrological chart.

What does that chart describe? On July 4, 1776, at 5:10 P.M. in Philadelphia, Sagittarius, ruled by Jupiter, was rising, the fixed star Regulus—which signified kings—was setting, and the Sun in Cancer conjunct the fortunate fixed star Sirius, which, by tradition, gave success in war and business as well as faithfulness and devotion, honor and renown. By accident or design, the moment Hancock picked up his pen was both stupendously propitious and symbolically apt, for it would notably link the revolution with energy and enterprise and the decline of royal power.

Our Founding Fathers, in truth, were not wholly averse to esoteric things. The Great Seal of the United States, of course, is replete with Masonic symbols and—with a bow to the thirteen colonies—abounds with implied tributes to the number 13: There are 13 stripes on the shield, 13 stars in the circle of glory above the eagle's head, 13 branches and 13 berries in the olive branch in the eagle's right talon, 13 arrows in the left, 13 layers of stone in the unfinished pyramid, even 13 letters in the two legends, *Annuit Coeptis* and *E Pluribus Unum*. More generally, the principal symbols on the seal—the bundle of arrows and the olive branch—were those of Manasseh, son of the biblical Joseph, whose progeny are said to have comprised Israel's 13th tribe. The Declaration of Independence, in fact, was authenticated by Hancock when the Sun in Cancer was at 13 degrees of its sign.

It is sometimes difficult to distinguish between a numerological conceit and the inner consistency of a cabalistic idea. Whether all this amounts to anything more than a conceit may be left to others to decide. American culture, however, was still primarily British, which means that astrology belonged to the fabric of colonial life. Long before America spawned its own professional class of adepts, astrology had taken root with the earliest settlers, who "planted by the signs," marked the benign or afflicted aspects of the planets, and tracked their maladies through the phases of the Moon. From the beginning, astrologers also did a brisk business at ports of call, where American ship captains, like their brethren in Bristol or Calais, sought to ascertain the fate of their ships and cargoes, and where relatives hoped to learn the fate of their loved ones on the main. From time to time, an astrologer might elect the best day for launching a ship or beginning a voyage to Barbados or Morocco, advise passengers about the risk of drowning or businessmen on whether their ship should be insured. Those involved in the slave trade, in particular, for some reason, often refused to sail without first having a horoscope cast.

Omens were also scanned from the skies. For example, prior to Bacon's Rebellion in Virginia, a precursor of the later great colonial revolt, a "horsetail" comet streamed across the heavens and seemed to presage turbulence or war.

For those who wished to explore the subject more deeply, texts on astrology could be readily had. Editions of Bonatti, Cardano, Lilly, Gadbury, and others were imported from England by the trunkload, while the first book to come off printer William Bradford's pioneering press in 1685 was *America's Messenger*, an astrological almanac. Others to appear about that time were *Holwell's Predictions: of Many Remarkable Things, which may Probably Come to Pass* and the *Monthly Observations and Predictions* of John Partridge for 1692. An astrological society was founded in Philadelphia—by Johannes Kelpius, who belonged to a Rosicrucian sect—and an astrological library and conservatory was set up on Wissahickon Creek. Enterprises were also occasionally undertaken at elected times—for example, construction on the Swedish Lutheran church at Wisaco, Pennsylvania, in 1698. The Revolutionary War by no means marked a divide. A steady stream of published or imported works continued up through and beyond those years, with the popular debut of *The New Book of Knowledge: Shewing the Effects of the Planets and other Astronomical Constellations,* for example, in 1789, or *The Predictions of John Nobles, Astrologer and Doctor* in 1793. College dissertations dealt now and then with astrology, and John Winthrop, Hollis Professor of Mathematics and Natural Philosophy at Harvard, was a devotee. He kept a book full of blank charts for his horary or decumbiture castings and prized for ready reference *A Table of the 12 Astrological Houses of Heaven* (published in London in 1654) and *Astrologica aphoristica Ptolemaei Hermetica (The Occult Astrological Aphorisms of Ptolemy,* published in Ulm in 1641). He also regarded comets as serving both a moral and a natural purpose, not without reference, as he put it, to "the minutes of divine justice" as emblazoned in the sky. The New England Calvinist Jonathan Edwards similarly considered comets "funeral torches to light Kings to their tombs." Thomas Jefferson owned John Gadbury's principal work, *Collection of Remarkable Nativities,* and Benjamin Franklin was acquainted with the art, though no one really knows what he thought of it deep down.

Franklin's *Poor Richard's Almanac* featured "Astrological Signs, Planets, and Aspects" on its cover, with the symbols attached, and the "Poor Richard" he proffered as its author was Richard Saunders, a well-known English astrologer (and author of the almanac *Apollo Anglicanus*) who had lived a century before. Written and published by Franklin over the course of twenty-five years, *Poor Richard's* was a potpourri of sayings and proverbs—"God helps them that helps themselves," "Fish and visitors smell in three days," and so on—interspersed with the rising and setting times for the Sun and Moon, the positions of the planets, and predictions of eclipses and other celestial events. He also sometimes boldly included long-term weather reports. But in so doing he framed a witty jest: "I am particularly pleased to understand that my Predictions of the Weather give such general Satisfaction; and indeed, such Care is taken in the Calculations, on which those Predictions are founded, that

I could almost venture to say, there's not a single One of them, promising Snow, Rain, Hail, Heat, Frost, Fogs, Wind, or Thunder, but what comes to pass punctually and precisely on the very Day, in some Place or other on this little diminutive Globe of ours." Perhaps he was just having fun.

More obscurely, the architecture of our nation's capital is star-spangled with zodiacs. There is a zodiac band on the bronze door entrance to the Library of Congress, a zodiac clock in its main rotunda, zodiacs in the southeast pavilion, and in the marble floor of the Great Hall. A bronze statue of Einstein outside the National Academy of Sciences shows him contemplating a star-spangled marble horoscope for April 22, 1979, his right foot resting on two stars—Boötes and Hercules. In the Academy itself, zodiac signs are built into the structure of the metal doors. In the adjacent building to the east—where the Federal Reserve Board meets—two other zodiacs were cut by the great Steuben Glass Company as decorative flanges for the lights.

All these may well be mere decorative motifs, though some ceremonies in Washington would seem (in a symbolic sense, at least) to have been astrologically timed. On Wednesday, September 18, 1793, for example, when George Washington as both president and Grand Master Mason of the United States laid the cornerstone for the new Capitol building, he did so with full Masonic honors—in concert with the Grand Lodge of Maryland and Lodge No. 22 from Alexandria, Virginia. He wore a Masonic apron for the occasion that had been given to him by Lafayette, which featured a number of the order's symbols—the radiant eye, or "Eye of Providence"; the Sun and Moon; the seven steps; open book, compass, square, and so on, along with the two legendary pillars referred to by Josephus, on which the sacred knowledge of astrology had been carved. The Sun was in Virgo, to represent the new nation's birth; the newly discovered planet Georgium Sidus (afterward called Uranus), named for the vanquished King George III, was exactly conjunct the fixed star Regulus, meaning the little king. As the ceremony proceeded, Georgium Sidus set in the west—just as Regulus had set at the Declaration's signing—to symbolize the end of royal power. Again, on Thursday, April 19, 1883, at 4 P.M., the unveiling of the bronze monument to the memory of Joseph Henry, the founding director of the Smithsonian Institution, was arranged so that Jupiter, the most auspicious of planets, would be directly overhead.

✶ ✶ ✶

ELSEWHERE, THROUGHOUT EUROPE, the aesthetic wonder of such cosmic arrangements—by which earthly and celestial events can be coordinated by design—continued to animate the imagination of artists, and in the work of Goethe, Schiller, Byron, Blake, and Sir Walter Scott, among others, astrological ideas abound. Scott conceived an astrological framework for his second Waverly novel, *Guy Mannering;* Blake collaborated with the painter and astrologer John Varley in seeking to correlate his own visionary drawings with the zodiac degree rising at the time; Goethe saw fit to open his autobiography with the astrological coordinates of his birth. In America, the poet Henry Wadsworth Longfellow immersed himself in the works of Abū Ma'shar.

Varley, not incidentally, took a special interest in Uranus, which had been discovered in his lifetime, in 1781. According to his son Albert, he "spent much time inserting the planet into his own and other people's charts, watching carefully to see what kinds of events it provoked." One morning he announced to his family that he was staying home that day because something very serious was about to happen, and he wanted to be there when it did. At noon, the house caught on fire. He sat down at once to write about it (as an example of Uranus in unfavorable aspect to Mars), and though he ultimately lost everything in the blaze, "he regarded that as a small matter compared with his discovery of the planet's power."

Goethe's autobiography began: "It was on the 28th of August 1749, at the stroke of twelve noon, that I came into the world at Frankfurt-on-Main. The constellation was auspicious: the Sun was in Virgo and at its culmination for the day. Jupiter and Venus looked amicably upon it, and Mercury was not hostile. Saturn and Mars maintained indifference. Only the Moon, just then becoming full, was in a position to exert adverse force, because its planetary hour had begun. It did, indeed, resist my birth, which did not take place until this hour had passed."

Byron was also an able astrologer. Like Dryden, he set up his own son's chart, and, again like Dryden, accurately predicted the main events in his life. In *Childe Harold's Pilgrimage,* we can hear his reverent voice:

Ye stars! which are the poetry of heaven!
If in your bright leaves we would read the fate
Of men and empires—'tis to be forgiven,
That in our aspirations to be great,
Our destinies o'erleap their mortal state,
And claim a kindred with you; for ye are
A beauty and a mystery, and create
In us such love, and reverence from afar,
That fortune, fame, power, life, have named themselves a star.

In his play *Sardanapalus,* the king is betrayed by an astrologer, who declares: "Thou sun that sinkest, and ye stars which rise / I have out-watched ye, reading ray by ray / The edicts of your orbs . . ."

Yet the common understanding of the tradition was slight. Scott remarked upon this in his introduction to *Guy Mannering:* "The scheme projected may be traced in the first three or four chapters of the work; but further consideration induced the author to lay his purpose aside. It appeared, on mature consideration, that astrology, though its influence was once received and admitted by Bacon himself, does not now retain influence over the general mind sufficient even to constitute the main-spring of a romance." Scott ultimately gave up on his original scheme of having the lives of his characters play out according to their charts. "It occurred to me," he wrote, "that to do justice to such a subject would have required, not only more talent than the author would be conscious of possessing, but also, involved doctrines and discussions of a nature too serious for his purpose and for the character of the narrative."

When those doctrines once again entered into the public discourse, they did so in somewhat adulterated form. In that sense, the popular revival of astrology may be said to have begun in England with the publication of various almanacs, such as Francis Moore's *Vox Stellarum,* or *Voice of the Stars,* which by 1803 had a huge print run of 200,000 copies. In 1824, a magazine called *The Straggling Astrologer* was the first to carry weekly prognostications—about love, marriage, business, travel, and so on—like today's astrology columns. This was followed by the *Prophetic Messenger,* which enjoyed even greater success and introduced forecasts for every day of the year. Other publications followed, and the appetite for astrology spread.

The leading apostle in this development was Robert Cross Smith, who called himself "Raphael." (For some reason, a number of 19th-century astrologers, in seeking to acquire an ethereal aura, called themselves after angels or sprites—Sepharial, Charubel, Aphorel, Raphael, Zadkiel, and so on—as if to confirm their link to the spirit world. Smith adopted "Raphael" as a pen name, which set the tone.) Born in Bristol, he had come to London in 1802; was introduced to astrology by G. W. Graham, a well-known professional balloonist; became the editor of *The Straggling Astrologer* in 1824; and subsequently of the *Prophetic Messenger,* where his success established him as "the founder of modern popular astrological journalism."

Smith was succeeded in this sphere by Richard James Morrison, also known as "Zadkiel." A man of genteel birth—his grandfather had been a sea captain, his father a gentleman pensioner of George III—Morrison was a former coast guard officer and navy lieutenant who had taken an early retirement upon his marriage to a niece of Sir Joshua Paul, an Irish baronet. In 1830, in emulation of Smith, he launched his own predictive almanac, which he first called *The Herald of Astrology,* and (after 1836) *Zadkiel's Almanac.* The first issue of the latter came out in January 1849 and included articles on astrology, phrenology, mesmerism, meteorology, and the newly discovered planet Neptune. Whereas *Old Moore's Almanac* (or *Vox Stellarum*) appealed to simpler folk, the almanacs that Smith and Morrison produced caught on with the middle class.

Morrison developed some skill and was notably successful in predicting earthquakes. In the 1852 issue of his almanac he surmised from the upcoming total eclipse of June 6, 1853, that since "Mars and Saturn are in Taurus in the precedent angle of the eclipse at Panama, I have no doubt there will be a fearful amount of earthquakes there and also about the isthmus of Darien, the shocks extending to Carthagena, along the northern coast of South America, to Honduras and California, Florida, etc., and the West Indies. These events may be looked for in July 1853, about the 16th day." It so happened that on the sixteenth day of July there was a terrific earthquake at Cumana on the Spanish Main. In 1864, Morrison also predicted that the American Civil War would be resolved by the following spring, and that "the peace would be associated with an eclipse of the Moon. This seemed borne out when on 10 April 1865, the day of the eclipse, Robert E. Lee surrendered to General Grant." In other work, Morrison published an abridged version of Lilly's *Christian Astrology* in 1835, his own *Grammar of Astrology,* and a *Handbook of Astrology* in two volumes. In 1844, he also founded the British Association for the Advancement of Astral and other Sciences (which inspired kindred organizations in England and abroad) and began (in the spirit of Dee) to experiment with scrying or divining with a crystal ball. The sessions at his house were attended by a number of prominent people, including the Countess of Erroll, Baron Bunsen (the Prussian ambassador), Admiral FitzClarence, the Marchioness of Ailesbury, the bishop of Litchfield, the Earl of Effingham, Lady Egerton of Tatton, the Earl of Wilton, and the novelist Sir Edward Bulwer Lytton.

Zadkiel's Almanac, meanwhile, continued to appear from year to year, making occasional waves with its predictions, and intriguing the popular imagination with intermittent allusions to members of the royal family as well as current events. These were usually innocuous, or harmlessly vague, but with his almanac for 1861 Morrison got into serious trouble over an implied warning to the prince consort about his health. That would probably not have attracted much attention if the prince consort had not unexpectedly died later that year at the age of forty-two. The *Daily Telegraph,* outraged at the almanac's success (which it ascribed to dumb luck), attacked it as "a farrago of wretched trash" and demanded that the author be prosecuted "as a rogue and a vagabond" under the Vagrancy Act. The next day Morrison's true identity was exposed in a letter to the paper from Rear Admiral Sir Edward Belcher, who had some grudge against him and denounced him as a fraud. Morrison demanded an apology, and, when Belcher would not retract, he sued. The following June, a trial was held before a special jury at the Court of Queen's Bench, Guildhall, with Sir Alexander Cockburn, the Lord Chief Justice, presiding. Belcher appeared on his own behalf without anyone to support him; Morrison had a distinguished witness list that included "various knights, lord and ladies, plus a bishop, an earl, a marchioness and various naval grandees." In the audience were also "numerous distinguished persons, nobility and gentry, who had had their nativities cast." The Lord Chief Justice, however, according to his own predilection, came down on Belcher's side. He scornfully referred to *Zadkiel's Almanac* as a "preposterous and mischievous" publication, and ridiculed astrology as a whole:

Of all the strange delusions that had ever misled the mind of man, the notion that our destinies are affected by the combinations of stars is perhaps the strangest. Ancient astronomers affixed, for convenience, certain names to certain stars, borrowing those names from heathen mythology; and then astrologers actually, in their ignorance, ascribed to the stars the character of those deities whose names they bore. Then, because one bright star was called Venus and another of a more fiery red was called Mars, they fancied that persons born "under" those stars had the characteristics of those particular heathen deities. Nothing could be more absurd, and such is the rubbish with which this almanac is filled. How people can be led to believe that planets named after heathen deities can have influence upon their birth and fortunes is indeed surprising.

After reading passages from the almanac aloud, he kept asking "What could be more absurd?" and concluded by inviting the jury "to consider the nature of the publication in your verdict as to the amount of damages you award." The jury took its cue from the judge, and after finding Belcher guilty on a technicality, fined him just twenty shillings, which was no fine at all. In short, Morrison won the case, but lost the public relations war.

Like *Raphael's Almanac,* which survived Smith's death, *Zadkiel's* likewise survived the death of its founder in 1875 and continued selling smartly until 1931. But Morrison himself never quite recovered from the trial.

The English Victorians had their American counterparts, of course, of whom the most prominent was probably Luke Broughton, an English émigré. Born in Leeds on April 29, 1828, Broughton belonged to a long line of astrologers, and was one of several siblings (all named after the evangelists) who had come to America in the 1850s to pursue their art. In 1854, he settled in Philadelphia, where two of his brothers, Matthew and Mark, had already established a modest practice and collaborated on a monthly periodical called the *Horoscope.* Broughton joined in their endeavors; worked on an ephemeris; and studied to become a doctor, obtaining his degree in 1858. In that year, he also began to publish his own *Monthly Planet Reader.* Subsequently, he wrote a manual called *The Elements of Astrology,* which helped to reestablish astrology in American life.

Broughton's astrology was mathematical and predictive. In his *Planet Reader* for January 1861 (when most expected the Civil War would be nasty, brutal, and short), he predicted that peace would not come "before the summer of 1865." One year later, he foresaw that General George McClellan would fail as the Union commander, and be replaced toward year's end, when Saturn in transit over his ascendant was squared "to its own place in the tenth house. His enemies will be rampant to have him removed, and the indications are that they will succeed." McClellan was removed from command on November 7. Broughton also seems to have glimpsed President Abraham Lincoln's fate. In October 1864, he correctly predicted Lincoln's reelection, but wrote: "Let the President be careful of secret enemies, and also of assassination, during this and the next few months." Ptolemy had written: "Should either of the luminaries be afflicted by Mars from cardinal signs, and Mars at the same time be elevated, the native will suffer a violent death." Lincoln had an elevated Mars in Libra, a cardinal sign, squared to his Moon. Broughton was simply going "by the book." As 1865 approached, he also saw that Lincoln's progressed Sun was in exact opposition to Mars, which he thought might trigger the event. In March 1865, he warned that in mid-April "some high official would die or be removed." On April 14, Lincoln was shot in the head in Ford's Theatre and expired the following day.

These forecasts raised his standing, but also made him an obvious target for those who viewed astrology as perverse. Broughton's home in Philadelphia was sacked by an enraged mob, and in 1866 he moved to New York City, where he acquired an assistant by the name of W. H. Chaney (or Cheney, according to the original spelling of his name), "a short, stocky man with a quick temper" and the father of the novelist Jack London. Something of a misfit, Chaney in his youth had been a wild rover, vagabond, and aspiring pirate; became a government surveyor; edited a newspaper; and after studying law, served as a district attorney in both Iowa and Maine. Though a fine mathematician (and an astrologer of proven skill), he lived on the edge, became embroiled in a number of scandals, some involving sexual liaisons, and engaged in a long and bitter feud with the Catholic Church.

In his *Primer of Astrology* and other books, Chaney laid great stress on "Primary Directions" or progressions (a degree for a year), and spent the last years of his life toiling away on a huge ephemeris that would accurately cover the 19th century as a whole. "No one," Broughton re-marked later of his colleague, "who does not understand astrology and mathematical astronomy, can have any idea of the amount of labor it takes to calculate back for fifty or a hundred years, the longitude, lati-tude, and declination of the Moon, Mercury, Venus, Mars, Jupiter, Sat-urn, and the planet Uranus, to the degree and minute for each day of the year; also the longitude, declination, and right ascension of the Sun, in degrees and minutes for each day for that length of time."

In 1866, Chaney had moved to New York City, and there in Octo-ber Broughton, who had a fairly wide clientele, had hired him to help with the mathematical calculations that had begun to devour his time. They set up shop together at 814 Broadway, where they presided over the Eclectic Medical University, a "New Age" school of sorts, where astrology, palmistry, herbalism, and other such disciplines were taught. Their relative success came to the unwelcome attention of James Gor-don Bennett, editor of the *New York Herald,* the mightiest of the country's papers at the time, who singled out the school for attack. At Bennett's instigation, the landlord of their building rented out the floor above them to an irascible band of Fenians (a group dedicated to freeing Ireland from British rule), who banged chairs on the floor and "drilled with crashing boots" while Broughton and Chaney tried to conduct their classes below.

The university disbanded but Chaney remained in New York until 1869, lecturing on astrology, phrenology, and natural theology, writing pamphlets, and casting charts. He had a hard time of it, however, and was often heckled at his lectures; and at one point was imprisoned for blasphemy in Ludlow Street Jail without bail for six months. Upon his release, he headed west, and from 1871 to 1872 lived in Salem, Ore-gon, where, he later recalled, "I enjoyed the friendship, 'in private,' of U.S. Senators, Congressmen, Governors, Judges of the Supreme and lower courts . . . I helped many a one to his position, working in se-cret, but they dared not reward me openly, although in private they were my best and truest friends." During this period, he also became a confidant of the mayor, a "spiritualist" himself, at whose home he first met Flora Wellman, the daughter of Marshall Wellman, a wheat ty-coon. Flora had been raised in her father's sprawling mansion in the "wheat city" of Massillon, Ohio (which her father helped build), but early on showed bohemian longings and at age sixteen ran away from home. She taught piano for a time, then became (or supposed herself) a medium and devotee of the spirit world. After she met Chaney, she became his lover and business partner, and in October 1873, the couple moved to San Francisco, where she conducted séances and he lectured on astrology, astronomy, "astro-theology," and the like, and their "love-child," Jack London, was born. "A very loose condition of society was fashionable in San Francisco in those days," Chaney wrote afterward, "& it was not thought disgraceful for two to live together without marriage. I mean the Spiritualists & those who claimed to be reformers."

Their union did not last long. On June 4, 1875, the *San Francisco Chronicle* carried an article entitled: "A Discarded Wife. Why Mrs. Chaney Twice Attempted Suicide. Driven from Home for Refusing to Destroy her Unborn Infant—A Chapter of Heartlessness and Domestic Misery." It told a lurid tale:

> Day before yesterday Mrs. Chaney, wife of "Professor" W. H. Chaney, the astrologer, attempted suicide by taking laudanum. Failing in the effort she yesterday shot herself with a pistol in the forehead. The ball glanced off, inflicting only a flesh wound, and friends interfered before she could accomplish her suicidal purpose.
>
> The incentive to the terrible act was domestic infelicity. Husband and wife have been known for a year past as the center of a little band of extreme Spiritualists, most of whom professed, if they did not practice, the offensive free-love doctrines of the licentious [Victoria] Woodhull.
>
> The married life of the couple is said to have been full of self-denial and devoted affection on the part of the wife and of harsh words and unkind treatment on the part of the husband. He practiced astrology, calculated horoscopes for a consideration, lectured on chemistry and astronomy, blasphemed the Christian religion, published a journal of hybrid doctrine called the Platomathean, and pretended to calculate "cheap nativities" on the transit of the planets for $10 each . . .
>
> The wife assisted him in the details of business, darned his hose, drudged at the wash-tub, took care of other people's children for hire, and generously gave him whatever money she

earned and could spare beyond her actual expenses. She never told her sorrows, not since her recent great trouble has she communicated them, except to intimate friends. She says that about three weeks ago she was *enceinte*. She told her husband, and asked to be relieved for two or three months of the care of the children by means of which she had been contributing to their mutual support. He refused to accede to the request, and some angry words followed.

Then he told her she had better destroy her unborn babe. This she indignantly declined to do . . .

He then left her, and shortly afterwards she made her first attempt at suicide . . .

Twenty-one years after this article appeared, Jack London was digging through old newspaper files in the San Francisco library for records of his birth, when he discovered who his true father was. In an issue of the *Chronicle* dated January 13, 1876, he read: "Births: Chaney—In this city, January 12, the 'wife' of W. H. Chaney, of a son." He searched backward for evidence of the divorce or separation, and found the article above recounting the sordid end of their affair.

In an ill-aspected hour, London decided to contact Chaney (then in Chicago) to learn more. On May 28, 1897, he wrote to ask him what kind of woman his mother had been in her youth, and whether Chaney in fact was his father. Chaney, who ran a "College of Astrology" out of his apartment at 2829 Calumet Avenue, at once disclaimed his paternity, out of fear he might have to pay some kind of support, and otherwise tried to put London off in a long (and somewhat twisted) reply. In a letter dated June 4, 1897, he acknowledged that he had lived with Flora for a year out of wedlock, but "I was impotent at the time, the result of hardship, privation & too much work." However, he suggested that Flora had slept around. For example, there was a fellow lodger she liked in the house that spring—"The weather was warm & windows were open. I went to my office early in the morning & he slept late"—and then there was "a very fine gentleman, a broker" (one of his own clients), whose "wife & daughter, a young lady, were then in Europe. I was told that Flora demanded ten thousand dollars hush money from him. He was then greatly embarrassed financially having stocks that required margins . . . Later I read in a San Francisco

paper that he had shot himself." But, said Chaney, he could not remember the names of either of these men, or be of help in tracking
them down. As for his own relationship with Flora, "There was a time
when I had a very tender affection for her; but there came a time when
I hated her with all the intensity of my intense nature, & even thought
of killing her & myself. . . . Time, however, has healed the wounds & I
feel no unkindness towards her, while for you I feel a warm sympathy,
for I can imagine what my emotions would be were I in your place."
Yet the bitterness of their breakup still ate at him from within: "The
trouble blighted ten years of my life," he told London. "[The false
charges against me] were copied & sent broadcast over the country.
My sisters in Maine read it & two of them became my enemies. One
died believing I was in the wrong. Another may be dead, I have not
heard from her these many years. All others of my kindred, except one
sister in Portland, Oregon are still my enemies & denounce me as a
disgrace." Indeed, despite a report prepared by a police detective
"showing that many of the slanders against me were false," he had
never been able to clear his name. "Then I gave up defending myself &
for years life was a burden. But reaction finally came and now I have a
few friends who think me respectable. I am past 76 & quite poor . . .
Yours truly, W.H. Chaney."

This appalling letter left London in a state of shock. It contained
more than he had wished to know, and less than he had hoped for, at
least with respect to his birth. In another ill-aspected hour, he wrote to
Chaney again, and an even more tortured reply came back. This time
Chaney explained that "the estrangement between him and Flora had
begun when she asked him whether he would be willing for her to
have a child by another man, since she wanted to enjoy the pleasures
of motherhood and he was too old to help her conceive." He said he
had reluctantly consented to this, provided the father of any such child
would pay for its support. Meanwhile, her infidelity, he recalled, had
already scandalized the boardinghouse where they lived. He recounted
her abject confession and plea for forgiveness, which he granted, and
the moral indignation with which the other lodgers had left. "Had I
followed my first impression," Chaney continued, "I should have left
her then myself, & it would have saved years of misery. But my own
life had been a broken one, & on reflection I forgave her." Not long afterward, she claimed to be pregnant by Chaney himself. He refused to

believe it and told her they were through. She became hysterical, went off to get a gun, and reappeared with "a wound on the left side of her forehead & the blood running over her face." However, he added, the suicide attempt had been faked, for "the pistol . . . had not been discharged" and there were no powder burns on her face.

That was the end of their affair. A year or two later, he said he had been told "that Flora had taken up with a widower who had some little children & was keeping a house of assignation." He also thought he had heard from someone somewhere that she had contracted a venereal disease. "My own life has been a very sad one," he wrote in closing, "more so than I think yours will be," but expressed sympathy for London's plight.

That was cold comfort to be sure. London later described his father vicariously in a story he wrote in 1899, as "a very learned man and a celebrated antiquarian, [who] gave no thought to his family, being constantly lost in the abstractions of his study . . . I can but say that he was the most abnormal specimen of cold-blooded cruelty I have ever seen." But London was also, of course, his father's son. Elsewhere he wrote, according to his own fatalistic streak: "Life's a skin-game. I never had half a chance . . . I was faked in my birth and flim-flammed with my mother's milk. The dice were loaded when she tossed the box, and I was born to prove the loss."

Chapter 12

THE SOMEWHAT SORRY SAGA of W. H. Chaney's unstable
life, together with the debased coinage of his talent, might
stand as a parable for the confused course astrology itself had
taken as it sought to find its footing in the modern world. Some would
lay the fault at one man's feet. That man was Alan Leo, "a traveling
sweets salesman from London," who in 1890 started *The Astrologer's
Magazine* on a lark to compete with *Old Moore's Almanac, Raphael's
Prophetic Almanac, Zadkiel's Almanac,* and other commercial ventures of
the kind. His magazine went nowhere, however, so he recast it as *Modern Astrology* to make it appear wholly different from its rivals and to attract a new clientele. At about the same time, he changed his name
from William Allen to Alan Leo "to accord with his Sun sign." Thus
Sun-sign astrology, in his own person and in the kind of astrology
he purveyed, became confused with astrology itself. Before long, he
would become the most famous and influential astrologer in the
world.

Born in London on August 7, 1860, Leo was the son of a soldier
who had returned from service in a Highland regiment in India to
work in a London dispensary, only to disappear mysteriously a few
years later without a trace. Leo was still a child at the time, and later
woefully recalled that he had been obliged to go to work almost at
once. By the time he was sixteen, he had been apprenticed to several
trades, and on his sixteenth birthday found himself homeless and penniless in Liverpool, where he slept on the street. Thereafter he had
dramatic ups and downs, being gainfully employed one year (as the

well-paid manager of a grocery store for example), down and out the next.

Meanwhile, Leo had been introduced to astrology by a herbalist in Manchester, through whom he met an astrologer who in turn introduced him to the linguistic scholar and cabalist Walter Gorn Old. Old belonged to several elite astronomical societies in Britain and France, and had studied a number of Oriental languages, including Coptic, Assyrian, Sanskrit, Chinese, and Hebrew, as well as Egyptian hieroglyphics. But his passion was numerology. While still in his twenties, he had become acquainted with Madame Blavatsky, the noble-born Russian occultist who had founded the Theosophical Society in 1875. Her *Isis Unveiled* (1877) and *The Secret Doctrine* (1888), notes one writer, "held that the world's religions were underpinned by the same divine wisdom, which had been passed down from the ancients. These ancient spiritual truths, hidden to escape persecution by the Western Church, were now re-emerging, along with the secrets of the East." As their medium, Blavatsky claimed to be "in touch with 'secret masters' who communicated to her astrally and via 'materialized' letters delivered through the ether from the masters' mountain fastness in Tibet." Before long, pilot "lodges" were established in countries throughout the world.

Walter Gorn Old was energetic on Theosophy's behalf and had become a member of Blavatsky's inner group of "Initiates," who affectionately dubbed him (due to his penchant for out-of-body experiences) "the astral tramp." Old (who later dubbed himself "Sepharial") brought Leo to the theosophical meetings that Blavatsky led, and Leo soon became a member of the society himself. A few years later, he married a fellow theosophist and "gave up his sales career to devote himself to astrology full-time." During the ensuing five years, his *Modern Astrology* proved wildly popular, based in part on his offer to subscribers of a "Test" or thumbnail horoscope for a shilling each. Leo was soon swamped with horoscope commissions. He enlarged his staff with clerks and mathematicians, started a correspondence course, founded a publishing house, and opened branch offices in Paris and New York.

Though his professional activities had many aspects, his relatively swift and remarkable success was largely founded upon the mass production of cheap horoscopes that treated all Sun, Moon, ascendant, and planetary-sign positions in generic terms. Before long, printouts were produced in stacks for collation and assembly, and upwards of fifteen thousand "personalized" horoscopes were sent out each year. At the same time, in order to make astrology accessible to the masses, "he made the Sun sign into the astrological catch-all it is today. For him," as Neil Spencer remarks, "the Sun was 'the universal principle,' the 'primal fount of existence,' and one to which all other planetary principles were subsidiary. This cut contrary to much previous astrological thinking, in which the Sun was one planetary energy among seven, and not necessarily beneficial, since it scorched adjacent planets, which were then 'combust.' "

That was one distinctive feature of the art to which he held. Yet Leo was also a convinced theosophist and oddly combined his elementary thinking and productions with an almost impenetrably abstruse idea of what astrology was. His brand of it was heavily suffused with Blavatsky's teaching, which "downgraded prediction and instead emphasized the craft's 'spiritual dimension.' " As a result, "Theosophy's teachings on reincarnation, karma and the evolution of the soul over many lifetimes—ideas that Blavatsky had taken from Hinduism and Buddhism—now became part of modern astrology's domain."

Leo reconceived the whole tradition in the light of this creed. In an article published in 1903, he wrote: "It is no exaggeration to say that all the astrologers I met at the commencement of my career were satisfied with the horoscope alone . . . it was clear that the meanings of the symbols found in the horoscope were not properly understood by them . . . it was not until the light of the Wisdom Religion [Theosophy] gave illumination to the ancient symbology that a few astrologers, turning the rays of that life upon astrological symbols, were able to penetrate behind the veil." The tradition was thereby redefined as a coded language that only theosophical teachings could construe. Instead of a method of prediction, "the horoscope could be regarded as a kind of esoteric document [in the theosophical sense] and its interpretation an occult exercise." In keeping with this conviction, Leo in 1915 founded the Astrological Lodge of the Theosophical Society, which became a beachhead for the spread of his ideas. He also wrote a number of hefty textbooks on the mathematics of astrology and interpretative technique: *Astrology for All, Casting the Horoscope, The Progressed Horoscope, The Key to Your Own Nativity, How to Judge a Nativity, The Art of Synthesis,* and

Esoteric Astrology. But their teachings were largely detached from the classic texts. The circumstances of his own life inclined him to distance them still further, for despite his own spiritualized conception of the craft, Leo was twice prosecuted for fortune-telling—first in May 1914, when he was acquitted on a legal technicality, and again in July 1917, when he was fined. As a result, he apparently decided "to revise all his publications on the advice of counsel to substitute psychological indications for forecasts of future events." After Leo, most astrologers shied away from prediction in order to keep out of court. And "psychological astrology" was born.

Whatever else may be said about Leo, however, he was intellectually engaging and, it seems, completely sincere. As one admirer put it, looking back: "He had the gift of drawing all kinds of people around him, and among these many of very high intellectual caliber, who were prepared to study astrology, which meant that they took up all phases of higher mathematics, spherical geometry and astronomy, in order that they might understand thoroughly what they were actually doing when they made horoscopes." That no doubt helped to keep part of the technique of it alive. But the art of the art was lost.

Indeed, the more arbitrary astrology became, as a grab bag of esoteric theory, the more charlatans emerged to claim it as their own. Typical in that regard was a certain Dr. Karr of *Dr. Karr's Guide to Success and Happiness,* published in 1900, which featured a portrait of the turbaned astrologer on the front, looking a little like Carmen Miranda in *The Road to Rio.* "The author, in offering this work to the public," it began, "has but one general purpose in view, viz., to be of the greatest possible service to the world and humanity." At the back of the twenty-four-page pamphlet were ads for "genuine magnetic lodestones" (guaranteed to attract wealth), magic roots, *The Book of Forbidden Knowledge* (explaining charms, signs, mesmerism, divination, "oriental electric psychology," color superstitions, the evil eye, lucky and unlucky days, and other "Hidden Secrets of the Ancients"), and ads for pamphlets that explained the meaning of dreams, palmistry, handwriting analysis, fortune-telling by cards, crystal gazing, and phrenology. The astrology in the *Guide* itself was of the generic type, and included obligatory lists of famous people born under the various signs—under Aries, for example: Thomas Jefferson, Otto von Bismarck, William Wordsworth, J. Pierpont Morgan, and Mary Pickford; under Pisces: George Washington, Victor

Hugo, Voltaire, Frederick Chopin, Copernicus, Geraldine Farrar, and "ALLA RAGAH, The Man Who Sees Tomorrow," who perhaps was Dr. Karr himself.

Meanwhile, though Sepharial (Walter Gorn Old) had introduced Leo to theosophy, Sepharial then fell away from it, believing that "the sooner we bring the science of astrology down from the clouds where the would-be esotericists have incontinently harried it, the sooner will it gain a proper recognition in the practical world." In particular, he became interested in whether astrology could be used for making money, and applied his theories about it to the stock exchange and commodity markets with some success. He also developed several systems—with catchy names like "Apex," "Golden Key," "Snapshot," "Flashlight," and "Spring Season"—for predicting how horse races and other contests would turn out. In his *Manual of Astrology,* he offered a programmatic prognosis for death based on the eighth-house planet and the position of Saturn in the signs:

> Fixed signs show death by blood (Aquarius), disorders of the throat (Taurus), heart affections (Leo), and generative system (Scorpio). Mutable signs show colds and affection of the lungs (Gemini), bowels (Virgo), and nervous system (Sagittarius and Pisces). Cardinal signs denote death by affections of the head, brain fever, etc. (Aries), stomach (Cancer), the kidneys, liver (Libra), also erysipelas and disorders of the skin (Capricorn). If Saturn be in the eighth, the death will be slow and tedious; Mars there, it will be quick and painful; Uranus, there, sudden and unexpected; Jupiter there, in comfort and order; Venus there, peaceful and without pang or throe of pain. The Moon and Mercury are passive, and act according to the nature of the planet in nearest aspect; but the Moon has a signification of death in the public streets, or in the presence of strangers, in public hospitals and other places of a public nature, when other testimonies point that way.

By this account, for example, someone who has Saturn in Leo with his Moon in the eighth house (sextile to Mars, by closest aspect) should die in a sharp, sudden manner in some public place of a heart attack.

Here is Sepharial on President James A. Garfield's untimely death. "In this horoscope we find Saturn in the twelfth, denoting secret enemies; Mars in Scorpio, opposing the Moon, and Uranus in the fifth in square to the Moon, the Moon being in parallel to Mars. Mars ruling the eighth is afflicted by Uranus and Jupiter from the fifth house, which rules theaters. He met his death while in a theater, and was shot in the bowels (Saturn in Virgo)." The horoscope of King George VI, on the other hand, provided

> the perfect example of a quiet, kindly type of ruler who made a perfect marriage and was very much in love with his charming Queen . . . He had the sign Libra rising, the outstanding zodiacal sign which rules love, companionship and marriage. Being the sign of the "balance" it indicated a character well fitted to preside over both marital and national destinies. Venus, the ruling planet, together with the Moon and Uranus were in the sign Scorpio denoting his strength of will and personal firmness, and the splendid trine from Jupiter in Leo to Mars in Sagittarius indicated his geniality, natural charm of manner, good humor and generosity. The Sun and Mercury in Sagittarius rendered His Majesty democratic and philosophical and showed his great interest in all forms of social and philanthropic work.

Finally, this is his read on Lord Byron's life: "Lord Byron was born on January 22, 1788, at 1:18 a.m. Venus is found in Aquarius, conjoined with Saturn; the Moon is with Mars in Cancer, opposed to Mercury; and the Sun in Aquarius is opposed by Uranus in Leo; four planets in fixed signs, and three in cardinal. The position of Venus, afflicted by Saturn, and the affliction of the Moon by Mars and Mercury, very aptly show the disappointment and trouble to which the Hellenic hero was born; the blighting of his love, the estrangement from his kindred and country, and the criticism to which he was subjected throughout his life, and even after his death."

✳   ✳   ✳

THE LEGAL PERSECUTION that had demoralized Alan Leo also touched America's shores. Astrology had been outlawed in England in 1736 under the Witchcraft Act and again in 1825 under the Vagrancy Act, which classified astrologers with "rogues and vagabonds." At the beginning of the 20th century, a related law enacted in New York State stamped them as "disorderly persons," along with "acrobatic performers, circus riders, and men who desert their wives." That law was decisively tested in 1914 when Evangeline Adams, a noted astrologer and claimed descendant of John Quincy Adams, the sixth president of the United States, was arrested by an undercover cop. Although Adams might have escaped lightly with a fine, she insisted on a trial. She came to court armed with reference books, expounded the principles of astrology to the judge at some length, and illustrated its practice by reading a blind chart that turned out to be that of the judge's son. The judge, John J. Freschi, was so impressed by her analysis that he ruled emphatically in her favor, concluding that she had "raised astrology to the dignity of an exact science" by her work.

Her lawyer, Clark L. Jordan, also proved superb on her behalf. In his brief, he had explained:

> Astrology is the science which describes the influence of the heavenly bodies upon mundane affairs and upon human character and life. It is a mathematical or exact science as it is based upon astronomy which describes the heavenly bodies and explains their motions, etc. It is an applied science in that it takes the established principles of astronomy as its guide in delineating human character, and all its judgments are based on mathematic calculations. It is an empirical science, because its deductions are based upon accurate data that have been gathered for thousands of years. Astrology is the oldest science in existence. It is not only pre-historic but pre-traditional, and must not be classed with fortune telling, or any of the many forms of demonology as practiced in ancient and modern times.

Judge Freschi agreed. In explaining his verdict, he noted that in her reading, Adams "went through an absolutely mechanical, mathematical process to get at her conclusions . . . Basing her horoscope on the well-known and fixed science of astronomy, she violated no law . . . The plain object of the statute is to protect the fool and the credulously weak from the knavery of those who claim wisdom and who resort to trickery and every device known to cunning as a means of gain in some form . . . I am satisfied that the element of fraud which we usually find accompanying the fortune-teller's case is absent here."

In her youth, Adams had been taught astrology by J. Heber Smith, a Sanskrit scholar and professor of medicine at Boston University, who was also one of New England's leading medical diagnosticians at the time. Smith often used astrology in connection with his medical work, and later became a specialist in analyzing the personalities of prison inmates from their charts. He had great confidence in Adams, who first set herself up in the Copley Hotel in Boston. But it was her foresight in connection with the horrific blaze that destroyed the Windsor Hotel in New York in 1899 that first established her name. Not long after the blaze occurred, the newspapers were full of astonished testimony from the hotel proprietor, who confessed to having ignored her warning of an impending calamity in which fire would play a part.

Adams eventually acquired a studio at Carnegie Hall, where she was consulted by the rich and famous, and visited by the likes of England's King Edward VII, Enrico Caruso, the naturalist John Burroughs; numerous theatrical celebrities (Geraldine Farrar, Lillian Russell, Tallulah Bankhead, Eva Le Gallienne, Charlie Chaplin, and Mary Pickford, among them); writers Eugene O'Neill and Anita Loos; a 21-year-old Joseph Campbell, the great mythologist, who received a life-reading that (he later declared) proved out; the head of the Pinkerton Detective Agency, O. M. Hanscom, who consulted her about his most difficult cases; and a famous New England surgeon, A. M. Thayer, who wanted to know the best days on which operations should be performed. Prominent businessmen and bankers also came, including Fritz Heinze, "the Montana copper king"; James T. Hill, President of the Great Northern Railroad; John P. Dryden, the head of Prudential Life; Charles Schwab (the elder); and J. P. Morgan, to whom she explained the general effects of the planets on business and politics. Morgan, though skeptical at first, became a true believer, and by his own account her forecasts helped to make him the rich man he became. Two presidents of the New York Stock Exchange, Seymour Cromwell and Jacob Stout, followed in his wake and professed themselves astounded at her skill. She even involved herself briefly in presidential politics. John W. Weeks of Massachusetts, a leading candidate for the Republican nomination in 1920, decided to consult her about his chances, only to drop out after she told him that Harding would prevail. She also predicted the election of Calvin Coolidge in 1924 and the unexpected victory of Al Smith as governor of New York in the same campaign.

In April 1930, Adams became the first astrologer to host a radio show, broadcasting three times a week. Listeners inundated her with requests for horoscopes (to which she could hardly reply), though her fame was solidified by several popular books—her autobiography *The Bowl of Heaven,* and three guidebooks, *Astrology for Everyone, Your Place in the Sun,* and *Your Place Among the Stars.* On the one hand, Adams took advantage of the appetite for Sun-sign astrology, which she once called "Solar Biology," with a touch of scorn; on the other, she seems to have practiced the art itself with considerable skill. As a horary astrologer, she developed her own unique methods that involved a consideration of the natal and horary charts together, and a horoscope that placed the natal planets in the houses of the horary chart. She also noted the degree rising at the moment her client arrived, which became the "Accidental Ascendant" of the figure she drew. An accomplished palmist—it was Adams who first discovered that palms smeared with printer's ink, rather than lampblack, made a better impression of the lines and mounts—she sometimes combined palmistry with her astrological readings, though she viewed it as a lesser skill.

Astrologers can sometimes be misled by their clients, of course, as Lilly had acknowledged; but Adams was not easily fooled. One day a woman brought her a horoscope that Adams assumed was that of a child. The woman asked her for a life reading (a character analysis and general prospects for the future); Adams examined the chart and told her that the individual had been born in a slum; would eventually live in style; would profit greatly from a connection with some woman; be clever; perhaps appear on stage; travel a lot; and live to a ripe old age. The woman then disdainfully revealed that the birth data belonged to a pet dog. However, it turned out that the dog belonged to a well-known actress who had bought it from a waif; that she had taught it tricks; that it had then appeared with her on stage in one of her successful plays; and had even accompanied her on tour. Moreover, it lived past the age of eighteen, which is a ripe old age for a dog.

Adams was a true believer. At the end of *The Bowl of Heaven,* she exclaims: "Astrology must be right. There can be no appeal from the Infinite." She died quietly in her sleep on November 10, 1932, at the age of sixty-four.

By then, newspaper astrology had also gotten off to a dashing start. In the summer of 1930, the *London Sunday Express* had invited the astrologer R. H. Naylor to cast and analyze the birth chart of the infant

Princess Margaret, born on August 21 to the future George VI. The article, published as a half-page feature, excited a good deal of attention, so the paper ran another column the following week. Soon Naylor was a regular, and circulation soared. Then on October 5 he wrote a column that proclaimed that British aircraft were at risk, and on that very day the warning was fulfilled when the great R-101 airship en route from Cardington to India crashed in France. The next Sunday, the *Express* reproduced its astrologer's prediction in bold headlines, and henceforth Naylor enjoyed the fame of a theatrical star. Other papers began to run their own astrology columns, and before long they were a feature of the popular press worldwide. Most of them were not as feckless as the columns of today, and a few (like Naylor's) sometimes hit the mark. In March 1936, when most astrologers were predicting peace, Naylor predicted the outbreak of World War II; in October 1938, Franklin Roosevelt's third term; and in 1944, that the United States would experience internal upheaval between 1966 and 1970 in connection with a war in Asia, which of course came true with Vietnam. The English astrologer Edward Lyndoe, who wrote a weekly column for *The People,* also had at least one notable success. After accurately predicting the death of George V, he announced that should the Prince of Wales (afterward Edward VIII and later the Duke of Windsor) succeed him, his reign would be short and end in abdication, which it did. Meanwhile, in India, in the October 1937 issue of the *Astrological Magazine,* B. V. Raman, today considered the father of 20th-century Indian astrology, had forecast the outbreak of war in Europe in 1939, followed by Hitler's relentless aggression and ultimate defeat.

★ ★ ★

UNDER THE INFLUENCE of Alan Leo and that of his principal disciples—Charles E. O. Carter, C. C. Zain, Marc Edmund Jones, and Dane Rudhyar—modern astrology took shape. All were theosophists of a sort, and enshrouded its doctrine and practice with related ideas. As astrology became mainly descriptive, its increasingly hazy character prepared the way for a psychologized version of it that essentially internalized the planets in a way that Ficino (sometimes invoked as the patron saint of this process) could scarcely have conceived. Meanwhile, under Annie Besant, Blavatsky's successor, the Theosophical Society issued a quarterly, *Astrology,* and gave courses with diplomas or

degrees. And in 1926 Llewellyn George founded the National Astrological Association, which became the American Federation of Astrologers. In England, a Faculty of Astrological Studies was founded in
1948.

After Alan Leo died, his thriving astrological lodge was taken over
by Charles Carter, who tended to eschew theosophical jargon in favor
of a more practical approach. Overall, however, Leo's ideas held sway.
In America, Leo's chief apostle was Marc Edmund Jones, a Presbyterian minister, editor, and educator, who had joined a clandestine Rosicrucian order in his youth and later developed his own "Sabian
System" of astrology as "my own name for the reconstruction of the
Chaldean or Babylonian science as far as I have been able to dig it out
through sources here and there. That is, it is a recovery on my part and
not an invention." Yet it was. The Sabian symbols for each degree of
the zodiac were psychically discerned, without provable reference to
any known ancient system, though a people known as the "Sabeans"
had inhabited ancient Yemen some eight hundred years before Christ.
In Jones's hands astrology became a means of exploring an individual's
potential for fulfillment, "a psychological method for charting or measuring experience," as he put it in *Astrology: How and Why It Works*. Under an epigraph from Cervantes, "Let us make hay while the Sun
shines," this is how Jones begins his page on Taurus: "Taurus is the
fixed sign in the earth group, and this means that it reveals, in every
horoscope, the characteristic way in which the native makes use of his
vicarious experience, to give himself a practical, everyday and external
steadiness in living. The key word for this mode of action is INTE-
GRATION. Here is the most definitely physical point of all in the
chart, showing the pattern of energy-release in the individual case, and
also measuring the particular indomitability or simple root constancy
of which an individual is capable."

Jones was also more interested in the patterns of the planets than
their natures. Thus, in his *Guide to Horoscope Interpretation* (1941), he
expounded an original theory of seven planetary groupings, or patterns, which he believed expressed (or imaged) the essential character
each horoscope possessed. These types or patterns were: the Splash,
Bundle, Locomotive, Bowl, Bucket, Seesaw, and Splay. Each was illustrated by the chart of some famous person, who was said to exemplify
the nature it described: for example, "the Splash" by Jacob Boehme;

the "Bundle" by Mussolini; the "Locomotive" by Isaac Newton; the "Bowl" by Oliver Cromwell; the "Bucket" by Napoleon; the "Seesaw" by Percy Bysshe Shelley; and the "Splay" by Henry VIII. This fanciful innovation (with names that sound like Motown hits) is now generally accepted by modern astrologers (who have an affection for such types), though its practical value is doubtful, since under the rubric of any one pattern may be found such disparate lives.

Nevertheless, Jones's work led to the "humanistic" astrology of Dane Rudhyar, who in 1936 wrote *The Astrology of Personality,* which made "the birth chart," in the words of one caustic observer, " 'a map of the psyche' that enabled the astrologer as therapist to probe into the murkiest recesses of the self, and to understand psychological crises in terms of planetary transits. No wonder you're feeling haunted by your relationship with your mother, declares the astro-shrink, Pluto is crossing your natal Moon." True, psychological astrology can be more interesting than that, but it is not uncommon to encounter in the work of its advocates a passage such as this: "The lunar symbol, the crescent, depicts the partial consciousness of the subconscious; the side of our-selves we know only through our dreams. This half-lit state is the life of our bodies and instincts before they reach the logic-centered realm of ego-consciousness and this state is also what Freud described as the id." By contrast, the somewhat stilted jargon of some of the older as-trologers has a resonantly clear ring.

Meanwhile, artists—especially those with a mystic bent—continued to be drawn to the subject, including the composers Arnold Schoen-berg, Aleksandr Scriabin, and Gustav Holst; the painters Paul Klee, Wassily Kandinsky, and Piet Mondrian; the architect Walter Gropius; and the poet William Butler Yeats (an expert astrologer himself, who elaborated a poetic view of history in which human life unfolded in great patterns analogous to the phases of the Moon). Henry Miller de-voted much of his book *The Colossus of Maroussi* to a discussion of the attributes of Saturn; the poet Louis MacNeice compiled an astrological textbook; H. G. Wells, T. E. Lawrence, Thomas Hardy, Edith Sitwell, Anaïs Nin, Aldous Huxley, William Carlos Williams, Lawrence Dur-rell, Katherine Anne Porter, and other writers made astrology part of their lives.

All sorts of theatrical celebrities, of course, have also paid attention to their stars, including (among the moderns) Susan Strasberg, Fanny

Brice, John Barrymore, Danny Kaye, Isadora Duncan, and Lillian Gish. If nothing else, such a list suggests its broad appeal. Barrymore kept track of the planetary hours and would only work a love scene when Venus reigned—or a fight scene under Mars. Margaret Mitchell based some of her characters in *Gone With the Wind* in part on the zodiac signs. Thus Aries, ruled by Mars, "the red planet," inspired Scarlett O'Hara, hence her name.

William Butler Yeats combined tarot card readings with astrological predictions (sometimes casting a horary chart for the moment the cards were read) and at the age of sixty-eight, five years before his death, was still calculating his own secondary progressions for the following year. Virginia Woolf noted in her diary: "[Yeats] believes entirely in horoscopes" and "will never do business with anyone" unless he first examines their chart. After his marriage in 1917 to Georgie Hyde-Lees, Yeats also became interested in automatic writing, but it is astrology that most permeates his work.

★   ★   ★

IF TRADITIONAL ASTROLOGICAL INTERPRETATION viewed the stars as objective forces determining our fate, psychological astrology, in a less deterministic way, soon came to regard them as representing psychological aspects of our makeup that incline us toward the ends we come to or the lives we lead. In this sense, its motto was "character is fate."

In time, it would be adopted by many psychologists, who found it useful to describe the various planetary influences (or the influences they symbolize) in psychological terms. The planets were regarded as entities in their own right, in somewhat the same way that Plato and the early Greeks viewed the Twelve Gods as archetypes—Saturn or Cronos as the principle of limitation and restriction, for example, Venus or Aphrodite of love and desire. These in turn were matched to the gods. "The Moon reflects the ego image of personal consciousness," we read in one description. "Mercury the principle embodied in the archetype of the trickster, Venus the maiden in dreams, Mars the heroic element in the psyche," and so on. The influence of the Swiss psychologist Carl Gustav Jung in all this was large.

Quite early on in his practice, Jung had begun to cast the horoscopes of patients to assist him in analysis. In a letter to Sigmund Freud, dated June 11, 1911, he wrote: "My evenings are taken up very largely with astrology. I make horoscope calculations in order to find a clue to the psychological truth. Some remarkable things have turned up which will certainly seem incredible to you . . . I dare say that one day we shall find in astrology a good deal of knowledge that has been intuitively projected into the heavens. For instance, it appears that the signs of the zodiac are character pictures." Freud, who had recently paid a visit to a psychic in Berlin, replied: "In matters of occultism I have grown humble . . . I promise to believe anything that can be made to look reasonable. I shall not do so gladly, that you know. But my hubris has been shattered." The two soon parted company, however, over Freud's theory of sexuality, and Freud decided that astrology was just a form of "thought transference"—as when a "fortune-teller," for example, becomes "receptive," like a medium, to a client's thoughts. How that could explain the blind reading of a chart at a distance Freud never ventured to explain.

Jung, meanwhile, continued to find insight in astrology, even to the end of his days. In 1947, he wrote to the Hindu astrologer B. V. Raman: "As a psychologist, I am chiefly interested in the particular light the horoscope sheds on certain complications in the character. In cases of difficult psychological diagnosis I usually get a horoscope in order to have a further point of view from an entirely different angle. I must say that I very often found that the astrological data elucidated certain points which I otherwise would have been unable to understand." In one experiment, he made a statistical study of the relationships between the horoscopes of 483 married couples and found that most of those who seemed to enjoy a good relationship had planetary aspects (Sun-Moon, Moon-ascendant, or Moon-Moon matches) indicating that they should.

There have always been different theories as to how astrology might "work"—some have thought by rays; others, by gravitation; still others by some electromagnetic force. The oldest notion, found in Genesis and elsewhere, is that the planets are signs, not causes—part of the Book of Nature or the Alphabet of God. Jung seemed to be saying something similar in 1930: "Whatever is born, or done, in this moment of time has the qualities of this moment of time." He called this "synchronicity" or "meaningful coincidence," which implied an acausal law of nature to explain the link. "Meaningful coincidence," however, is a contradiction in terms, and the classic texts held, rather, that everything happens where and when it does because of the patterned coordinates of time and space (or place). Then there is a sort of matrix theory, as set forth by a former editor of *Science and Astrology* magazine:

> In the moments prior to birth the child has no separate existence but consists of a collection of living cells which may well be considered part of the mother's organism. But, at the moment of drawing its first breath, a vital change takes place. For it is at that critical moment that self-animation comes to the child—the product of the life-forces which enter it. And, at that same moment, the lines of force of the existing magnetic field surrounding the Earth (product or resultant of the composite magnetic fields belonging to the Sun, Moon, Planets and Stars) run through the child's body, setting the paths of the electrons within the atoms of the matter composing the child's body into a fixed and definite pattern which appears to exist for the duration of its life.

Jung might have been intrigued by that. He would almost certainly have been appalled at the way psychological astrology has sometimes been applied to his own life. Here is one serendipitous account of his years following the fall of 1909, when he set up a private practice in the States:

> Jung was on the move, spiritually and intellectually, and in particular his relationships with women began to change. As Neptune made its second transit of his natal Venus, Jung joked to Freud that the "prerequisite for a good marriage [was] the license to be unfaithful." At the turn of the year, his progressed Venus moved into Virgo and Toni Wolff, a Virgoan, became his patient. She was to become his lover and remain the "other woman" throughout his life . . . Jung was certainly a ladies' man; he had a fiery Mars in promiscuous Sagittarius, a sweet-talking Mercury-Venus conjunction and a 7th house Sun in Leo. By 1911, there were already four women vying for his affections . . . and with his Sun square to Neptune, strict adherence to analytic boundaries were not his forte.

One of those on the cusp of these new, dubious developments was Grant Lewi, a hugely popular astrologer who taught English literature at Dartmouth College and the University of North Dakota before exchanging his cap and gown for the astrologer's cloak. In the 1930s and '40s he helped edit *Horoscope Magazine,* moved to Arizona, founded his own periodical, which he called *The Astrologer,* wrote a couple of novels—*Star of Empire* and *The Gods Arrive*—and two best-selling books, *Astrology for the Millions* and *Heaven Knows What.* These expounded his own system of interpretation, based on sign-house equations, and his exclusive use of transits in predicting future events. He also developed a theory concerning Saturn returns (which, on average, occur every twenty-eight years) as pivotal moments in a person's life. Lewi once remarked with a certain Pythagorean wit that he believed in astrology for the same reason other people believe in the multiplication tables—because they work. It is said that he deliberately refused to take out a life insurance policy until he saw his end was near, then, at the age of forty-nine (despite the appearance of perfect health) bought one, "paid one premium, and died."

Lewi's conception of astrology was of its time. "The chief use of the horoscope," he wrote, "is its aid in self-discovery . . . This is, among other things, The Age of Psychiatry." In *Heaven Knows What,* published in 1937, he wrote about Sun/Moon combinations in an intimate, chatty manner that in some respects set the tone for the advice columns of today. This was his take on someone whose Sun and Moon were both in Aries:

You belong to the positive or executive group. . . . Whatever anyone else can say about you, it will never be that you don't know your own mind. In fact, it's far more likely that they will say you know your own mind too well and are too ready to express whatever comes into it; for independence is the outstanding trait of your nature—independence in both intellectual and material matters. You are an individualist, first, last, and always; you believe in Yourself, and in your own ability to overcome the forces of heaven and earth, man and God, in order to arrive at that goal of personal ambition which is so clearly defined in your mind as to be already an actuality. You are likely to be somewhat intolerant of divergent opinions on the part of others. Not only are you intolerant of them, but you may just plain not understand them, for Understanding is not your strong point. The troubles of other people are not especially interesting to you, although by a paradoxical twist you frequently expect other people to listen to your troubles at some considerable length.

Yet, don't despair, he said, for "this combination leads far upward, once you have mastered the more self-centered leanings of your nature and learned to objectify." Grant Lewi was not without his gifts, but such preposterous generalizations—from so little, so much—became fairly standard fare.

Lewi's *Astrology for the Millions,* published three years later, included chapters on "Why I Believe in Astrology," "The Everlasting Quest: Man's Search for Himself," "The Pattern of Life: How Astrology Provides It in the Vitasphere," "Fate or Free Will? Astrology's Answer," two sections on "Strong Men of Destiny" (with astrological portraits of Lincoln, Woodrow Wilson, Hitler, Mussolini, Stalin, and F.D.R., among others); a chapter on "Self-Destruction"; and paragraphs on the planets and their meaning in the signs. His page on the Moon begins: "While the Sun's position by sign determines what motives and urges dominate your life as it meets the naked eye, the sign position of the Moon tells the desire of your heart which may or may not be expressed or realized in your life. When you 'know what you mean but can't say it,' it is your Moon that knows it and your Sun that can't say it. 'Thoughts that do often lie too deep for tears' are the thoughts of your Moon's nature. The wordless ecstasy, the mute sorrow, the secret dream, the esoteric picture of yourself that you can't get across to the world, or which the world doesn't comprehend or value—these are the products of the Moon in your Vitasphere."

Lewi's style was enormously influential and echoes through most astrological handbooks written since. Here, from a recent one, is an account of the "Sun in Pisces": "Pisces is multifaceted and multitalented. It is the ultimate sign . . . intuitive, imaginative, and possesses an insight into human nature that bestows compassion and an 'others before me' attitude. Pisces is empathetic and people oriented. As a theoretical sign, it displays, feeds, and expresses its ego through other people." We also learn that Pisces "picks up the vibrations of others," which can be "mentally and emotionally draining"; "needs solitude," yet "fears being alone"; sometimes wants to please others "for fear of being rejected";

can be "lazy, vague, and lack concentration"; will "hang on to imaginary hurts"; yet "with a good education, self-awareness, and fortitude, can be creative, artistic, and productive." In an obligatory list of "Famous Pisceans," the usual suspects are rounded up: "Elizabeth Taylor, Albert Einstein, Michelangelo, Frederic Chopin, Victor Hugo, Rudolf Nureyev, George Washington, and John Steinbeck"—given so the Piscean reader can vicariously bask in their light.

⋆ ⋆ ⋆

SOME ASTROLOGERS ARE MORE EQUAL THAN OTHERS, of course, and various threads of the tradition were always being spun. Mundane astrology, unfettered by legal restraint, also claimed its own predictive sphere. It continued to flourish, in and out of newspaper columns, and the lore of comets and eclipses held their honored place. In a sense, "official" astrology—that is, astrology as used by public or state officials—had never quite died out. Otto von Bismarck and his chief general, Helmuth von Moltke, had reportedly consulted with an astrologer before embarking on the Franco-Prussian War. Its subsequent course went so well for them that when they reached Versailles, they sent the man a handsome reward. "A statesman cannot create anything himself," Bismarck once remarked. "He must wait and listen until he hears the steps of God sounding through events; then leap up and grasp the hem of his garment." After another one of his triumphs, he professed: "I am content when I see where the Lord wishes to go and can stumble after him." In 1904, Japanese astrologers helped determine the timing of military actions that led to the unexpected defeat of the Russians in the Russo-Japanese War. Among omens (going back in time) the comet of 1704 had been seen to coincide with the dissolution of Poland and the capture of Gibraltar by the English in the war of the Spanish Succession; the comet of 1807, with Napoleon's invasion of Portugal; that of 1812, with the French invasion of Russia and war between the United States and Britain; that of 1858 (Donati's comet), with the troubles in Kansas preceding the Civil War.

Astrologers had also been engaged by the omens leading up to World War I. Throughout 1914, the Sun had followed Mars, as if to dog his tracks, until the two came into conjunction on Christmas Day. That event had been preceded four months before by a solar eclipse on August 21, which fell in the last decanate of Leo, "with the great fixed star Regulus in the shoulder of the Lion, a signification of terrifying aspect to emperors, kings, princes and autocracies," in the words of Katherine Craig. According to tradition, an eclipse affected those countries from which it was "visible or total, or nearly so," and when falling in a fixed sign as well as a royal one, its effects were lasting upon the countries and dynasties concerned. Those consulting Raphael's little book *Mundane Astrology* would have read, more specifically, that an eclipse in the last decan of Leo means: "Captivity, the ransacking and besieging of towns, the sacking and profanation of holy places . . . the exile, imprisonment or murder of some king, notable person or great ruler. Much discontent among the people. Movements of armies, fighting, fires, fevers, pestilence and scarcity of the fruits of the earth."

In August 1914, the central path of the eclipse had extended from the Baltic Sea through Eastern Europe, the Asian subcontinent, Persia, Africa, and the North Atlantic states. During the eclipse, the Sun held the exact place held by Mars at the first new moon of the summer solstice on June 22. Then on June 26, Mercury, the Moon, the Sun, and Jupiter had all joined with Uranus in the same sign. As Craig construed it, with uncommon eloquence:

> Once before in history a similar planetary configuration had stirred the degeneracy of old Rome into a ferment that rose to the surface in Sulla, in Marius and in Catiline, in the assassination of Julius Caesar, in the down fall of the Roman Republic; and from Rome, the great heart of civilization, the virus of anarchy and of rebellion had permeated the veins of the world. Mithridates was conquered; Asia fell to pieces; and Greece died in all but name; Spain, Britain and Gaul in turn bowed to Rome; old Egypt met its Battle of Actium and the Priests of Isis were betrayed at a harlot's whim. Then, the turmoil quieted, the cycle of Mars passed and the light followed. . . . The defeat of Britain, Gaul and Spain had paved the way for newer and stronger nations; Rome gloried in her Augustan Age, and from a manger in manacled Judea beamed the light that was to redeem the world.

Astrological prediction at the time was almost universal in foreshadowing humiliation and defeat for the Kaiser, in part because in his own chart Mars and Saturn were in a square. Under that same aspect, which formed in the heavens in the late summer of 1914, the Kaiser had begun the war, and by it his defeat would be sealed.

Astrology figured, in turn, among the Allied and Axis powers of World War II. In the 1930s there were several serious German periodicals on the subject, but Hitler promptly suppressed them when he came to power. As Heinrich Himmler explained: "We cannot allow astrologers to follow their calling unless they are working for us. In the National Socialist state, astrology must remain a *privilegium singulorum*. It is not for the masses."

Hitler had reason to be wary. In 1923, soon after he became head of the Nazi Party, the astrologer Frau Elsbeth Ebertin, who edited a popular annual called *Ein Blick in die Zukunft (A Glimpse into the Future)*, had publicly warned him against undertaking "anything of importance" that November. In the issue published in July 1923, she wrote: "A man of action born on 20 April 1889, the Sun in 29 degrees Aries [she was off by a few degrees] at the time of his birth, can expose himself to personal danger by excessively incautious action and could very likely trigger off an uncontrollable crisis. His constellations show that this man is to be taken very seriously indeed; he is destined to play a 'Fuhrer-role' in future battles . . . [and] sacrifice himself for the German nation." The reference to Hitler was unmistakeable, but when her article was called to his attention, he exclaimed: "What on earth have women and the stars to do with me?" On November 8, he undertook his abortive Munich Beer Hall Putsch, after which he was arrested and imprisoned for nine months.

Years later, in 1939, after Hitler had been appointed chancellor, a Swiss astrologer, Karl Ernst Krafft (said to have successfully predicted trends in the commodities market) warned in a letter to the German Head Office for State Security, dated November 2, that Hitler's life would be in danger "from explosive material" that month between the 7th and the 10th. Such are the coordinates of fate that three months before, a Swiss clockmaker by the name of George Elser had quietly begun to assemble various tools and gadgets needed for planting a time-bomb in a pillar of the Munich beer hall to which Hitler returned each year on November 8 to commemorate his Putsch. On August 5, Elser had arrived in Munich, secured work as a carpenter in the crew assembled to renovate the hall, and got to work. In his toolbox, he had an assortment of planes, hammers, squares, shears, engraving tools, a pad saw, a precision ruler, scissors, pliers, wood clamps, and several rasps and fine wood files. Another kit contained fifty kilograms of high explosives, six clock movements, insulated wire, and a six-volt battery, all neatly packed into a 180-mm brass artillery encasing he had somehow obtained. Night after night, he returned to the building through an obscure passageway and hid in the gallery until the last person had left, then "working by the weak beam of a flashlight shrouded with a blue handkerchief," carefully pried off a small piece of molding from the column, inserted a special cabinet-maker's saw into a tiny hole he had drilled in a corner of the panel, and for three to four hours at a stretch, sawed ever so slowly through a few millimeters of wood. Eventually, with hammers and drills exactly suited to his task, he managed to create a cavity of the requisite size in the column. Meanwhile, he had also begun converting two fifteen-day alarm clocks, with backup triggering devices, detonators, and charges, cogwheels and levers, into a bomb that could be set 144 hours in advance. On the night of November 4, it was wired and set for 9:20 P.M., November 8. On that evening, Hitler arrived at the beer hall at 8:10, mounted the stage in front of the column—draped with an enormous swastika-emblazoned flag—and began his speech. It was scheduled to last for at least an hour and a half. But at 9:12 he suddenly "wrapped it up, gave a quick Nazi salute, and trotted out of the hall." Eight minutes later, Elser's bomb exploded, killing everyone on stage.

Krafft was promptly arrested on suspicion of complicity in a plot, then compelled to work for the German propaganda machine. He did so reluctantly—and paid for his candor with his life. Meanwhile, Winston Churchill had enlisted the services of Louis de Wohl, an astrologer who had "accurately predicted the outbreak of the war and the invasion of Poland at a dinner party attended by Lord Halifax" in 1931. "It was clear to me, as to every student of astrology who knew Hitler's horoscope," de Wohl wrote afterward, "that he would launch his great attack against the West when Jupiter was in conjunction with his Sun, in May 1940."

When the British learned about Krafft's service, they mistakenly assumed he was Hitler's personal astrologer. De Wohl was familiar with Krafft's working methods, and also knew the other adepts supposedly in Hitler's employ; so he was called upon not only to make predictions, but to predict what they would predict, or advise. In his first memo to the British War Office in 1940, he professed himself sure that the Germans would not invade England if Hitler heeded their advice. Later, when Hitler learned that some astrologers thought his invasion of Russia would end (like that of Napoleon) in disaster, he was so enraged he had a number of them killed. Meanwhile, Rudolf Hess, on the advice of his own astrologer—Ernst Schulte-Strathaus—had fled to Scotland in a tiny plane, and before the war came to an end, Frau Elsbeth Ebertin would be killed by an Allied bomber when her house suffered a direct hit. Krafft would die on his way to Buchenwald.

Hitler, in fact, hated astrologers, and would have been loathe to consult them. According to Hugh Trevor-Roper (in *Hitler's Table Talk*) he once complained to some generals—on July 19, 1942—at a dinner party that all it took was one lucky guess and the record of all their failed predictions was expunged.

Meanwhile, back at the British War Office, de Wohl had been assigned to a secret department of British Intelligence called the Political Warfare Executive, which specialized in disinformation campaigns. In a sense, he was constitutionally suited to such work, for he had an inscrutable manner, crocodile-like eyes, and the odd gift of being able to write simultaneously with both hands, spelling the same words backward with his left that he was writing in normal fashion with his right. One of his more successful projects involved the production of a bogus German astrology magazine called *Der Zenit (The Zenith),* of which half a dozen issues were produced at regular intervals in 1942–43. All contained troubling predictions for Hitler and his forces, and being dated before, but issued after, the events, appeared incredibly accurate and spread alarm. Indeed, *Der Zenit* apparently caught on with some members of the German high command. After the British learned that Admiral Karl Dönitz was paying attention to the advice given in a previous issue about "unfavorable days," copies were distributed with strategically chosen dates to dissuade him from launching his U-boats when allied ships were on the seas. The readers of the March 1943 issue, for example, were asked to "consider the sad case of U-boat 335. This vessel had a good horoscope when launched but that of its commander, Kapitan-Leutnant H. Pelkner, was bad. As a consequence, his boat was sunk after four days at sea—on 3 August 1942—when the Moon in transit was square to the Sun and Pluto in his chart." A new table showing favorable and unfavorable days for

U-boat operations was then produced. Needless to say, not many days in March 1943 were considered auspicious for German patrols.

Other such projects of the Political Warfare Executive included the production of bogus ration coupons; leaflets that explained to German soldiers how to get out of the army by faking some illness; articles about rampant infidelity among wives on the home front; and other items designed to demoralize the German troops. Forged documents were also used to spread rumors "about the deterioration of Hitler's physical and mental health." One, for example, claimed that one of Hitler's personal physicians had been treating him for prostate problems with a cocaine-based drug called Cycloform and "massive doses of ovarial hormones." It implied that the drugs had produced a manic state, and that Hitler, as a result, had made a number of disastrous military decisions, including his dismissal of some twenty top generals and his sudden resolve to assume personal command over the Russian front.

Meanwhile, in his role as Allied seer, de Wohl correctly predicted the outcome of a number of battles, including that of El Alamein in 1942 (after comparing the horoscopes of Rommel and Montgomery), even as Joseph Goebbels despaired over the gloomy forecasts that cluttered his files. Finally, in April 1945 Goebbels unearthed two charts that had correctly foreseen the victories and reversals the Reich had sustained, but which also predicted peace by August 1945. He decided to present them to Hitler in the context of an historical argument that favored their desperate case. When Prussia was on the verge of defeat at the end of the Seven Years' War, he explained, the czarina of Russia had unexpectedly died and Prussia had escaped her apparent fate. Just so, said Goebbels, the two horoscopes he had examined indicated a similar denouement. The death of a major Allied leader was imminent, and the tide would turn. Sure enough, Roosevelt promptly died, and Goebbels telephoned Hitler to tell him it all seemed to be working out as they hoped. Hitler must have marveled at that, since he was already sequestered in the concrete bunker that would become his tomb. "My Fuhrer, I congratulate you," said Goebbels. "Roosevelt is dead. It is written in the stars that the second half of April will be the turning point for us. This is Friday, April 13. It is the turning point."

The stars were right; but Goebbels's read on them was wrong. Peace came that August, but under the Allies' heel.

Rightly or wrongly, most astrologers today accept the role of the outer planets, in which Uranus (discovered in 1781), Neptune (discovered in 1846), and Pluto (discovered in 1930) represent, respectively, the force, or principle, of unpredictable or revolutionary change; the nebulous and the occult; and the unleashing of tremendous subterranean powers. Uranus has been the most studied of the three and is said to indicate, in material and social matters, violent changes, sudden loss and gain, impulsive and unaccountable acts, and so on. In mundane astrology, its transits are often said to indicate a period of travail. One astrologer, in retroactively considering the astrology of World War II, has noted that transits of Uranus coincided with a number of crises in the Nazi state. Taking the birth of that state to be January 30, 1933, at 11 A.M., when Adolf Hitler was appointed chancellor, he finds that Uranus squared Mercury (May 1935 to April 1936) when Hitler invaded Austria and occupied the Rhineland; squared Saturn (May 1936 to April 1937) when he established a totalitarian state; squared the Sun (May 1937 to March 1938) when he laid out his long-term plans for war; and was conjunct the ascendant (May 1938 to May 1940) when he attacked Poland and World War II began. Then during the 1940s, Pluto stood in opposition to the same cluster of planets that Uranus had previously squared. The march of aspects proceeded in lockstep with events. With Pluto in opposition to Mercury, Hitler invaded Russia; with Pluto in opposition to Saturn (October 1942 to June 1945), Hitler's forces suffered a series of catastrophic reversals in the field.

★   ★   ★

ALL NATIONS, whether they realize it or not, in fact have their own horoscopes, based on the time of their founding, and some, including Thailand, Myanmar (formerly Burma), and Sri Lanka (formerly Ceylon) were deliberately created at astrologically chosen times. The twelve signs of the zodiac are notably stamped on the reverse side of Israel's official state seal, and the Israeli secret service (known as Mossad) reportedly uses a horoscope based on the data of the nation's founding to help predict security threats to the state. The late Indian prime minister Indira Gandhi subscribed to astrology and apparently knew of the danger of her assassination in advance; Theodore Roosevelt took note of his horoscope—particularly the opposition of his seventh-house Moon to his first-house Mars—and "kept a weather eye on his chart." That chart was etched and mounted on a board that stood on a table in his drawing room. The inauguration of Ronald Reagan as governor of California took place, for strictly astrological reasons, in the middle of the night—at 12:16 A.M., to be precise, when Jupiter was high in the sky; and his subsequent campaign for the White House was similarly framed. According to Donald Regan, President Reagan's chief of staff, "virtually every major move and decision Reagan made" as president was cleared in advance with astrologers, including the timing of his announcement to seek reelection, his invasion of Grenada, his attack on Libya, and his delicate negotiations over disarmament with Mikhail Gorbachev.

The so-called twenty-year cycle governing the deaths of American presidents in office—Harrison, Lincoln, Garfield, McKinley, Harding, F.D.R., and Kennedy—not incidentally, applies only to those elected at the time of a Saturn-Jupiter conjunction in an Earth sign. By the time Kennedy ran for president, the apparent coincidence was sufficiently publicized to come to his attention. But he brushed it aside with the remark that he would be the first to "break the jinx." The connection (for astrologers) is not a coincidence but explained through the influence of the conjunction on the national horoscope, in which Jupiter and Saturn are squared.

Reagan's supposed reliance on astrology has given rise to some dispute, in part because the first lady downplayed its role. But among political leaders, he was certainly not alone. In August 1944, Charles de Gaulle met a regimental band leader and astrologer by the name of Maurice Vasset in Toulon and from then on consulted him until the end of his career. François Mitterand, in turn, carried on a seven-year dialogue with the French astrologer Elizabeth Teissier, while during the same period Boris Yeltsin had recourse to the Academy of Astrology in Moscow for counsel and advice.

Whether the counsel any of them received had much merit is a question. The record is simply too obscure.

★   ★   ★

SOME MODERN STUDENTS OF ASTROLOGY have sought to test it by statistical methods in the spirit of some of the late 17th-century reformers who hoped to correlate astrological data with character traits. The first statistical studies of any consequence were done in France by

Paul Choisnard, a late-19th-century artillery officer and engineer, who claimed that Sun-Mars aspects were often conspicuous in the charts of those who suffered violent deaths; that Mercury-Moon aspects strongly marked the charts of philosophers and visionary thinkers; and that Sun-Moon aspects were featured in the charts of theatrical stars. A half century later, two other French scientists, Michel and Françoise Gauquelin, associated with the Sorbonne, assembled over the course of twenty years the birth data of thousands of individuals from registrars all over Europe, and studied the angular positions of the planets in relation to career. With a statistical frequency far beyond what chance would allow (in fact, by as much as 1,000,000:1), sports champions tended to have Mars prominent—that is, rising (near the ascendant) or culminating (near Midheaven)—in their charts, scientists Saturn, writers Mercury, and so on. These results were replicated by a highly skeptical group of Belgian scientists who spent four years trying to discredit what their French counterparts had done. In the end, their own data yielded the same results. Meanwhile, the Gauquelins pressed forward with their work, and eventually published the results of their findings in thirteen volumes in 1976. Those findings, however imposing, are less a proof of the art than of their own paradigm.

Even so, the tremendous efflorescence of astrology, with all its innovations, might seem to demonstrate its protean variety and strength. It might also be viewed, with some justice, as a kind of Hydra-headed monster, with no coherent doctrine, and with nothing "established" from which it does not also diverge. Modern astrology includes Jungian psychology; "cosmobiology" (the technical study of the impact of the cosmos on individuals as reflected in their "cosmograms"); the French symbolist school (according to which an astrological symbol "is the only available means of expressing a complex emotional reality that cannot be clearly conceptualized"); humanist astrology, with its emphasis on the analysis of personality; sidereal astrology; the heliocentric zodiac; the influence of a hypothetical group of as-yet undiscovered transneptunian planets; the vibrations of various asteroids (such as "Lilith," an obscure cloud of small dust particles that is said to orbit Earth like a second moon); new sign rulerships, as given to the outer planets (Uranus for Aquarius, Neptune for Pisces, Pluto for Scorpio); midpoint theory; various configurations such as T-squares (in which two planets are in opposition, with a third at their midpoint); Yods (in which two planets are in sextile, with a third planet at inverse midpoint, forming quincunxes to both ends); and the Grand or Cosmic Cross (formed by four or more planets, in which each planet successively makes a square to the one that precedes it). There are also new, minor aspects, such as the sesquisquare (135 degrees), semisextile (30 degrees), biquintile (144 degrees), and decile (36 degrees).

Chaos has ensued, along with a hodgepodge of Iranian, Indian, Vedic, Aztec, Chinese and other astrological material commingled with esoteric theosophical and New Age ideas. Out of this mix have also come arbitrary rules and reforms. In the ancient system, the question asked was: how strong is the planet in a sign? In the 19th century, astrologers began to "free associate" a bit and equate the nature of the sign with the planet that ruled it, which is an altogether different concept and today holds sway. The result has been what J. Lee Lehman calls a "skewing of the meaning of a planet in dignity," and the loss of the meaning of the five essential dignities themselves of sign, exaltation, triplicity, term, and face. Moreover, as she notes, "because of the modern practice of equating planet, house, and sign, many astrologers have even lost the ability to distinguish between these," too. Again, tradition held that each planet had its own orb. Alan Leo introduced the concept that aspects had orbs, which was a very different thing.

Modern astrology, which takes its limited cue from modern psychology, regards planets and planets in signs as types of energy, or even as "symbolizing" energy, which may be so. However, that has little basis in the tradition, which has an intelligible structure governed by mathematical rules. Modern astrology, by contrast, tends to work by associations, is more impressionistic, and adheres to the notion of an "inner" or "subjective sky." That allows for a certain arbitrary latitude in analysis, so that even on its own terms, a sharp psychological portrait can seldom be drawn.

There is and always has been, of course, a place for psychological astrology (or astrological psychology), which identifies not only types, but reveals the contradictions within the human personality in all its variety, with far greater sophistication and nuance, one could claim, than the crudities of psychoanalysis and its sweepingly useless categories of superego, ego, and id. The warrant for this is ample in tradition, especially among those who wrote and practiced in the Renaissance, when the idea of the "interior" or subconscious self began to be born. One can follow it in the development of Shakespeare's plays, with the debut of the interior monologue, and in the nascent thinking of Renaissance astrologers such as Marsilio Ficino, who wrote: "These celestial bodies are not to be sought by us outside in some other place; for the heavens in their entirety are within us, in whom the light of life and the origin of heaven dwell." Valentin Wiegel, in *Astrology Theologized,* a 16th-century work, also set forth the idea that the heavens at birth were an objective mirror of the self. "The saying that a wise man shall rule his stars," he wrote, "is not only to be understood of the external stars in the heaven or Firmament of the great world, but of the internal stars. . . . The stars over which we ought to rule, if we will be wise, are all the cogitations, speculations, cupidities, affections, etc., ascending by imagination out of our hearts . . . and tending by free will and reason to pleasure and abuse." Neither meant, however, that the stars themselves were psychological phenomena, or "projections," but objective phenomena patterned in us.

✶   ✶   ✶

IF ASTROLOGY STILL OCCASIONALLY ATTEMPTS to predict the future, so does meteorology, or medicine, or economics, among other disciplines. A meteorologist forecasts the weather (not always correctly) before it arrives, an economist (not always correctly) economic trends, a physician (not always correctly) the course of a disease. Jess Stearn's best-seller *A Time for Astrology* contained the following anecdote:

> It was a clear day in June 1970 in sunny California. Kate Lyman, sitting out in her patio and drinking in the heavenly sunshine, was idly scanning her personal horoscope when she suddenly sat up with a start. In the chart, which she had progressed ahead in time, she saw coming up an aspect that she found unbelievable as she looked around the wooded serenity of her beloved Malibu hills.
>
> There, inescapably, in the fourth house of her chart, the house of the home, in the sign of sorrow and one's undoing—there the planets

Mars, standing for fires and accidents, and Saturn, standing for restriction and loss, were in exact conjunction with one another, merging in the same degree to give added force to what they both stood for.

As a keen student of astrology . . . Kate Lyman had no question in her mind as to what Saturn and Mars together in the fourth house meant . . . [She] immediately put aside her chart, walked into her house, reached for the telephone, and called her insurance agent. "I want to increase my fire insurance," she said.

Three months later, as flames, fanned by searing winds from the desert, raced through the Malibu hills, Mrs. Lyman grabbed up her six-year-old twins, her small menagerie of cats, and a few odds and ends, and beat the flames to Pacific Coast Highway on her way to safety.

As she had foreseen, the affliction was to her fourth house, and not to the first house of her person or to the fifth of her children. Her home in Malibu's beautiful Sierra Retreat, overlooking the Pacific, had been leveled, and all its furnishings and her possessions destroyed . . . Wiping away the suspicion of a tear as she surveyed the smoldering ruins . . . she marched off, chin up, to gather the charred pieces of her life and plan for the new home she would need—and the extra insurance money that her devotion to astrology had ensured her.

Actually, the aspects that alarmed her could have meant a number of things, but one can't fault her educated guess.

Where does prediction end and analysis begin? In the old astrology, they were divided only by the question the astrologer brought to the chart. The procrustean bed of the modern approach substitutes one for the other, as may be seen in the astral portraits set forth in most celebrity horoscopes. Astrology books abound in these. The late Princess Diana was a perennial favorite, until her sudden death took the trade by surprise. Since then her drastic end has been flexibly explained in a hundred different ways. Another example is Adolf Hitler, whose oft-psychologized chart, taken as a touchstone, proves a rocky shoal on which most moderns come to grief.

"By their fruits shall ye know them." That biblical adage is as good as any by which to discriminate between those who know what they are doing and those who don't. One adept who makes this case with ascerbic brilliance is John Frawley, an English astrologer who has done much to resurrect the teachings of the past. As a comparative test of the capacity of ancient and modern techniques to sort out what matters in a chart—and what the things that matter mean—he looks at how a blind reading of Hitler's horoscope by each of them might read.

Born on April 20, 1889, Hitler had Libra rising; the Sun, Venus, Mercury, and Mars in Taurus; the Moon and Jupiter in Capricorn; and Saturn in Leo, high in the tenth house. In aspect, Saturn was in double square to Venus and Mars. A typical modern guidebook, notes Frawley, tells us that with a Libra ascendant, "our native 'is an easy-going, charming and kind personality who is diplomatic, cooperative and will do anything for peace and harmony. He is intellectually intelligent but inclined to indecision and easily influenced by others.' " With the ascendant ruler, Venus, falling in Taurus and the seventh house, he is also "pleasant, kind, faithful, appreciates good manners; has a love of luxury, especially the home; good voice; good taste; interested in gardening; can be very possessive" but has a problem with close relationships. The Moon in Capricorn gives a "reserved and cautious nature; . . . ambitious and hard worker . . . a drive for success," while, being in the third house, it shows "thinking strongly governed by emotions; prone to day-dreaming," and one who "soon tires of monotonous routine." At the same time, his Saturn-Venus square indicates a craving for "attention and respect." And so on. This brew of personality traits is then, in Frawley's words, synthesized according to the astrologer's taste.

The traditional astrologer, on the other hand (in this case basing his own blind reading on Ptolemy) would first determine temperament—"the cloth from which the person is cut." That temperament (as assessed by traditional methods) turns out to be "choleric," not "earthy" as moderns would claim (by merely counting the number of planets in the signs). How, asks Frawley, is this choleric fire going to find its outlet? "A strong, well-placed Mars [best suited to express it] . . . would suggest that the native can successfully integrate his nature into society . . . Hitler's Mars, however [by dignity and aspect] is in a dreadful state." Moreover, by antiscion (an ancient mathematical calculation that "gives . . . the shadow of each planet") Saturn "falls exactly" on the same Venus/Mars conjunction that it squares. Saturn, the great malefic, is thus their shadow-self. That ominous development is compounded by the fact that in Hitler's chart there is no strong planet . . . Jupiter is in its fall; the Moon and Mars are in their detriment; Mercury and the Sun are peregrine. This is a strong indication of the degeneracy of the nature. Even Venus, which has strength by virtue of falling in its own sign, Taurus, is grievously handicapped by being retrograde and by its immediate contact with the two malefics, Mars and Saturn. This contact is all the more serious because both the malefics are weak, and the weaker the malefics are, the worse their effects. Occurring in fixed signs, this gives an unshakeable malaise. With most of the planets above the horizon (the horizontal axis of the chart) and in angular houses, this will find its outlet in the world: were the planets hidden below the horizon, Hitler would have spent his life thinking dreadful thoughts rather than acting them out . . . [But] with all these planets severely afflicted, the outcome is not so favorable. Ptolemy suggests that in these cases, it "makes his subjects robbers, pirates, adulterers, submissive to disgraceful treatment (we might recall Hitler's sexual predilections), takers of base profits, godless, without affection, insulting, crafty, thieves, perjurers, murderers, poisoners, impious, robbers of temples and of tombs, and utterly depraved." All this would manifest through the degenerately choleric temperament. We might feel that Ptolemy has given us a picture somewhat closer to the mark than the peace-loving gardener that our modern blind reading has produced—not bad for someone writing almost 2,000 years before Hitler's birth.

The fixed-star lore of Ptolemaic tradition then fills the portrait out. The Sun, Moon, and Midheaven (or career cusp) of Hitler's chart, notes Frawley, were all conjunct malevolent fixed stars. These, by turns, were associated with "piled-up corpses," "destruction by fire or war," "unscrupulous defeat," and "a violent death." In short, modern astrology gives us an earthy, genial soul, ambitious, with some crowd-pleasing needs; Ptolemy, a complete degenerate whose career will lead to wholesale carnage, "piled-up corpses," "unscrupulous defeat," and "a violent death."

With a bow to one of the outer planets, we might also note that Uranus was conjunct Hitler's ascendant, in opposition to Mercury, indicating a violently hysterical nature and an erratic mind.

Chapter 13

O N MAY 21, 2000, Saturn moved into conjunction with Jupiter and—in the time-honored view of some astrologers— the world was about to change. The forecast of what this might mean was doom-laden from the start because Saturn was also approaching its own conjunction with Aldebaran, a fixed star known for causing famine, devastation, and war. That conjunction was effected in 2002, repeating a celestial aspect that had last coincided in 1914–18 with the First World War. A number of astrologers had predicted a decade or so ago that the next world war would be between the West and the Islamic world. And so it may prove.

The cardinal event in that conflict so far was doubtless the September 11, 2001, suicide attack on America engineered by Osama bin Laden, a wealthy Saudi fundamentalist with an implacable hatred of the United States. Retroactively, that attack has received a good deal of confident analysis. A month after the fact, it was noted in *Today's Astrologer* (October 16, 2001) that "an extremely powerful solar eclipse" had taken place on July 1, 2000, exactly conjunct the planet Mars, which was in its fall in Cancer, which the Moon rules. The eclipse lasted two hours and fifty-one minutes, which indicated (by tradition) that its effects would be felt for two years and ten months. The Moon at the time was in perigee, or at its closest position relative to the Earth, and since it rules the public in general, in mundane astrology its conjunction with Mars meant violence and war. Since the eclipse also occurred in the ninth house of foreign affairs, "with Mars activating, it made it likely that before those two years and ten months were up, the

United States would suffer some attack from abroad, and in turn attack foreign countries and powers." And this, of course, is what took place. Terrorist agents of Al-Qaeda, originating overseas, attacked the United States mainland. The United States in turn attacked Afghanistan, and in March 2003 invaded and occupied Iraq, all within the time frame established by the length of the eclipse.

In the same issue of *Today's Astrologer,* another wrote:

All the benefic aspects of Mercury show the success of the attacks by the terrorists. Air flight is indicated by Mercury conjunct the Ascendant in an air sign; it is in the twelfth house of secret enemies. Mercury also rules the ninth house of travel and foreigners. The Moon is in Gemini in the ninth house opposite Mars in Capricorn in the third house of short travel, with Mars ruling the seventh house of open enemies. The Sun is conjunct the United States natal Midheaven. Saturn in the eighth house in Gemini indicates death from the air. The fact that transiting Mercury is the ruler of the natal seventh house and is in partile conjunction with Saturn in the tenth house reveals the intent of the enemy was to destroy the government and the president (Virgo ruling the tenth house) from the air.

Regardless of which July 4 chart one accepts—aside from the one by Sibly, there are others cast for different times that day—the transits of September 11, 2001, to the national horoscope "showed three very important aspects: Transiting Saturn conjunct Mars/Uranus, explosive violence and frustration; transiting Pluto in opposition to Mars, which emphasized this; and transiting Uranus conjunct the Moon." The lunar aspect indicated "an emotional shock to the country's inhabitants." Since the Moon also rules the chart of New York City (founded as New Amsterdam on May 16, 1626, 8:52 A.M., 74W00, 40N43), with Cancer rising, "this powerful transit brought unexpected emotional turmoil and disruption."

All this seems plausible, if ingenious. It is also after the fact. Retroactively, anything can be wrenched into line.

Predictively, however, one astrologer seems to have got it just about right. The Sanhedrin (or Teaching of the Council of Hebrew Elders), dating back to ancient times, tells us that "an astrologer can determine by his calculations under which planet, in which month, and on which particular day any given nation will undergo an attack." Such astrologers, if they exist, may be said to be rare, but those few who embrace the traditional (and mathematically rigorous and demanding) techniques cultivated by the Greeks and Arabs, as epitomized in the work of Guido Bonatti, Girolamo Cardano, William Lilly, and others, seem able to make a go of it, at least from time to time. It so happens that the astrologer Robert Zoller, a Latin scholar, medievalist, and (aptly enough) a specialist in Arabic astrology, saw the attack of September 11 coming almost as clear as day. His prediction unfolded in a series of forecasts, each one more sharply focused than the last. In the July 1999 issue of his monthly four-page newsletter, *Nuntius,* devoted to "mundane astrological forecasts and brief articles on medieval philosophy and modern occultism," he wrote: "If the U.S. does not stop acting incompetently, it will invite the depradations of adventurers such as Osama bin Laden, Saddam Hussein, Slobodan Milosevic [not yet toppled from power] and others. This is a wake-up call. Our way of life and cultural values (& our lives) are at stake." The dangers, to his mind, were signified by an upcoming solar eclipse, which occurred on August 11, 1999, and, according to Ptolemaic principles, would last, with respect to its effects, for about two and a half years. One year later, in the August 2000 issue of *Nuntius,* he wrote: "There is an increasing threat to the U.S. citizens and this is particularly so on the Eastern seaboard." And the following month: "I again draw attention to the increasing threat of Islamic terrorism and that *it will be felt on the U.S. mainland. The greatest period of danger is in September 2001 . . . I am looking at Islamic terrorism that is rooted in Islamic fundamentalism and directed at everyday citizens going about their daily business in our own cities. The destruction and loss of life will shock us all.* [emphasis added] I repeat my warning now for the third time that unless the U.S. remains vigilant it will be caught unprepared and we will be rocked to our very core. It will be an act of war but unlike any other in our history. Our culture and way of life and lives are at risk." In a postscript, he noted that in September 2001 the stock market would also, as a consequence, "suffer major disruption followed by a steep decline." This prediction was made, of course, exactly one year before the attack on the United States took place.

It would appear that a lowly but capable astrological scholar poring over a handful of tell-tale charts—using the mundane predictive methods of Ptolemy, Masha'allah, Abū Ma'shar and Bonatti—knew more about what was likely to happen that day than all the intelligence experts combined in the FBI and CIA.

Perhaps, after all, "traditional astrology" has some application for our time.

It may not be out of place to point out that the United States chose to attack Iraq in March 2003, when Saturn (the great malefic) was conjunct Mars in the seventh house (open enemies) of our national horoscope. Both in turn were conjunct or aligned with a martial "fixed double star" known as El Nath—from Al Natih, in Arabic, meaning "the Butting One"—at the tip of the Bull's North Horn. This notable triple conjunction signified a formidable military posture and determined might. Mars, moreover, was exalted in Capricorn (which might be thought good for a campaign of "shock and awe") but failed to make any benefic aspect with either Jupiter or Venus "to calm it down." Hence, an astrologer might say—a drawn-out war. Mercury, ruler of the seventh house (of foreign affairs), was also in Pisces (its detriment) in aspect to Caput Algol, the fixed star of beheadings— which might help to explain why the conflict began with a much-trumpeted, if hastily-arranged, "decapitation attack" (as the Pentagon boastfully called it) on the Iraqi regime, and why the awful spectacle of beheadings has since become such a conspicuous feature of the war.

Again, opposite Antares, one of the four Royal Stars of Persia and known as "The Watcher of the West," stands another "Watcher of the Heavens," the brightest star in Taurus, Aldebaran, from an Arabic word which means "the following," because it follows upon the Pleiades. Ptolemy noted two thousand years ago that any astrological aspect between Aldebaran and Mars meant military action, fraught with danger. The two were opposite, with Mars exactly conjunct Antares on January 31, 2003, when, we are told, the Bush administration made its decision to go to war.

The timing was not the best. An astrologer might ask (with Chaucer, in "The Man of Law's Tale"), "Was there no astrologer [to elect a better time] in all thy town . . . ? "

Unfortunately, the horoscope of President George W. Bush would seem allied to dread events. When the ascendant falls on a powerful fixed star it can palpably affect a person's fate. Salvador Dali's ascendant was on the fixed star Propus, which gives eminence; Freud's on

Pollux, which gives a subtle mind; Einstein's on Sirius, which gives honor and renown; that of Bush on Praesaepe, one of the most malevolent fixed-star clusters in the sky. We have already met it. "Of the nature of Mars and the Moon, it causes adventurism, insolence, wantonness, brutality, and disgrace." Associated also with blindness (physical and otherwise), it indicates a person who cannot see clearly the consequences of his acts. That does not mean that the president is not, in some sense, a "nice man" personally, or even "well-meaning," however coarse (Praesaepe and coarseness are inseparably linked); it does mean, astrologically, that he tends to bring horror in his wake. There is hope in the conjunction of his Sun with Canopus (a white star in one of the oars of Argus), which gives "piety and conservatism," with the potential of "turning evil to good"; on the other hand, his Moon, conjunct Seginus (a small star situated on the left shoulder of Boötes), gives "illegitimate preferment" often "followed by ruin and disgrace." Or so an astrologer might say. I am not one; and these are not my own, willful judgments, but Ptolemaic lore. "Time governs princes; princes govern men," as Cardano put it. "Look for the end to time": History will tell his tale.

That tale, in a sense, is already inscribed. Here is a political prediction made in March 2004 by horary means for the outcome of the presidential race between John Kerry and Bush. The question, posed by the author as an experiment, was submitted on March 17 (when Kerry was even or ahead in the polls) to an astrologer in Scotland, who had been expertly trained in traditional techniques. The chart was drawn for 8:33 A.M. Greenwich Mean Time, which gave it an ascendant of 12 degrees 19 minutes Gemini, with a structure thus construed: Bush, being the incumbent president, was signified by the tenth house; Kerry, as the challenger and party out of power, by the fourth. The second house signified the economy; the seventh, foreign affairs. (This is all according to the rules.) Since the sign on the cusp of the tenth was Capricorn, Bush himself was represented in the chart by Saturn (which rules that sign); since that on the cusp of the fourth was Cancer, Kerry was represented by the Moon. The astrologer here quoted an aphorism from Bonatti: "Saturn obtains kingdoms or supremacy of power by labor, fraud, and infamy." Anyway, that is what the tradition says.

At first glance the chart seemed to represent the end of a tenure, which looked good for Kerry, since the fourth- and tenth-house cusps were in the last degree of their respective signs. But Capricorn was followed by Aquarius—also, by tradition, ruled by Saturn—"so the rulership remains the same." That in itself suggested that Bush would remain in power. Moreover, the Moon (Kerry) was "weak" or "peregrine" (that is, in an undefined state), though Bush's Saturn also "lacked strength" (being in Cancer, the sign of its detriment), which showed him to be "incompetent" (I am quoting the astrologer here, again according to tradition), but the Moon (which also signifies "the public" in any horary chart) was under Saturn's dominion, being in Aquarius, that planet's sign. So Bush would win the popular vote. Now, it so happens that the two competing planets were also in each other's signs, and so were joined by "mutual reception." That indicated a general equality in the competition, and showed that Bush was "worried by Kerry" while Kerry was "not at all confident that he would prevail." The chart also suggested that smear tactics would be used in the campaign, that the race would be close, but that Bush would win on the issue of terror and a stabilized economy—as shown by Saturn in the second house (the economy) in sextile to Jupiter the ruler of the seventh (foreign affairs). This verdict was then confirmed by a comparison of the solar return charts for each man. Kerry's revealed nothing to indicate a triumph; Bush, on the other hand, seemed favored by fate. Mars in his return fell on the cusp of his natal ascendant, but also ("a strong sign of a win") ruled the tenth-house cusp of both his return and natal charts. It seemed there was no way he could lose.

Most "modern" astrologers, not incidentally, were predicting a Kerry triumph based on "auspicious transits." Again, however one may wish to explain it, everything predicted by this "traditional" astrologer in March, seven and a half months before the election, hit its mark.

★   ★   ★

ALDOUS HUXLEY WAS ONCE ASKED what modern scientists would say if a prominent member of their group announced that he believed in astrology. He replied: "They would say, 'Here is a great scientist with a foible.' " The immensely popular astronomer Carl Sagan had little use for astrology but declined to sign a manifesto against it that appeared above the signatures of 192 scientists, including nineteen

Nobel Laureates, in the September/October 1975 issue of *The Humanist*. In a letter to the editor he explained:

> I find myself unable to endorse the 'Objections to Astrology' statement, not because I feel that astrology has any validity whatever, but because I felt and still feel that the tone of the statement is authoritarian. The fundamental point is not that the origins of astrology are shrouded in superstition. This is true as well for chemistry, medicine and astronomy, to mention only three. To discuss the psychological motivation of those who believe in astrology seems to me quite peripheral to the issue of its validity. That we can think of no mechanism for astrology is relevant but unconvincing. No mechanism was known, for example, for continental drift when it was proposed by [Alfred] Wegner. Nevertheless, we see that Wegner was right, and those who objected on the grounds of an unavailable mechanism were wrong.

In the practice of astrology, some predictions have come true, some not. Jacques Gaffarel, astrologer to one of the wiliest of French statesmen, Cardinal Richelieu, called it "the handwriting on the wall of heaven"; more cautiously, Louis de Wohl observed: "Astrology is not prophecy. It deals with tendencies, not certainties. It has a fairly wide margin of error—but it works." Similarly, Evangeline Adams wrote: "The horoscope does not pronounce sentence . . . It gives warning," but added, "Astrology must be right. There can be no appeal from the Infinite." Its veracity may be argued, in any case. As Sir Isaac Newton reportedly once said to Edmund Halley, who disparaged astrology in an offhand remark: "I, Sir, have studied the subject, and you have not." An apocryphal story, perhaps, but with truth at its heart.

There has been prescience enough, to be sure, as we have seen. In the Hundred Years' War between England and France, French astrologers correctly forecast the outcome of the Battle of Crécy in 1346; the capture of John II of France at the Battle of Poitiers on September 18, 1356, and his deliverance from the hands of the English four years later; the death of Edward, the Black Prince, in 1376; and the deposition of Richard II in 1399. In 1552, the English bishop of St. Andrews consulted the famed astrologer-mathematician Girolamo Cardano about a difficult illness, and after Cardano erected the bishop's horoscope (which resulted in a correct diagnosis and treatment) he took his leave with these words: "I have been able to cure you of your sickness, but I cannot change your destiny, nor prevent you from being hanged." Eighteen years later the bishop was arraigned for treason by the queen regent of Scotland and hanged by her palace gate. Johannes Kepler correctly predicted the death of his patron Count Albrecht Wallenstein for late February–early March 1634 (the horoscope for this prediction still exists); in 1648, William Lilly, England's most renowned astrologer, predicted the approximate date for the outbreak of the Great Fire of London in 1666. Lilly also made the following judgment on a horoscope he cast in 1647 for Charles I: "Luna is with Antares, a violent fixed star which is said to denote violent death, and Mars is approaching Capul Algol, which is said to denote beheading." Two years later the king's head fell on the block. In the issue of the *Berliner Auskunftsbogen,* published in December 1958, a German astrologer correctly predicted the death of Pope John XXIII for 1963—when, he explained, Mars would be in opposition to both Saturn and Uranus in the pope's horoscope, and Pluto in square to his Sun.

However we may regard, or disregard, the subject's allure, it is possible, after all, for an astrologer to say that he stands with Brahe, Galileo, Kepler, and Newton in astronomy; with Dante, Chaucer, Shakespeare, Dryden, Goethe, and Byron in literature; with Plato and Ficino in philosophy; with Aquinas and Bonaventura in theology; with J. P. Morgan and various other tycoons in finance; with Queen Elizabeth I, Charles de Gaulle, and Ronald Reagan in politics; with Botticelli, Tintoretto, Albrecht Dürer, and Hieronymus Bosch in art; with Carl Jung in psychology; and with Alexander the Great, Darius of Persia, Hadrian, al-Rashid, Lorenzo de'Medici, and numerous other generals and conquerors in war. That is a daunting list.

A number of developments in the history of science have helped astrology reclaim a measure of its former place. The "New Physics" has presented a formidable challenge to the doctrine of scientific materialism; the social sciences have reinstated human behavior as a fit subject for research; psychoanalysis has suggested hidden, "occult" forces in the structure of the human psyche that guide human behavior independent of the conscious will; and most importantly of all, perhaps, the development of genetics has begun to describe, in semi-deterministic terms, the biological map or blueprint with which we are born. Its analogy with astrology turns out to be strangely precise. Recent statistical research has shown that astrological patterns in the charts of parents are often replicated in those of their offspring, with a frequency that follows the Mendelian paradigm for inheritable traits. In short, as Kepler had insisted, "the child who is about to enter the world carries on an obscure dialogue with the planetary gods," and seems to exhibit a hereditary tendency to be born under the same cosmic conditions that prevailed at its parents' birth.

The analogy, moreover, is strengthened on the astrological side by the phenomenon of "astral twins." One such famous case involved an English subject and his king. On June 4, 1738, in the parish of St. Martins-in-the-Fields, two boys were born less than a minute apart. One was William Frederick, later crowned George III, King of England; the other, James Hemmings, an ironmonger's son. Widely separated by class, yet bound to a parallel fate, these two men, each in his own social sphere, lived out the edict of his stars. In October 1760, when George III succeeded his father on the throne, thereby fulfilling the purpose to which he was born, Hemmings took over his father's business. Both men were married on September 8, 1761, fathered the same number of children (even, weirdly, the same number of boys and girls), suffered the same accidents, succumbed to the same diseases, and died within less than an hour of each other on Saturday, January 29, 1820.

Another famous prince-and-pauper story yoked together a chimney sweep with the Prince of Wales (later King George IV). E. Oakes-Smith, in *The Shadow Land, or the Seer* (1852), wrote:

Of the career of the Prince of Wales it is unnecessary to speak—his vices, his follies, his perjuries were all royal, and his fellow, the sweep, was not a jot behind him. The broom and scraper were as ill-adapted to the hands of one as the scepter to the hands of the other. The parents of the sweep, tired and ashamed of his profligacy, finally established him as a tallow chandler. On the same date, George IV was put on a royal allowance and both men embarked on separate but similarly notorious careers as gamblers, philanderers and spendthrifts—but on entirely different financial and social levels. The commoner acquired a stable of asses and ran the best donkey races of the day. George IV kept the best blooded ponies and ran the finest horse races in the country. On the day that "Prince George" was kicked in the hip by a donkey, George IV was kicked in the ribs by a horse. Both were injured and incapacitated for the same amount of time. When the Prince of Wales lost everything and went bankrupt, so did the commoner "prince." And when all the King's horses were sold by the royal horse-seller, the ex-chimney sweep lost his asses under the hammer of the auctioneer.

As above, so below—and, it seems, within. The biblical idea that man is made in the image of God (which is good Judeo-Christian doctrine) is but a version of that still more ancient notion that man is a micro-cosm or miniature image of the cosmos, by which the power of God is revealed. For the celestial pattern is itself an image of God.

In the realm of prediction, all things, we may say—as all sacred teachings teach—are known to him. To his eternal mind, past, present, and future form one continuous round. Fate and free will in this sense are not at odds, for God knows in advance not only all things "fated" but all the free choices we will make. By that foreknowledge, the fu-ture already exists, and indeed has happened, in his mind. The universe is a visible expression of that mind, and the capacity to read it—by astrological means—is a window into eternity, a portal to the throne.

★   ★   ★

THE BEGINNINGS OF A NEW, blended astral science have begun to emerge. In weather forecasting, for example, planetary conjunctions and oppositions appear to affect the development of tornadoes (the Aviation Weather Center in Kansas City uses such data as a critical part of its base), while disturbances of the Sun's all-embracing magnetic field (as affected by different planetary groupings) have been shown to permeate to the upper layers of the Earth's atmosphere and affect con-ditions on Earth. The Electronics Research Center at NASA uses planetary aspects (under the name of "gravitational vectoring") to pre-dict magnetic storms, as does NATO to prevent its radio frequencies from being interrupted when they occur. In November 1969, NASA reportedly invited ten members of the American Federation of As-trologers to Cape Kennedy for the memorable moon shot of Apollo XII. Two days before the launching, Frances Sakoian, a Boston as-trologer, announced that the launch chart indicated trouble with the fuel. And to the great consternation of officials, just before liftoff, one of the fuel tanks had to be replaced. "Mars," an astrologer explained afterward, "was in the first house square Saturn in the third house of communications and short journeys and square Venus in the ninth house of long trips."

In the realm of finance, astrologers have also been relatively successful in correlating market fluctuations with planetary cycles, and a number of London banks now have "astroeconomists" on their staff. Astroeconomics is as old as astrology itself. "Almost 4,000 years ago," as Nicholas Campion and Steve Eddy tell us, "the Babylonians used the planets to anticipate good and bad harvests, which eventually meant higher or lower profits for merchants. In ancient Rome and medieval and Renaissance Europe, most educated people consulted the stars over their financial prospects. The Muslims used points in the sky to examine individual commodities." On the whole, solar, lunar, and planetary returns are said to be more accurate than directions in the predictive readings of a business chart. One of the classic statements in this regard was made by Clement Hay in *Today's Astrologer,* October 20, 1941:

In a business solar return chart, the first, fourth, seventh, and tenth houses are the clue to the prosperity within that year. All of the planets in the four houses are to be weighted without regard to the signs. If those houses are occupied by malefics with only good aspects to them, the firm will prosper. If the four major houses are occupied by malefics with only bad aspects, there will be loss and adversity. If there are benefics there afflicted, there will be no gain; but if benefics have only good aspects in the four major houses, then prosperity will be great. If Jupiter, Venus, and Pluto should all be in the major angles (and no others there) and well aspected, then it will be a bonanza year.

Opinions differ as to the best chart for the New York Stock Exchange to use. One foundation chart gives a default time of 12 noon LMT for May 17, 1792, in New York, "when a group of traders," Graham Bates writes, "concluded an agreement to regulate their activities." Another, said to be "more reliable," is drawn for May 11, 1869, "when the various New York Exchanges were reorganized into one exchange, with trading beginning at 10:25 EST." Perhaps the most successful trader in Wall Street's history was an "astroeconomist" of sorts by the name of

W. D. Gann, who used astrological methods in part to acquire his stupendous wealth.

Born on June 6, 1878, on a farm outside Lufkin, Texas, Gann was the son of a dirt-poor cotton farmer, with scarcely a prospect in life outside his own determined drive. He sold newspapers and cigars on trains as a boy, met and engaged businessmen in conversation, talked himself into a job in a brokerage firm in Texarkana, and attended business school at night. In 1903, he moved to New York City, and by his early thirties had begun to make his name. In October of that year, "The Ticker and Investment Digest" (a house organ of the New York Exchange) reported that he had 286 winning, and 22 losing, trades that month with "a win rate of 92%." This astonishing success, viewed at first as a fluke, continued unabated, and in time stretched from months into years. In 1919, he began publishing a daily market digest of his own, *The Supply and Demand Letter,* which covered both stocks and commodities and provided its readers with annual forecasts, which were usually right. Needless to say, he foresaw the stock market crash of 1929. At the beginning of that year, he predicted that the market would hit new highs until early April, experience a sharp break, resume with new highs until September 3—then collapse. Meanwhile, in 1927 he had written a prophetic work of fiction entitled *Tunnel Through the Air,* which predicted a surprise attack on the United States by Japan and an air war between the two powers.

Throughout the 1930s, Gann's portfolio grew: he acquired seats on various commodities exchanges; traded on his own account; wrote the "Wall Street Stock Selector," the "New Stock Trend Detector," and other house organs of the perennial "market watch." But his methods remain obscure. Some of them relied heavily on numerical relationships, such as the Square of Nine, the Square of 144, and swing charts in combination with planetary aspects—in particular, the Jupiter-Saturn cycle. They also used some of the astrological formulas for determining the outcome of contests that Sepharial had worked out. By the time he retired—in splendor, and bought the first private jet—he had made over $50 million (equivalent to about twice that today).

Others, with their own methods, have followed in his wake.

★ ★ ★

OBVIOUSLY, ALL ORGANIC LIFE is influenced to some degree by the rotation of the Earth, its path and orbit around the Sun, the change in seasons, the waxing and waning of the Moon. After all, we live in the Milky Way. Our celestial and terrestrial environments—linked by light, gravity, even rhythmic pattern, in the biological sense—are part of the same whole. This means we are demonstrably tied to lunar and solar cycles by an internal clock. The metabolic rates of plants and animals have been shown to have a correlation to the solar and lunar cycles; chemical precipitates (bismuth oxychloride, for example) are affected by solar activity, and "naphthalene solidifies fastest at the New Moon." This is perfectly in line with the alchemical idea that planetary positions and chemical reactions are linked.

Thomas Hardy has a charming poem, addressed to a spring song-bird, about the instinctual connection between nature and the stars:

> How do you know that the pilgrim track
> Along the belting zodiac
> Swept by the sun in his seeming rounds
> Is traced by now to the Fishes' bounds
> And into the Ram, when weeks of cloud
> Have wrapped the sky in a clammy shroud,
> And never as yet a tinct of spring
> Has shown in the Earth's appareling:
>   O vespering bird, how do you know,
>   How do you know?

Everything in the world has a pattern. Why wouldn't you? or I? We do, of course, in the form of DNA and other genetic markers; but the timing of events, as well as the span allotted to each life, also belongs to time itself. Time as we know it is measured out by the revolutions of the Sun and Moon, so why wouldn't the length and pattern of our lives also be calibrated by the great celestial clock? Of course it is (an astrologer would say), because it must be. How could we be exempt? Everything has a predetermined span, just as the planets, stars, and the two great lights of our world, the Sun and Moon, make their predetermined rounds. Kepler would have agreed with the poet George Meredith, who gave lyrical expression to a kindred idea, with respect to the Music of the Spheres:

So may we read, and little find them cold:
Not frosty lamps illuminating dead space,
Not distant aliens, not senseless Powers.
The fire is in them whereof we are born;
The music of their motion may be ours.

Perhaps this is what Paracelsus meant when he said: "In the heavens you can see man, each part for itself; for man is made of heaven. And the matter out of which man was created also shows the pattern after which he was formed."

One recent writer on the stars assures us: "The desire to construct a zodiac is motivated by the natural tendency we all have to seek the familiar in the unfamiliar, to express the unknown in terms of what we know, in this case what we all can plainly see in the sky. We lend worldly attributes to the unseen forces of nature. We invent the cosmos after ourselves." Yet I would have thought the cosmos invented us. The ideas we have about it come from our own partaking, and the structure we give it—like the inborn structure we give to language—reflects the thing of which we are a part. As the naturalist John Burroughs put it: "We are just as much in the heavens now as we ever can be. If a man was in the Pleiades, he would be no more in the heavens than [a drunk] in the ditch."

GLOSSARY

accidental dignity Applied to planets that fall on an angle, or in angular houses, and to the most elevated planet in a horoscope.

affliction Adverse aspect. A planet is afflicted when it is parallel, in conjunction with, in square, or opposition, to Mars or Saturn, or in square or opposition to any of the other planets.

air signs Gemini, Libra, and Aquarius.

anaretic From the Greek, meaning "destroyer." The term is applied to a malefic that occupies an anaretic place (in the fourth, sixth, eighth, or twelfth house) or an anaretic degree (the final degree of any sign) and afflicts the Hyleg. See also APHETA, HYLEG.

angles The first, fourth, seventh, and tenth houses. Also the ascendant, descendant, Midheaven, and Imum Coeli—the four cardinal points in a horoscope.

antiscion From the Greek, meaning "opposite shadow." A degree and its antiscion are equidistant from the summer-winter solstice axis. Planets related by antiscion have the force of a conjunction.

aphelion The point of a planet's orbit that is most distant from the Sun. Compare PERIHELION.

apheta Literally, "the giver of life." Synonymous with HYLEG. A well-aspected benefic.

apogee The point in a planet's orbit that is most distant from the Earth. Compare PERIGEE.

apply When a swifter planet approaches a slower one to form an aspect. The Moon, being the swiftest, applies to all the other planets in turn each month as it passes around the zodiac. However, if the slower-moving planet is retrograde both planets may be said to apply. The influence of the planets is stronger when they are applying than when separating. See DIRECT and RETROGRADE.

articulate or voiced signs Gemini, Virgo, Libra, Sagittarius, and Aquarius. Also called "human signs."

ascendant The eastern angle; the sign and degree of the zodiac rising on the eastern horizon for the time and place a chart is drawn. A new degree rises every four minutes, a new sign about every two hours, and the twelve signs every twenty-four hours for every place on Earth. Whatever sign is on the ascendant is called the rising sign. See HYLEG.

ascension There are signs of long ascension, short ascension, right ascension, and oblique ascension. The signs of long ascension are Cancer, Leo, Virgo, Libra, Scorpio, and Sagittarius. Those of short ascension are Capricorn, Aquarius, Pisces, Aries, Taurus, and Gemini. Due to the obliquity of the ecliptic, signs of long ascension rise slowly in northern latitudes, taking longer than the two-hour average for all twelve signs rising each day at a uniform rate. In the Southern Hemisphere the signs of short ascension become signs of long ascension, and vice versa. Longitude is measured on the ecliptic or Sun's path from the first point of Aries; right ascension is measured on the equinoctial or celestial equator.

aspect The angular relationship between two planets or a planet and an angle or sensitive point. Aspects are determined by celestial longitude; parallels and contraparallels are based on declination.

axis The Earth's axis is an imaginary line on which the planet rotates, producing the phenomena of day and night. The axis always points to the North Star in the constellation Ursa Minor (the Great Bear).

barren or unfruitful signs Gemini, Leo, and Virgo.

benefic Planets: Jupiter and Venus; aspects: sextile and trine.

bestial signs Aries, Taurus, Leo, and Capricorn.

bicorporeal signs Gemini, Sagittarius, and Pisces.

cadent The third, sixth, ninth, and twelfth houses, and the planets that occupy them.

cardinal signs Aries, Cancer, Libra, and Capricorn, which fall naturally at the cardinal (angular) points of a horoscope.

celestial equator The extension of the Earth's equator into space, perpendicular to the Earth's axis.

celestial sphere The apparent sphere, with the Earth at the center, to which the heavenly bodies appear to be attached.

combust Any planet within several degrees of the Sun is combust, or burned up, by the Sun's rays. Its power is thereby consumed.

common or mutable signs Gemini, Virgo, Sagittarius, and Pisces.

conjunction When exact, an aspect formed by two planets that share the same sign and degree of celestial longitude.

constellation A group of fixed stars.

contraparallel The angular relationship between two planets that occupy the same degree of declination, one north of the celestial equator, the other south. See DECLINATION.

critical degrees Degrees that mark the approximate end of the Moon's daily journey (according to its average daily motion) through the twelve signs. The Moon takes about 27½ days to complete its passage around the zodiac, averaging about 13 degrees each day. Thus, starting with the first degree of Aries, the first day's travel will end at the 13th degree, the second day's travel at the 26th degree, and so on. The critical degrees are therefore the 1st, 13th, and 26th of the cardinal signs, the 9th and 21st of the fixed signs, and the 4th and 17th of the common or mutable signs.

culminate A planet culminates when it arrives at the Midheaven.

cusp The first degree of a house or sign.

debilitated A planet in the sign of its detriment or fall.

decan The subdivision of a sign into thirds, each of 10 degrees.

declination The angular distance of a planet north or south of the celestial equator. The maximum declination of the Sun is reached at the summer solstice (north declination) and winter solstice (south declination), 23 degrees, 27 minutes. Mars, Mercury, and the Moon reach declinations of 27 degrees north, and on rare occasions Venus attains 28 degrees, but Jupiter and Saturn have approximately the same declination as the Sun. Planets that occupy the same degree and direction of declination are parallel; those that occupy the same degree in opposite directions are contraparallel.

decumbiture   In medical astrology, a chart cast for the time a patient first consults a doctor or takes to bed with an illness.

degree   A degree is one 360th of a circle. There are 30 degrees in each of the twelve signs of the zodiac, and the motion of the planets through these signs is stated in degrees and minutes of longitude, beginning with the first degree of Aries. The circle of the Sun is called the ecliptic, and is taken as the standard line of celestial motion so far as our solar system is concerned. The planets zigzag along that line according to their celestial latitude. A degree in this sense is used as a unit of measurement to fix the position of the planets on the celestial sphere containing the fixed stars.

descendant   The point opposite the ascendant; the sign and degree setting on the western horizon when the same degree and opposite sign is rising in the east.

detriment   A planet in the sign opposite its sign of rulership or dignity.

direct   Describes the motion of a planet forward through the signs. Opposite to RETROGRADE.

directions   Any of a number of methods that advance the planets and angular house cusps of a natal chart to a particular time after birth, usually according to systems that reckon either a day or a degree for a year of life. Also called "progressions."

dodecatemorion   A division of each sign into twelfths; a twelfth of that sign.

Dragon's Head   From the Latin *caput draconis,* the Moon's north node. The nodes are points in the orbit of a planet where it crosses the ecliptic, or the Sun's apparent path around the Earth. The point of crossing from south to north is called the ascending or north node; that from north to south, the descending or south node, also known as:

Dragon's Tail From the Latin *cauda draconis,* the Moon's south node.

Earth signs Taurus, Virgo, and Capricorn.

eclipse See LUNATIONS.

ecliptic The Sun's apparent path among the constellations of the zodiac.

election The process of determining the most advantageous time, astrologically, for undertaking some action or enterprise.

elevate The zenith occupied by the Sun at noon is the highest point in the heavens. The nearer to the zenith or Midheaven a planet is, the more elevated it is said to be.

ephemeris A guide to the longitudes and declinations of the planets for a given year.

equator An imaginary line in a plane at right angles to the axis of the Earth, midway between the north and south poles. It divides the Earth into two hemispheres, north and south. The celestial equator is the line of the Earth's equator projected into space.

equinox A time when day and night are of equal length. This occurs twice each year when the Sun crosses the equator: once at the vernal equinox, when it enters 0 degrees Aries, and again at the autumnal equinox, when it enters 0 degrees Libra.

essential dignity A planet in its rulership by sign, exaltation, triplicity, term, or face.

exaltation A planet's sign, other than that which it rules, in which it is powerfully strengthened.

face One of the three equal subdivisions of a sign, each with its own planetary ruler. Synonymous with decan.

fall A planet's sign opposite that of its exaltation.

feminine signs The Earth and water signs: Taurus, Virgo, and Capricorn; Cancer, Scorpio, and Pisces, with reference to gender. Also called "negative" signs. (Earth and water are "negative" and inert, but are acted upon by the "positive element." Thus, "the winds stir the waters of the ocean and volcanic fires shake the Earth." Therefore the fiery and airy signs are called "masculine.") See also MASCULINE SIGNS.

fire signs Aries, Leo, and Sagittarius.

fixed signs Taurus, Leo, Scorpio, and Aquarius.

fixed stars The stars and star clusters composing the constellations of the zodiac. By their immense distance from the Earth, and, therefore, lack of parallax, they seem to preserve the same position relative to one another, as if they were fixed in space.

figure The horoscope or map of the heavens cast by an astrologer.

fruitful signs The water signs: Cancer, Scorpio, and Pisces.

heliacal The heliacal rising of a star or planet is its first emergence from invisibility as it separates from a conjunction with the Sun.

horary astrology The science of judging how some matter will turn out from a figure drawn for the moment a question is presented to an astrologer who can answer it.

horizon In astrology, the birthplace is always considered the highest point on Earth, and the principal circle seen from there is the horizon. The sensible horizon is the circle that bounds our view, where heaven and earth appear to meet.

horoscope The chart or symbolic figure of the heavens drawn for the birth of anything—including a person, a nation, or a question; also, such a figure drawn for a time deemed best for some act or enterprise.

house The houses are divisions of the heavens relative to the birthplace; the signs are divisions of the heavens relative to the vernal equinox. The zodiac appears to rotate at the rate of one degree every four minutes, but the houses are considered stationary relative to the birthplace in the scheme. From the point of the birthplace, two imaginary lines are drawn to the four cardinal points: north, east, south, and west. One line is the meridian. It extends to the point directly overhead, where the Sun is at noon and where it forms the cusp of the tenth house (or Midheaven); directly opposite is the cusp of the fourth house. A line drawn at right angles across the meridian from east to west forms the cusps of the first and seventh houses.

The things signified by each of the twelve houses are, in brief: *First house:* The self. *Second house:* Movable assets. *Third house:* Brothers and sisters; short journeys. *Fourth house:* Property, such as land; the father. *Fifth house:* Children; pleasure. *Sixth house:* Ill health; servants. *Seventh house:* Partners—in marriage, business, or other relationships; open enemies. *Eighth house:* Death. *Ninth house:* Religion, learning; long journeys. *Tenth house:* Social standing, career. *Eleventh house:* Friends; hopes and wishes. *Twelfth house:* Self-undoing; prisons, hospitals; secret enemies.

human signs Also called "voiced" signs: Gemini, Virgo, and the first 15 degrees of Sagittarius and Aquarius.

Hyleg A term used by ancient Arab astrologers for the Sun, Moon, and ascendant as the principal foci of vitality and health in the horoscope.

Imum Coeli From the Latin, meaning "bottom of the heavens." The point opposite the Medium Coeli or Midheaven and forming the cusp of the fourth house.

inarticulate signs Aries, Taurus, Leo, and Capricorn.

infortunes Mars and Saturn (modern astrologers say also Uranus, Neptune, and Pluto). Also called *malefics.*

ingress The entry of a planet into a sign. The Sun's ingress into the cardinal signs (Aries, Cancer, Libra, and Capricorn) marks the beginning of each season of the year.

intercepted Describes a sign that does not appear on a house cusp but is wholly contained within that house. Owing to the spherical shape of the Earth and the inclination of the Earth's axis, some of the mundane houses in the higher northern latitudes are only 12 or 15 degrees long, while others are 40, 50, or 60 degrees. The signs of the zodiac, on the other hand, are 30 degrees each. In cases where a mundane house is long, one or even two whole signs may be included within its cusps. When a whole sign is included in a house, the sign is said to be intercepted. When a sign is intercepted, the sign opposite is also intercepted in the opposite house.

latitude In astronomy, the distance of a planet north or south of the ecliptic.

lights The Sun and Moon.

longitude In astronomy, the position of a planet or point along the ecliptic. When the position is reckoned on the celestial equator, it is called right ascension.

lord A planet that rules a sign, such as Mars for Aries and Scorpio; Venus for Taurus and Libra.

lunation A conjunction of the Sun and Moon at "new Moon." During her monthly course the Moon zigzags across the ecliptic, and at the conjunction, or new Moon, is generally a number of degrees away from the ecliptic. Under such conditions, we have an ordinary new Moon. For a total solar eclipse to occur, the Moon must be directly in the Sun's path as seen from the Earth, and the declination of the Sun and Moon must have practically no latitude.

There are never less than two solar eclipses in a year, and never more than seven; the usual number of all eclipses (Sun and Moon) is four: two solar and two lunar, which typically come in pairs six months apart. The full Moon preceding or following a solar eclipse is usually a lunar eclipse.

lunar mansions A division of the zodiac (or horoscope) into twenty-eight houses or mansions based upon the Moon's average daily motion. See also CRITICAL DEGREES.

malefics See INFORTUNES.

masculine signs The fire and air signs: Aries, Gemini, Leo; Libra, Sagittarius, and Aquarius, without reference to gender. Also called *positive signs*. See also FEMININE SIGNS.

Medium Coeli From the Latin, meaning "middle of the heavens." The Midheaven or culminating degree of the ecliptic.

meridian An imaginary circle drawn through the north and south poles over the face of the Earth. Therefore, a line of longitude.

movable signs The same as cardinal signs (Aries, Cancer, Libra, and Capricorn).

mundane astrology A branch of astrology that deals with events in the public sphere.

mutable signs The same as common signs (Gemini, Virgo, Sagittarius, and Pisces).

mute signs Cancer, Scorpio, and Pisces.

mutual reception When two planets are in each other's sign of essential dignity—for example, Mars in Pisces (traditionally ruled by Jupiter) and Jupiter in Scorpio (traditionally ruled by Mars)—they are in mutual reception. As such, they support each other and are harmoniously linked.

nativity A natal *horoscope* or *figure;* a map of the heavens cast for the moment of birth.

negative signs The earth and water signs: Taurus, Virgo, and Capricorn; Cancer, Scorpio, and Pisces.

nodes See DRAGON'S HEAD.

opposition When exact, an aspect formed between two planets that occupy the same degree of celestial longitude 180 degrees apart.

parallel An aspect formed between two planets when they are in the same degree of declination, either north or south of the celestial equator.

Part of Fortune An arithmetically derived point of positive significance on the ecliptic, and the most commonly used of the so-called "Arabian parts." It is found by adding the position of the ascendant to that of the Moon, and subtracting that of the Sun.

peregrine Said of a planet that does not occupy a sign of essential dignity or debility. A peregrine planet is like a wanderer or homeless vagabond and in essence lacks any clear power to act.

perigee The point in a planet's orbit that lies closest to the Earth. Compare APOGEE.

perihelion The point in a planet's orbit that lies closest to the Sun. Compare APHELION.

planetary hours In addition to ruling the days of the week, the planets have dominion over the hours of the day, and the underlying system, order, and connection between the rulership of the days and hours becomes apparent when it is noted that the planet for which any day is

named rules the first hour after sunrise on that day. The order of the
planets is: Saturn, Jupiter, Mars, Sun, Venus, Mercury, Moon, in an endless round. Starting with the hour of sunrise on Sunday, which is ruled
by the Sun, the next hour is therefore allotted to Venus, the third to
Mercury, and so on. Thus, the Sun rules the first hour of Sunday, the
Moon the first hour of Monday (which is the twenty-fifth from the hour
of the Sun that ruled Sunday morning), Mars the first hour of Tuesday
(which is the twenty-fifth hour from the hour of the Moon that ruled
Monday morning), and so on, through the other days of the week.

positive signs The fire and air signs: Aries, Leo, and Sagittarius; Gemini, Libra, and Aquarius.

precession The apparent, gradual backward (or westward) movement
of the vernal equinox through the zodiac at the rate of 50 seconds of arc
per year. Precession is due to the wobble of the Earth on its axis, which
causes the pole of the equator to revolve around the pole of the ecliptic.

promittor The slower moving of two planets in aspect.

quadruplicity The cardinal, fixed, and mutable qualities, each of which
contain four zodiacal signs.

radical Pertaining to the horoscope at birth.

rectification Method of ascertaining a true birth time through studying
the planetary correspondences at the time of important events.

retrograde Describes the apparent backward motion of a planet when it
appears, as observed from Earth, to reverse its natural direction of travel
and move backward in the horoscope. Compare DIRECT.

right ascension Measurement along the celestial equator eastward from
zero degrees Aries that describes planetary positions in terms of degrees,
minutes, and seconds.

rising sign The ascendant.

ruler The ruler of a horoscope is the planet that has greatest dominion
and influence over the life, and to which the native most readily re-
sponds. All things being equal, it is the lord of the ascendant, unless another planet is stronger with respect to elevation, dignity, exaltation,
angular position, and aspect. The ruler of a sign is the planet that rules it

in essential dignity; the ruler of a house is the lord of the sign on the cusp. Where there is an intercepted sign, its lord has partial rulership over the house, although less so than the planet which rules the sign on the cusp.

separating When a planet that has been in aspect with another moves onward and thus dissolves the aspect, it is said to be separating from that aspect.

sextile When exact, an aspect formed by two planets that occupy the same degree of celestial longitude 60 degrees apart.

short ascension The spring and winter signs in the northern hemisphere, which take less time than the other six to rise above the horizon—Aries, Taurus, Gemini, Capricorn, Aquarius, and Pisces.

sidereal day The time that elapses between two successive passages of a fixed star over the meridian of a given place.

sidereal year The time that elapses between a conjunction of the Sun with any fixed star and its return again to the same conjunction.

sidereal zodiac See TROPICAL ZODIAC.

significator The faster moving of two planets in aspect, as opposed to the "promittor."

solar return or revolution A horoscope erected for the exact time in a given year when the transiting Sun reaches the same position by sign and degree it held in the natal horoscope.

square When exact, an aspect formed by two planets that occupy the same degree of celestial longitude 90 degrees apart.

stationary The apparent lack of motion of a planet when it changes direction from direct to retrograde.

succedent houses Houses two, five, eight, and eleven, which succeed or follow the angular houses.

table of houses A table calculated to show what signs and degrees of the zodiac are on each of the cusps of the twelve mundane houses at any given time, day or night, in the year. A table of houses is always the same for a certain degree of latitude.

term One of five unequal divisions of a sign with its own subruler.

transit The position and movement of a planet on a given day.

trine When exact, an aspect formed by two planets that occupy the same degree of celestial longitude 120 degrees apart.

triplicity A group of three signs belonging to the same element: fire (Aries, Leo, Sagittarius); earth (Taurus, Virgo, Capricorn); air (Gemini, Libra, Aquarius); and water (Cancer, Scorpio, Pisces).

tropical zodiac The tropical zodiac invariably starts at the equinoctial point, or zero degrees Aries, and is based on the seasons. It consists of twelve equal signs and is wholly unaffected by precession, or the apparent movement of the equinoctial point backward through the constellations. The twelve unequal fixed-star constellations, on the other hand, which lie on either side of the ecliptic or the Sun's apparent path among the stars, make up the natural or sidereal zodiac.

void of course Describes the Moon when the planets are so placed that it makes no major aspect with them before leaving the sign it is in.

NOTES

Epigraph to Part One:

Page

1 "The universe, eternity": *The Heart of Burroughs's Journals*, January 13, 1882, p. 85.

Chapter One

Page

4 "*All astronomers are agreed*": Wedel, *The Medieval Attitude Toward Astrology*, p. 20; Geneva, *Astrology and the Seventeenth Century Mind*, p. 133.

4 "*These vast realms*": Abbott, *Makers of American History: Christopher Columbus*, p. 19.

5 "*waiting impatiently*": Irving, *The Life and Voyages of Christopher Columbus*, p. 63.

5 "*the voice of prayer*": Ibid., p. 288.

5 "*Come see the people*": Ferris, *Coming of Age in the Milky Way*, p. 57.

5 "*God-given*": Watts, "Prophecy and Discovery," p. 100; Geneva, *Astrology and the Seventeenth Century Mind*, p. 133.

6 "*Astrology is astronomy*": MacNeice, *Astrology*, p. 66.

7 "*separate the gems*": Negus, "Kepler's Astrology"; West and Toonder, *The Case for Astrology*, p. 87.

7 "*believe in astrology*": Brau et al., *Larousse Encyclopedia of Astrology*, p. v.

8 "*lights in the firmament*": Genesis 1:14.

8 "*whether we understand*": McCaffery, *An Astrological Key to Biblical Symbolism*, p. 80.

9 "*each one paying homage*": Hall, *The Story of Astrology*, p. 123.

9 "*were celebrated in the temple*": Ibid.

9 "*coopted by the Church*": Spencer, *True as the Stars Above*, p. 119.

10 "*the solar deity*": Ibid., p. 127.

10 "*appeared as the Fisher of Men*": Ibid.

11 "*Jove and Latona's son*": quoted in McCaffery, *Astrology: Its History and Influence*, p. 53.

11 "*every scene on earth*": Ibid.

11 "*What do you think*": quoted in Grafton, *Cardano's Cosmos: The Worlds and Works of a Renaissance Astrologer*, p. 131.

12 "*whose life unfolded*": McCaffery, *Astrology: Its History and Influence*, p. 38.

12 "*belonging to the Sun*": Dobin, *Kabbalistic Astrology*, p. 109.

12 "*We feel that when all possible*": quoted in Wigzell, *Reading Russian Fortunes*, p. 14.

12 "*entered into the councils*": from J. M. Ashmand, Introduction to Ptolemy, *Tetrabiblos*, p. xi.

13 "*Do you believe*": quoted in Krupp, *Skywatchers, Shamans & Kings: Astronomy and the Archaeology of Power*, p. 223.

13 *an intellectual environment favorable:* Thomas, *Religion and the Decline of Magic,* pp. 643–44.

13 *"all the more dangerous":* Roosevelt, *History As Literature,* p. 247.

14 *"There are in Astrologie":* McCaffery, *Astrology,* p. 276.

Chapter Two

Page

15 *"Very deep is the well":* Mann, *Joseph and His Brothers,* p. 1.

16 *"light of the astrologers":* Dobin, *Kabbalistic Astrology,* p. 151.

16 *"prophet":* McCaffery, *Astrology,* p. 20.

16 *"When Venus appears":* Stewart, *Astrology: What's Really in the Stars,* p. 36.

16 *"When the fiery light":* Sachs, *The Astrology File,* p. 26.

16 *"If a halo":* Stone, *The United States: Wheel of Destiny,* p. 26.

16 *"When Jupiter stands":* Krupp, *Skywatchers,* p. 229.

17 *"approximately determined":* Cumont, *Astrology and Religion Among the Greeks and Romans,* p. 7.

17 *"accurately fix":* Ibid., p. 8.

17 *"stars in the path":* Stewart, *Astrology,* p. 50.

17 *the temporary resting places:* Cornelius and Devereaux, *The Secret Language of the Stars,* p. 50.

17 *"[In ancient times] the farmers":* Kepler, *Report on the Fiery Triplicity,* quoted in Negus, "Kepler's Astrology," p. 2.

18 *"directly or indirectly":* Cumont, *Astrology and Religion,* p. 16.

18 *"all those who professed":* Ibid.

18 *"The star which the Greeks":* Ibid., p. 28.

18 *"not only common":* McCaffery, *Astrology,* p. 46.

18 *"when Socrates":* Cramer, *Astrology in Roman Law and Politics,* p. 5.

19 *"the Egyptians had learned":* Campion, *Mundane Astrology,* p. 48.

19 *"darkness falling":* McCaffery, *Astrology,* p. 54.

20 *"Finding that the lords":* Gleadow, *The Origin of the Zodiac,* p. 69.

21 *"at once lost":* Ibid., p. 72.

23 *"Overhead, the glorious":* Maunder, *The Astronomy of the Bible,* p. 256.

23 *"of unquestioned authority":* quoted in Seward, *The Zodiac and Its Mysteries,* p. 5.

24 *"A Sumerian priest":* Stewart, *Astrology,* p. 135.

Chapter Three

Page

28 *"Despite the warnings":* quoted in Cramer, *Astrology in Roman Law and Politics,* p. 75.

29 *"Were all those who perished":* quoted in Wedel, *The Medieval Attitude,* pp. 12–13.

29 *"a star that brings":* Cicero, *Nine Orations and The Dream of Scipio,* p. 300.

29 *"planet that bodes ill to men":* Ibid.

29 *"based on empirical observations":* On Divination: see McCaffery, *Astrology,* p. 97.

29 *"his hair long and disheveled":* Cramer, *Astrology in Roman Law and Politics,* p. 78.

30 *"a great and almost incredible":* Suetonius, *The Twelve Caesars,* p. 103.

30 *"If Jupiter in his own house":* Firmicus Maternus, quoted in Molnar, *The Star of Bethlehem: The Legacy of the Magi,* p. 6.

31 *"He will wield":* Manilius, *Astronomica,* Book 4, 11. 549–54; p. 267.

31 *"ushered in":* Cramer, *Astrology in Roman Law and Politics,* p. 78.

31 *"astronomical calculations":* Vitruvius, *The Ten Books on Architecture,* p. 269.

31 *"You were born to be":* quoted in Cramer, *Astrology in Roman Law and Politics,* p. 89.

32 *"If fortune will, she may":* in Temple Hungad, *A Brief History of Astrology,* p. 3, and reprinted in Lewis, *The Beginnings of Astrology in America.*

32 *"old whores":* Barton, *Ancient Astrology*, p. 173.

32 *"Give me the ways":* John Dryden's translation, quoted in Annabella Kitson, "Some Varieties of Electional Astrology," in Kitson, ed., *History and Astrology: Clio and Urania Confer*, p. 175.

33 *"It seems my own life":* quoted in Tacitus, *The Annals of Imperial Rome*, p. 205.

34 *"You, too, sonny":* quoted in Cramer, *Astrology in Roman Law and Politics*, p. 131.

34 *"had as much chance":* Suetonius, *The Twelve Caesars*, p. 158.

35 *"unfinished by nature":* quoted in Scramuzza, *The Emperor Claudius*, p. 35.

35 *"incapable of acting":* quoted in ibid., p. 36.

35 *"slobbered horribly":* Suetonius, *The Twelve Caesars*, p. 199.

35 *"unclear":* Scramuzza, *The Emperor Claudius*, p. 37.

36 *"His health":* Suetonius, *The Twelve Caesars*, p. 200.

36 *"Oh, let him kill me":* quoted in Cramer, *Astrology in Roman Law and Politics*, p. 116.

37 *"Imagine I were coming":* quoted in ibid., p. 117.

37 *"Whatever happens":* quoted in Schechner Genuth, *Comets, Popular Culture, and the Birth of Modern Cosmology*, p. 21.

38 *"where slave minions":* Dill, *Roman Society from Nero to Marcus Aurelius*, p. 15.

38 *"although the comet portended":* Cramer, *Astrology in Roman Law and Politics*, p. 118.

38 *"a substitute":* Ibid.

40 *"was so tranquil":* Ibid., p. 138.

41 *"the common people":* Suetonius, *The Twelve Caesars*, p. 293.

42 *"the very year, day, hour":* Cramer, *Astrology in Roman Law and Politics*, p. 140.

42 *"Oh, come on, lad":* Suetonius, *The Twelve Caesars*, p. 305.

42 *"a dark if distant":* Berlinski, *The Secrets of the Vaulted Sky*, p. 73.

42 *"There will be blood":* Suetonius, *The Twelve Caesars*, p. 306. See also Michael Molnar, "Blood on the Moon in Aquarius: The Assassination of Domitian."

43 *"had not failed":* Cramer, *Astrology in Roman Law and Politics*, p. 155.

44 *"traveled through one province":* Cassius Dio, *Roman History*, Book 69, ll. 1–4.

45 *"In this chart Saturn is the lord":* Cramer, *Astrology in Roman Law and Politics*, p. 164.

46 *"was born to become":* Ibid., pp. 176–77.

49 *"The emperor's fears":* Dill, *Roman Society*, pp. 45–46.

49 *"If [Sirius] rises when the Moon":* Stewart, *Astrology*, p. 10.

49 *"Year 27 of Caesar [Augustus]":* Ibid.

50 *"a comprehensive idea":* Campion, "Dorotheus of Sidon: His Life and Significance," p. 4.

51 *"surrounded by exotic gardens":* Berlinski, *Vaulted Sky*, p. 42.

51 *"after settling into":* Ibid.

51 *"In conformity with nature":* Ptolemy, *Tetrabiblos*, pp. 77–81.

52 *"[Morally] Monstrous":* Ibid., p. 85.

53 *"we apprehend the movements":* Berlinski, *Vaulted Sky*, p. 43.

53 *"the changes that the [heavens]":* Ibid.

53 *"It is not possible that particular forms":* McCaffery, *Astrology*, p. 81.

54 *"Fate rules the world":* Manilius, *Astronomica*, Book 4, l. 14; p. 223.

55 *"Astrology, it is true":* quoted in McIntosh, *Astrology: The Stars and Human Life*, p. 20.

55 *"Look at Venus where it is":* quoted in Holden, *A History of Horoscopic Astrology*, p. 55.

55 *"It is the sign that is turned away":* Ibid., p. 83.

56 *"The first 8":* Ibid., p. 52.

57 *"cadent from the Ascendant":* Plant, "Vettius Valens: An Ancient Judgement of Wealth," p. 2.

57 *"to eavesdrop on cosmic":* Bram, Preface to Maternus, *Ancient Astrology: Theory and Practice*, p. 3.

57 *"a degree's movement":* Barton, *Ancient Astrology*, p. 161.

57 *"If, at birth, the Moon":* Ibid., p. 161.

57 *"if the Moon is void":* Ibid., p. 162.

57 *"Venus in Capricorn"*: Ibid., p. 165.
57 *Study and pursue:* Maternus, *Ancient Astrology: Theory and Practice*, pp. 68–69.
58 *"It shows the quality of life"*: Ibid., pp. 136–37.
59 *"The decline of Rome"*: Gibbon, *The Decline and Fall of the Roman Empire*, p. 800.

Chapter Four

Page

61 *"in the general decline"*: Wedel, *The Medieval Attitude Toward Astrology*, p. 25.
63 *"lamb cooked over a spit"*: Berlinski, *Vaulted Sky*, p. 82.
64 *"he seemed to them"*: Notker, "Life of Charlemagne," in *Two Lives of Charlemagne*, p. 143.
65 *"the Arabs were not able"*: Ibid.
65 *"a royal house"*: Ogg, *A Source Book of Medieval History*, p. 127.
65 *"that they rose"*: Notker, "Life of Charlemagne," in *Two Lives of Charlemagne*, p. 144.
65 *"The envoys were more merry"*: Ibid., pp. 145–46.
66 *"They seemed to have despoiled"*: Ibid., p. 146.
66 *"specially chosen"*: Ibid., p. 147.
67 *"Harun al-Rashid, Commander"*: "Harun al-Rashid," *Wikipedia*, http://en.wikipedia.org/wiki/Harun_al-Rashid.
67 *were coined by the Arabs:* Kenton, *Astrology: The Celestial Mirror*, p. 17.
68 *how astrology could:* Geneva, *Astrology*, p. 222.
68 *"violence befalls him"*: quoted in ibid., p. 127.
69 *You have to know that reception:* quoted in Holden, *A History of Horoscopic Astrology*, pp. 104–6.
70 *"the teacher of the people"*: Abu Maʿshar, *The Abbreviation of the Introduction to Astrology*, p. iii.
70 *"conditions of the planets"*: Ibid., pp. 20–28.
71 *"Once with some travelers"*: Holden, *A History of Horoscopic Astrology*, p. 113. (My abridgement and paraphrase.—Benson Bobrick)
72 *"The Native's Wealth"*: Al-Khayyat, *The Judgments of Nativities*, pp. 28–30.
74 *"There are certain signs"*: Al-Bīrūni, *The Book of Instruction in the Elements of the Art of Astrology*, p. 256.
74 *"Prosperity is associated"*: Ibid., p. 100.
75 *"like an ingot of silver"*: Al-Masu'di, *The Book of Golden Meadows*, p. 100.
75 *"Rakkah"*: Ibid.
75 *"O thou whose reign"*: Ibid.
76 *"beaten to death in a sack"*: Huxley, *From an Antique Land*, p. 188.

Chapter Five

Page

77 *"Now when Jesus was born"*: Matthew 2:1–2, 7–12, King James Version.
78 *"saw the star reflected"*: Kidger, *The Star of Bethlehem: An Astronomer's View*, p. 34.
78 *"Its light was unspeakable"*: quoted in ibid., pp. 17–18.
78 *"a new star unlike"*: quoted in ibid., p. 18.
78 *"the star was not a star"*: quoted in George, "Manuel I Komnenos and Michael Glycas: A Twelfth-Century Defence and Refutation of Astrology," Part 1, "History and Background," p. 12.
78 *"The star which the Magi saw"*: quoted in Tester, *A History of Western Astrology*, p. 112.
79 *"He [God] appointed"*: quoted in Hyde, *Jung and Astrology*, p. 15.
79 *"the birth of a Hebrew king"*: Molnar, *The Star of Bethlehem: The Legacy of the Magi*, p. 7.
79 *"Herod died after a lunar"*: quoted in Krupp, *Beyond the Blue Horizon: Myths and Legends of the Sun, Moon, Stars, and Planets*, p. 307.

80 *"a massing of nearly all"*: Kidger, *The Star of Bethlehem*, p. 88.

80 *"during their journey"*: Ibid., p. 262.

80 *"All stars except the ones"*: Ibid., p. 26.

81 *"in the first light of dawn"*: Ibid., p. 28.

82 *"sent to the Magi"*: George, "Manuel I Komnenos and Michael Glycas," Part 1, p. 11.

83 *"pretended to divine"* and *"casters of spells"*: Brau et al., *Larousse Encyclopedia of Astrology*, p. 184.

83 *"learned in astronomia"*: Carey, *Courting Disaster: Astrology at the English Court and University in the Later Middle Ages*, p. 28.

83 *"every Carolingian lord"*: Campion, *Mundane Astrology*, p. 53.

84 *"astrology was even more popular"*: Reston, *Galileo: A Life*, p. 219.

85 *"a theologian of great"*: Grafton, *Cardano's Cosmos*, p. 102.

85 *"a sacrilegious heretic"*: Ibid., p. 101.

85 *"whatever moves in the heavens"*: quoted in Berlinski, *Vaulted Sky*, p. 60.

86 *"cupola of Islam"* and *"minaret of piety"*: Sedgwick, *A Short History of Spain*, pp. 38–39.

86 *"the astrolabe of Ptolemy"*: quoted in Introduction to Maternus, *Ancient Astrology: Theory and Practice*, p. 6.

89 *"Mathematicians are those"*: quoted in McCaffery, *Astrology*, p. 193.

89 *"describes the whole form"*: quoted in Wedel, *The Medieval Attitude*, pp. 49–50.

89 *"analyze the prospects"*: George, "Manuel I Komnenos and Michael Glycas," Part 1, p. 11.

90 *"From morning to evening"*: Vasiliev, *History of the Byzantine Empire*, Vol. 2, p. 702.

91 *"the influence of the planets"*: Smoller, *History, Prophecy, and the Stars: The Christian Astrology of Pierre d'Ailly, 1350–1420*, p. 30.

92 *"one could have predicted"*: Ibid., p. 78.

92 *"a tree under the shadow"*: quoted in *The Jewish Encyclopedia*, p. 245.

92 *"The real science"*: Epstein, Meira B. *The Correspondence*, pp. 14–15.

93 *"rash and working against"*: Smoller, *History, Prophecy, and the Stars*, p. 30.

93 *"When you want to make"*: Gleadow, *The Origin of the Zodiac*, p. 57.

94 *"engraving astrological images"*: Smoller, *History, Prophecy, and the Stars*, p. 30.

94 *"operate in fractions"* and *"in millionths"*: McCaffery, *Astrology*, p. 137.

95 *"if someone were to forget"*: Ibid.

95 *"As to what they"*: St. Augustine, *The City of God*, p. 145.

96 *"it does not follow"*: Ibid.

96 *"beginning with the shortest"*: Nicholas Campion, "Astrological Historiography in the Renaissance" in Kitson, ed., *History and Astrology*, p. 114.

96 *"So if you want to know"*: Geneva, *Astrology and the Seventeenth Century Mind*, p. 124.

96 *"contributed to an atmosphere"*: Smoller, *History, Prophecy, and the Stars*, p. 155.

97 *"If the world shall last"*: quoted in ibid., pp. 155–56.

97 *"it followed that the world"*: Ibid., p. 62.

98 *"For when the sun is directly opposite"*: quoted in Berlinski, *Vaulted Sky*, pp. 156–57.

98 *"portend certain changes"*: quoted in Grafton, *Cardano's Cosmos*, p. 53.

98 *"all rode in"*: Cramer, *Astrology in Roman Law and Politics*, p. 117.

99 *"like a spear"*: Ryan, *The Bathhouse at Midnight*, p. 375.

99 *"blood-stained and threatening"*: Geneva, *Astrology and the Seventeenth Century Mind*, p. 88.

100 *"to accelerate the manifestation"*: Brau et al., *Larousse Encyclopedia of Astrology*, p. 74.

100 *"On the 16th day"*: quoted in Geneva, *Astrology and the Seventeenth Century Mind*, p. 98.

100 *"On the 20th day"*: Ibid.

100 *"For I believe"*: Act I, Scene 3, ll. 31–32.

100 *"visible or total, or nearly so"*: Brau et al., *Larousse Encyclopedia of Astrology*, p. 74.

100 *"guest stars"* and *"odd appearances"*: Krupp, *Skywatchers*, p. 232.

101 *"the last eclipse"*: Brau et al., *Larousse Encyclopedia of Astrology*, p. 101.

101 *"was on the Midheaven"*: Ibid.

101 *"the path of"*: Ibid.
101 *"The year that Mithradates"*: quoted in *Larousse Encyclopedia of Astrology*, p. 73.
102 *"Everything depended"*: Heilbron, *The Sun in the Church*, p. 3.
104 *"the Eildon hills"* and *"bridle the [river] Tweed"*: Canto 2, Stanza 13, ll. 9–10.
105 *"God damn you"*: Burckhardt, *The Civilization of the Renaissance*, p. 325.
105 *"All things are known"*: quoted in Wedel, *The Medieval Attitude*, p. 79.
106 *"probably the most important"*: Robert Hand in Introduction to Bonatti, *Liber Astronomiae*, Part 1, p. i.
106 *"The 111th is"*: Lilly, ed., *Anima Astrologiae*, p. 48.
106 *"The 120th Consideration is"*: Ibid., p. 51.
107 *"Look at the first house"*: quoted in Holden, *A History of Horoscopic Astrology*, p. 133.
107 *"Note in the Nativities of Kings"*: Lilly, ed., *Anima Astrologiae*, p. 61.
108 *"horary question and elections"*: Laird, "Christine de Pizan and Controversy Concerning Star-Study in the Court of Charles V," p. 41.
108 *"regarded himself"*: Carey, *Courting Disaster*, p. 107.
108 *"an ever-increasing"*: Ibid., p. 23.
108 Pedro Alphonso, physician: "Astrology in Medieval Europe," www.meta-religion.com.
109 *"What astrologer could predict"*: quoted in Carey, *Courting Disaster*, p. 83.
110 *"Should I marry?"*: Ibid., p. 105.
110 *"had consulted Bolingbroke"*: Whitfield, *Astrology: A History*, p. 126.
110 *"commissioned two"*: Ibid., p. 127.
111 *"up into the leads"*: Allen, *The Star-Crossed Renaissance*, p. 102.
111 *"pander to his whims"*: Thomas, *Religion and the Decline of Magic*, p. 287.
111 *"friend and counselor"*: Grafton, *Cardano's Cosmos*, p. 104.
112 *"fantastical hodge-podge"*: Larkey, *Astrology and Politics in the First Years of Elizabeth's Reign*, p. 181.
112 *"our dispositions are caused"*: Parker, *Familiar to All: William Lilly and Astrology in the Seventeenth Century*, p. 56.
112 *"went about to dissuade"*: quoted in Allen, *The Star-Crossed Renaissance*, p. 180.
112 *"I am astrologically induced"*: quoted in Woolley, *The Queen's Conjurer: The Science and Magic of Dr. John Dee, Adviser to Queen Elizabeth I*, p. 141.
112 *"in the Moon's own house"*: Geneva, *Astrology and the Seventeenth-Century Mind*, p. 137.
113 *"a calculated effort"*: Annabella Kitson, "Some Varieties of Electional Astrology," in Kitson, ed., *History and Astrology*, p. 174.
113 *fortune-telling*: Ibid., pp. 220–21.
114 *"counsels a patience"*: McCaffery, *Graphic Astrology*, pp. 298–99.
114 *"The hour was morning's prime"*: Dante, *Inferno*, Canto 11, l. 113.
114 *"cradling the Earth"*: Ovason, *The Secret Architecture of Our Nation's Capital*, p. 358.
114 *"Our life, and also the life"*: Kay, *Dante's Christian Astrology*, p. 4.
114 *"according to its stars direct every seed"*: Dante, *Purgatorio*, Canto 21, ll. 109–11; Kay, *Dante's Christian Astrology*, p. 4.
115 *"Alas! alas!"*: Chaucer, Prologue to "Wife of Bath's Tale," ll. 614–16, in *The Canterbury Tales*, in *The Works of Geoffrey Chaucer*.
115 *"For Goddes love"*: "Knight's Tale," ll. 1084–91.
116 *"Was there no philosopher"*: "Man of Law's Tale," l. 310.
116 *"fall"*: Ibid., l. 303.
116 *"feeble Moon"* and *"disastrous fashion"*: Ibid., l. 306.
116 *"slain"*: Ibid., l. 301.
116 *"For in the sterres"*: Ibid., ll. 194–96.
116 *"unsophisticated, farmyard"*: Whitfield, *Astrology*, p. 102.
116 *"the timely arrival"*: Ibid., p. 102.
116 *"calculing"*: Chaucer, *Troilus and Creseyde*, Book 1, l. 71, in *Works*.
116 *"in good plyt"*: Ibid., Book 2, l. 74.
117 *"badde aspects"*: Ibid., Book 3, l. 716.

117 *"Venus combust":* Ibid., 1. 717.
117 *"the cursed constellation":* Ibid., Book 4, 1. 745.
117 *"increasing in light":* Annabella Kitson, "Some Varieties of Electional Astrology," in Kitson ed., *History and Astrology,* p. 176.

Chapter Six

Page
121 *"the golden age":* Brau et al., *Larousse Encyclopedia of Astrology,* p. 240.
121 *"Like the patient":* Grafton, *Cardano's Cosmos,* p. 23.
121 *"recorded on a form":* Ibid.
122 *"In all the better families":* Burckhardt, *The Civilization of the Renaissance in Italy,* p. 325.
123 *"he labored mightily":* Russell, *History of Astrology & Prediction,* p. 56.
123 *"Petrarch was left":* Ibid., pp. 57–58.
123 *"Whence it comes I know not":* quoted in Schechner Genuth, *Comets,* p. 83.
124 *"Some of the greatest scientific minds":* Allen, *The Star-Crossed Renaissance,* p. 100.
124 *"not a divine oracle":* quoted in ibid., p. 78.
125 *"cheerful and festive":* Clydesdale, "Marsilio Ficino's Holistic Astrology," p. 25.
125 *"coined the phrase":* Ibid.
125 *"untroubled course":* quoted in ibid., p. 28.
126 *"We may think of the stars":* Voss, "The Astrology of Marsilio Ficino: Divination or Science?" p. 36.
126 *"Saturn seems to have impressed":* quoted in Clydesdale, "Ficino's Holistic Astrology," p. 26.
126 *"The heavens will promote":* Ibid., p. 94.
126 *"could . . . only be justified":* Voss, "The Astrology of Marsilio Ficino," p. 40.
126 *"a material action":* Ibid., p. 38.
129 *"The blood of the just":* quoted in Russell, *History of Astrology & Prediction,* p. 133.
129 *"In thrice one hundred years":* in Craig, *Stars of Destiny: The Ancient Science of Astrology and How to Make Use of It To-Day,* p. 85.
129 *"with an accuracy":* Balzac, *Catherine de Medici,* p. 101.
129 *"And this proved":* Ibid.
130 *"straightway Ennio Verulano":* Thorndike, *A History of Magic and Experimental Science,* Vol. 5, p. 100.
130 *"during single combat":* Greene, *The Astrology of Fate,* p. 143.
131 *"Le Lyon jeune":* Ibid., p. 144.
131 *"was instantly recognizable":* Ibid.
133 *"both luminaries":* Cardan [Cardano], *The Book of My Life,* p. 4.
133 *"tormented by a tragic passion":* Ibid., p. 25.
134 *"small, white, restless":* Morley, *Jerome Cardan,* Vol. 1, p. 120.
134 *"like the beating":* Ibid., p. 121.
134 *"at the festival of the birth":* Ibid., p. 77.
135 *"friends and acquaintances":* Ibid., p. 93.
135 *"as true a leveler":* Ibid.
135 *"the doctor who does not understand":* quoted in McCaffery, *Astrology,* p. 58.
135 *"A doctor without astrology":* Ibid., p. 252.
135 *Aries-Calcination:* Christianson, *This Wild Abyss,* p. 110.
136 *"This is a case of":* quoted in Morley, *Jerome Cardan,* Vol. 1, p. 165.
136 *"lauded his discernment":* Ibid.
136 *"tall, frank-looking":* Ibid., Vol. 2, p. 72.
136 *"Whether or not it is true":* quoted in Nancy Siraisi, "Girolamo Cardan and Medical Astrology," in Grafton and Newman, eds., *Secrets of Nature: Astrology and Alchemy in Early Modern Europe,* p. 91.

136 *"medicine is a many-sided art"*: Cardan [Cardano], *The Book of My Life,* pp. 180–81.

137 *"From the start he had"*: Grafton and Newman, eds., *Secrets of Nature,* p. 70.

138 *"I have been able to cure you"*: Forman, *The Story of Prophecy,* p. 120; Allen, *The Star-Crossed Renaissance,* p. 52.

138 *"about insignificant things"*: quoted in Grafton, *Cardano's Cosmos,* p. 19.

138 *"Looking at it, I said"*: Ibid., pp. 94–95.

139 *"When he showed me"*: quoted in Grafton and Newman, *Secrets of Nature,* p. 94.

139 *"When Jupiter shall be"*: Cardano, "Aphorisms," in Lilly, ed., *Anima Astrologiae,* p. 83.

139 *"When Venus and Jupiter are in"*: Ibid.

139 *"When Venus is with Saturn"*: Ibid.

139 *"Mars is seldom joined"*: Ibid., p. 89.

140 *"amba hi libris corrupti"*: quoted in Morley, *Jerome Cardan,* Vol. 1, p. 76.

140 *"because I knew"*: Ibid., pp. 223–24.

140 *"with slobbering hands"*: Ibid., p. 192.

140 *"Men in this worlde"*: Ibid., pp. 196–97.

141 *"Make me of ten four"*: Ibid., p. 234.

141 *"Two persons were in company"*: Ibid.

141 *"Where did you ever find"*: Ibid., pp. 231–32.

141 *"so weak"*: Ibid., p. 238.

141 *"having all their terms"*: Ibid., p. 275.

142 *"in a manner before"*: Ibid.

142 *"mathematics is its own explanation"*: Cardan [Cardano], *The Book of My Life,* p. 246.

142 *"Considering the proposition"*: Ibid.

142 *"There is no apparent reason"*: Frawley, *The Real Astrology,* p. 119.

142 *"no other man"*: Morley, *Jerome Cardan,* Vol. 1, p. 90.

142 *"a magnificent moth"*: Ibid., pp. 90–91.

142 *"the science [of astrology]"*: Ptolemy, *Tetrabiblos,* p. 5.

143 *"so vehemently bent"*: French, *John Dee,* p. 2.

143 *"speak and confer"*: Ibid., p. 25.

144 *"divided by the Mediterranean"*: Ibid., p. 26.

144 *"observations"*: quoted in Woolley, *The Queen's Conjuror,* p. 18.

144 *"a mad fighting fellow"*: Ibid., p. 22.

145 *"due maintenance"*: Ibid., p. 27.

145 *"the top of the 7"*: Holden, *A History of Horoscopic Astrology,* p. 172.

145 *"to counteract any harm"*: French, *John Dee,* p. 7.

146 *"Master Key"*: quoted in Woolley, *The Queen's Conjuror,* p. 116.

146 *"watch-clock"*: Ibid., p. 85.

146 *"with five or six"*: Ibid.

146 *"the image of a solar eclipse"*: Ibid., p. 83.

146 *"enjoyed almost universal esteem"*: French, *John Dee,* p. 4.

147 *"performed 'marveilous Actes' "*: Ibid., p. 8.

147 *" 'Naturally, Mathematically' "*: Ibid., p. 7.

147 *"calculing"*: Ibid.

147 *"diabolical"*: Thomas, *Religion and the Decline of Magic,* p. 362.

147 *"those dark times"*: Aubrey, *Aubrey's Brief Lives,* p. 5.

147 *"met the Spirits"*: Ibid.

147 *"great security"*: quoted in French, *John Dee,* p. 7.

147 *"characters imbued"*: Allen, *The Star-Crossed Renaissance,* p. 53.

147 *"an Art Mathematical"*: quoted in Allen, *The Star-Crossed Renaissance,* p. 53.

147 *"a star is often a sign"*: McCaffery, *Astrology,* p. 61.

148 *"a Pythagorean art"*: Zoller, *The Arabic Parts in Astrology,* p. 2.

148 *"is not the physical movements"*: Ibid.

148 *"arithmetically derived points"*: Ibid.

148 *"like an artist's gown"*: Aubrey, *Aubrey's Brief Lives,* p. 89.

148 *"white as milk"*: Ibid.
149 *"a considerable number"*: Parker, *Familiar To All*, p. 186.

Chapter Seven

Page

150 *"What purpose then"*: quoted in Cowling, *Isaac Newton and Astrology*, p. 4.
151 *"In the middle of all"*: quoted in Geneva, *Astrology and the Seventeenth Century Mind*, p. 178.
151 *"the choral dance"*: Thorndike, *A History of Magic*, Vol. 5, p. 406.
151 *"giant of a man"*: Ferris, *Coming of Age in the Milky Way*, p. 74.
152 *"something divine"*: Christianson, *This Wild Abyss*, p. 133.
152 *"bright star which appeared"*: quoted in Gleiser, *The Prophet and the Astronomer*, p. 63.
152 *"It surpassed all the other stars"*: Ibid.
153 *"a gleaming brass"*: Ferris, *Coming of Age*, p. 73.
153 *"a matchless series"*: Berlinski, *Vaulted Sky*, p. 179.
154 *"private game preserves"*: Ferris, *Coming of Age*, p. 73.
154 *received visitors*: Encyclopedia Britannica, Vol. 2, p. 460.
154 *"We cannot deny"*: Dreyer, *Tycho Brahe*, p. 100; Christianson, *This Wild Abyss*, p. 144; Lynch, ed., *The Coffee Table Book of Astrology*, p. 31.
154 *"all Europe stood at gaze"*: Woolley, *The Queen's Conjuror*, p. 141.
154 *"Such unnatural births"*: quoted in Schechner Genuth, *Comets*, p. 51.
154 *"just above the Tropic of Capricorn"*: Woolley, *The Queen's Conjuror*, p. 143.
155 *"When beggars die"*: Shakespeare, *Julius Caesar*, Act 2, Scene 2, 1. 130.
155 *"generally speaking, kings"*: quoted in Dreyer, *Tycho Brahe*, p. 55.
157 *"pleasant, comely and voluptuous"*: Christianson, *This Wild Abyss*, pp. 150–52.
158 *"after a pleasant day"*: Ibid., p. 194.
158 *"Let me not seem to have lived"*: quoted in Christianson, *This Wild Abyss*, p. 224.
159 *"had been raised by an aunt"*: Ferris, *Coming of Age*, p. 74.
159 *"Yet this was the man"*: Ibid., p. 75.
159 *"the most acute thinker"*: Ibid.
160 *got through Latin*: Koestler, *The Sleepwalkers*, p. 240.
160 *"unusual subjects"*: Ibid.
160 *"the sight of Atlantis"*: Ibid.
160 *"to the minutest detail"*: Ibid.
160 *"argued with men"*: Ibid.
161 *"Look how the floor of heaven"*: Merchant of Venice, Act 5, Scene 1, 11. 66–71.
161 *"nesting"* and *"the five Platonic solids"*: Nick Kollerstrom, "Kepler's Belief in Astrology," in Kitson, ed., *History and Astrology*, p. 156.
161 *"It happened on July 19th"*: Berlinski, *Vaulted Sky*, p. 181.
162 *"celestial choir"*: Christianson, *This Wild Abyss*, p. 246.
162 *"So far the almanac's"*: quoted in Casper, *Kepler*, p. 63.
163 *"suffered from multiple"*: Ibid., p. 369.
163 *"small, thin, swarthy"*: Koestler, *The Sleepwalkers*, p. 227.
163 *"On New Year's Eve"*: Ibid., p. 232.
163 *"he had run away"*: Ibid., p. 230.
165 *"If two stones were placed"*: quoted in Koestler, *The Sleepwalkers*, p. 337.
165 *"if the Earth and the Moon"*: Ibid., p. 228.
165 *"It influences a human being"*: Ibid., p. 152.
166 *"The whole business of crises"*: in Kitson, ed., *History and Astrology*, p. 159.
166 *"the belief in the effect of"*: quoted in Koestler, *The Sleepwalkers*, pp. 243–44.
166 *"intermediary light reflected"*: Christianson, *This Wild Abyss*, p. 234.
167 *"Regard this as certain"*: quoted in Nick Kollerstrom, "Kepler's Belief in Astrology," in Kitson, ed., *History and Astrology*, p. 166.

167 *"Philosophy, and therefore"*: translated in Negus, "Kepler's Astrology," p. 3.
167 *"Although the sky is constantly making"*: Ibid.
167 *"A human being's nature"*: Ibid.
167 *"The sun should be directed"*: Ibid.
168 *"This character [of the heavens]"*: Ibid.
168 *"The natural soul"*: Kitson, ed., *History and Astrology*, p. 154; Negus, "Kepler's Astrology," p. 3.
168 *"I believe that at the birth of children"*: Negus, "Kepler's Astrology," p. 2.
169 *"Look at the relationship"*: quoted in Gauquelin, *Cosmic Influences on Human Behavior*, pp. 249–50.
169 *"the wanton little daughter"*: quoted in West and Toonder, *The Case for Astrology*, p. 87.
169 *"a hard-working hen"*: Negus, "Kepler's Astrology," p. 2.
169 *"no one should regard it"*: J. Allen Hynek, Foreword to Gauquelin, *Cosmic Influences*, p. 12.
169 *"can devote as much industry"*: Negus, "Kepler's Astrology," p. 3.
170 *"when the storm rages"*: quoted in Casper, *Kepler*, pp. 348–49.
171 *"an unnamed lord"*: Ibid., p. 340.
171 *"I might truthfully"*: Ibid.
172 *"horrible event"*: Ibid., p. 342.
172 *"Almost every motion of the body or soul"*: quoted in Nick Kollerstrom, "Kepler's Belief in Astrology," in Kitson, ed., *History and Astrology*, p. 160.
173 *"all the documents pertaining"*: Casper, *Kepler*, p. 357.
173 *"a day fittingly followed"*: quoted in Casper, *Kepler*, p. 358.
173 *"the prince of astronomy"*: Ibid., p. 359.
173 *"Mensus eram coelos"*: Ibid.
173 *"My soul being from heaven"*: author's own translation.
174 *"the lynx-eyed astrologer"*: Reston, *Galileo*, p. 18.
174 *"the most benign star"*: quoted in Sobel, *Galileo's Daughter*, p. 34.
174 *Isaac Newton's birth*: Ferris, *Coming of Age*, p. 101.

Chapter Eight

Page

175 *"Napier's bones"*: McCaffery, *Astrology*, p. 266.
176 *"Sickness and health"*: quoted in Aveni, *Conversing with the Planets*, p. 129.
176 *"Good days to buy"*: Ibid.
176 *"do not occur"*: Ibid.
176 *"There never was any great"*: quoted in Thomas, *Religion and the Decline of Magic*, p. 298.
177 *"Is there a single cardinal"*: quoted in Yates, *Theatre of the World*, p. 63.
177 *"Most Illustrious Prince!"*: *Modern History Sourcebook*, on-line. http://www.fordham.edu
177 *"ravened, embraced"*: Thomas, *Religion and the Decline of Magic*, p. 290.
177 *"Join then and keep"*: quoted in Geneva, *Astrology and the Seventeenth Century Mind*, p. 266.
178 *"garbled astrological nostrum"*: Stone, *The United States, Wheel of Destiny*, p. 28.
178 *"laid the foundations"*: Wright, *Middle-Class Culture in Elizabethan England*, p. 602.
178 *"lived close to the land"*: Curry, *Prophecy and Power*, p. 97.
178 *"Who is there"*: quoted in Whitfield, *Astrology*, p. 176.
178 *"Saturn and Venus this year"*: Shakespeare, *Henry IV, Part 2*, Act 2, Scene 4, 11. 286–87; in *Complete Works*. Other Shakespeare references herein are also to this edition of his works.
178 *"If thou want'st"*: quoted in Thomas, *Religion and the Decline of Magic*, p. 331.
178 *monastery wall*: Wigzell, *Reading Russian Fortunes*, p. 41.
180 *"In sooth, thou wast"*: *Twelfth Night*, Act 1, Scene 3, 11. 146–49.

180 *"that the Dragon's"*: Johnstone Parr, "Shakespeare's Artistic Use of Astrology," in Lynch, *The Coffee Table Book of Astrology,* p. 259.

180 *"meteors fright the fixed stars"*: *Richard II,* Act 2, Scene 2, 1. 8.

180 *"a portent of broached"*: *Henry IV, Part 1*: Act 5, Scene 1, 1. 9.

180 *"Hung be the heavens"*: *Henry VI, Part 1*: Act 1, Scene 1, 11. 3–7.

181 *"When beggars die"*: *Julius Caesar,* Act 2, Scene 2, 1. 30.

181 *"When stars wear locks"*: Tourneur, *The Revenger's Tragedy,* Act 5, Scene 3, 1. 23.

181 *"sick almost to doomsday"*: *Hamlet,* Act 1, Scene 1, 1. 132.

181 *"when the planets"*: *Troilus and Cressida,* Act 1, Scene 3, 11. 85–92.

181 *"ire"* and *"angry stars"*: *Pericles,* Act 2, Scene 1, 1. 9.

181 *"star-cross'd"*: *Romeo and Juliet,* Prologue, 1. 6.

181 *"some ill planet reigns"*: *The Winter's Tale,* Act 2, Scene 1, 1. 107.

181 *"these late eclipses"*: *King Lear,* Act 1, Scene 2, 11. 98–131.

182 *"It is the stars"*: Ibid., Act 4, Scene 3, 11. 32–35.

182 *"What's in a name"*: *Romeo and Juliet,* Act 2, Scene 2, 11. 46–47.

182 *"All of the astrological"*: Parr, *Tamburlaine's Malady,* p. 62.

183 *"I would entreat you"*: Byron's *Conspiracy,* Act 3, Scene 3, 11. 38–127.

184 *"al Ghul"*: Robson, *The Fixed Stars & Constellations in Astrology,* p. 124.

184 *"If Mars be in imperfect signs"*: Ptolemy, *Tetrabiblos,* p. 136.

184 *"Men at some time are masters"*: *Julius Caesar,* Act 1, Scene 2, 11. 139–41.

184 *"Wonder not at the grievous"*: quoted in Geneva, *Astrology and the Seventeenth Century Mind,* p. 86.

185 *"I find my zenith"*: *The Tempest,* Act 1, Scene 2, 11. 181–84.

185 *"There is a tide"*: *Julius Caesar,* Act 4, Scene 3, 11. 248–52.

185 *"bodies high reign on the low"*: Sidney, *Astrophel and Stella,* Sonnet 26, 1. 11, in Bender, *Five Courtier Poets of the English Renaissance,* p. 345.

185 *"in the house of agonies"*: quoted in Allen, *The Star-Crossed Renaissance,* p. 174.

185 *"have complete power"*: quoted in Allen, *The Star-Crossed Renaissance,* p. 154.

185 *"Our remedies oft in ourselves"*: *All's Well That Ends Well,* Act 1, Scene 1, 11. 231–34.

185 *"Court any woman"*: quoted in Allen, *The Star-Crossed Renaissance,* p. 158.

186 *"Y'are much inclined"*: quoted in Allen, *The Star-Crossed Renaissance,* p. 174.

186 *"The Lord of the first house"*: quoted in Allen, *The Star-Crossed Renaissance,* p. 177.

186 *"Ah: how falls your question?"*: Webster, *The Duchess of Malfi,* Act 2, Scene 2, 11. 29–30.

186 *"the notorious astrological physician"*: Traister, *The Notorious Astrological Physician of London,* passim.

186 *"good yeoman stock"*: Rowse, *Sex and Society in Shakespeare's Age: Simon Forman the Astrologer,* p. 5.

187 *"bold and impudent"*: as quoted in Traister, *The Notorious,* p. 3.

187 *"by what authority [he] meddled"*: quoted in Rowse, *Sex and Society,* p. 49.

187 *"learned his skill"*: Ibid.

187 *"pisspot physic"*: Ibid., p. 33.

187 *"he had the right of it"*: Ibid.

188 *"Throughout the world"*: Traister, *The Notorious,* p. 23.

188 *"a small face"*: Rowse, *Sex and Society,* p. 93.

188 *"lend out sums"*: Ibid.

189 *"If thou wilt"*: quoted in Traister, *The Notorious,* p. 37.

189 *"They will sing and rhyme"*: Ibid., p. 49.

189 *"Those that have the scurvy"*: Ibid., p. 58.

190 *"the sexual appetite"*: Ibid., p. 5.

190 *"when she found"*: Ibid., p. 150.

190 *"She was begotten"*: Ibid., p. 151.

191 *return safely*: Rowse, *Sex and Society,* p. 164.

191 *"could hardly make a move"*: Ibid., p. 120.

191 *"proud and inconstant"*: Ibid., p. 128.
191 *"very fair"*: Ibid.
192 *"10 past 2 p.m."*: Ibid.
192 *"he shall not be bishop"*: Ibid., p. 151.
192 *"best go home"*: Ibid.
193 *"She will prove a whore"*: Ibid., p. 191.
193 *"There seems to be"*: quoted in Rowse, *Sex and Society*, p. 220.
193 *"particularly those trading"*: Ibid., p. 161.
194 *"beyond Cripplegate"*: Ibid., p. 38.
194 *"have their crystals"*: *The Devil Is an Ass*, Act 1, Scene 2, 11. 44–48.
194 *"this Doctor . . . he is the Faustus"*: *The Alchemist*, Act 2, Scene 3, 1. 127.
195 *"to compel Carr's love"*: Rowse, *Sex and Society*, p. 11.
195 *"a jury of matrons"*: Ibid., p. 255.
195 *"was dragged up"* and *"his memory held up"*: Ibid., p. 260.
195 *"The fact that some"*: Ibid.
196 *"This is the book of the life"*: quoted in Rowse, *Sex and Society*, p. 267.
196 *"I shall die ere"*: Ibid., p. 259.
196 *"to magical practices"*: Rowse, *Sex and Society*, p. 40.
197 *"physick"*: *A Briefe Description of the Notorious Life of John Lambe*, p. 2.
197 *"evil Diabolical"*: Ibid., p. 4.
197 *"to disable, make"*: Ibid., pp. 6–7.
197 *"the High Sheriffe"*: Ibid., p. 14.
198 *"duke's devil"*: Amundsen, "The Duke's Devil and Doctor Lambe's Darling," p. 49.
198 *"Let Charles and George"*: Ibid., p. 51.

Chapter Nine

Page
200 *"They have learned little"*: quoted in Warren-Davis, "Culpeper: Herbalist of the People," p. 5.
200 *"an art which teachest"*: Ibid.
200 *"physic without astrology"*: Ibid., p. 10.
200 *"he treated all comers"*: Kitney, "Culpeper's Herbal Medicine," p. 12.
200 *"If you do but consider the whole"*: quoted in Thomas, *Religion and the Decline of Magic*, p. 332.
201 *"First consider what Planet"*: quoted in Parker, *Familiar to All*, pp. 241–43.
201 *"Why should we rob"*: quoted in Thomas, *Religion and the Decline of Magic*, p. 333.
202 *"had the virtue . . . of"*: Frawley, *The Real Astrology*, p. 69.
202 *"in each case meticulously"*: Kenton, *Astrology: The Celestial Mirror*, p. 27.
202 *"In each decumbiture"*: Warren-Davis, *Astrology and Health*, p. 76.
203 *"my mother always intending"*: Lilly and Ashmole, *Lives of Those Eminent Antiquaries Elias Ashmole and William Lilly, Written by Themselves*, p. 5.
203 *"could make extempore"*: Ibid., p. 8.
204 *"If any scholars from remote"*: Ibid.
204 *"country labor"*: Ibid., p. 5.
204 *"no hope by plain"*: Ibid.
204 *"an upper servant"*: Ibid., p. 33.
204 *"ceremoniously walk before him"*: Ibid., p. 34.
204 *"She had many suitors"*: Lilly and Ashmole, *Lives*, p. 27.
205 *"reveled in the opportunity"*: Parker, *Familiar to All*, p. 41.
205 *"suffering from a monumental"*: Ibid., p. 45.
205 *"a squat little man"*: Lilly and Ashmole, *Lives*, p. 31.
205 *"could set a figure"*: Ibid., p. 30.
205 *"a comely old man"*: Ibid., p. 35.

206 *"Captain Bubb":* Ibid., p. 36.

206 *"a nibbler":* Ibid., p. 38.

206 *"a little smattering":* Ibid., p. 37.

206 *"which he well understood":* Ibid., p. 44.

206 *"had the most piercing":* Ibid., p. 31.

207 *"I standing by all the while":* Ibid., p. 34.

207 *"Be humane, courteous":* "To the Student in Astrology," foreword to *Christian Astrology,* n.p.

208 *"who had earned":* Parker, *Familiar to All,* p. 89.

208 *"made many impertinent":* Lilly and Ashmole, *Lives,* p. 64.

208 *"Lo, hear what Lilly saith":* quoted in Geneva, *Astrology and the Seventeenth Century Mind,* p. 64.

208 *"You are yet in your ABC's":* quoted in McCaffery, *Astrology,* p. 292.

208 *"God is on our side":* quoted in Geneva, *Astrology and the Seventeenth Century Mind,* p. 64.

209 *"The Stars are now":* Ibid.

209 *"he would rather eat":* Parker, *Familiar to All,* p. 153.

209 *"know nothing"* and *"was lawful":* Ibid., p. 148.

209 *"in what quarter":* Lilly and Ashmole, *Lives,* p. 88.

209 *"I told him to go":* Ibid., p. 89.

209 *"representing his Majesty":* quoted in Geneva, *Astrology and the Seventeenth Century Mind,* p. 182.

210 *"Dazzle mine eyes":* Henry VI, Part 3, Act 2, Scene 1, 1. 25.

210 *"Three glorious suns":* Ibid., 11. 26–32.

210 *"when mock Suns":* quoted in Geneva, *Astrology and the Seventeenth Century Mind,* p. 100.

210 *"Luna is with Antares":* Lynch, *Astrology,* p. 32.

211 *"Whether one absent will return":* Lilly, *Christian Astrology,* p. 406.

211 *"Whether a Damsel":* Ibid., p. 312.

211 *"being well-dignified":* Ibid., pp. 57–80.

212 *"Question: Money lost":* Ibid., p. 395.

212 *"as Luna applied":* Ibid., p. 396.

213 *"I only wrote 'senes' ":* quoted in McCaffery, *Astrology,* p. 305.

213 *"is nothing appertaining":* Lilly, *Christian Astrology,* p. 50.

213 *"Whether Presbytery shall stand?":* Ibid., p. 439.

213 *"the general significator":* Ibid., p. 440.

213 *"peregrine, occidental":* quoted in Campion et al. *Mundane Astrology,* p. 360.

213 *"within three years":* Ibid.

213 *"a more amenable":* Ibid.

214 *"unlawfully given":* Parker, *Familiar to All,* p. 189.

214 *"The astrologers have made":* quoted in Geneva, *Astrology and the Seventeenth Century Mind,* p. 260.

215 *"In the year 1665":* quoted in Innes, *Horoscopes,* p. 75.

215 *"representing a great sickness":* quoted in McCaffery, *Astrology,* p. 307.

215 *"after the coffins":* "The Great Plague 1665," http://www.historic-uk.com/HistoryUK/England-History/GreatPlague.htm

215 *"Lord have mercy":* Ibid.

215 *"Bring out your dead":* Ibid.

216 *"A woman might piss":* quoted in *Cities Guide, London.* www.economist.com/cities.

216 *"we beheld that dismal":* www.pepys.info/fire.html.

217 *"all the sky was afire":* Ibid.

217 *"The Committee seemed":* quoted in Craig, *Stars of Destiny,* p. 87.

217 *"most convenient to signify":* quoted in Whitfield, *Astrology,* p. 82.

217 *"the finger of God":* Ibid.

217 *"a rash, ravenous":* quoted in Geneva, *Astrology and the Seventeenth Century Mind,* p. 248.

218 *"The Fixed Stars are slow"*: Hazelrigg, "Metaphysical Astrology," p. 14; in Lewis, *The Beginnings of Astrology in America.*

218 *"English Atlas"*: Ashmole, quoted in McCaffery, *Astrology,* p. 306.

Chapter Ten

Page

219 *"some technical gossip"*: Parker, *Familiar to All,* p. 164.

220 *"addicted to astrology"*: McCaffery, *Astrology,* p. 296.

220 *"By Elections we may"*: Annabella Kitson, "Some Varieties of Electional Astrology," in Kitson, ed., *History and Astrology,* p. 171.

220 *"About 10 after noon"*: Ibid., p. 186.

220 *"a sign associated"*: Ibid.

220 *"in that condition"*: Lilly and Ashmole, *Lives,* p. 304.

221 *"uncomfortably early hour"*: Kitson, ed., *History and Astrology,* p. 187.

221 *"good aspects between"*: Ibid.

221 *"high reputation for"*: Ibid., p. 189.

221 *"for increase of honor"*: Thomas, *Religion and the Decline of Magic,* p. 635.

221 *"for use as contraceptives"*: Ibid.

221 *"the fruit of prolonged"*: Sutherland, *English Literature of the Late Seventeenth Century,* p. 285.

222 *"those who take this course"*: quoted in McCaffery, *Astrology,* p. 232.

222 *"The Planets and Stars are"* and *"Expect not that all"*: quoted in Parker, *Familiar to All,* p. 98.

222 *"Ptolemy may be something"*: quoted in Curry, *Prophecy and Power,* p. 96.

222 *"multitude of Pretenders"*: quoted in McCaffery, *Astrology,* p. 296.

222 *"scorne and contempt"*: Ibid.

222 *"Trust not to all"*: Ibid.

223 *"seditious"*: Butler, *Hudibras,* Part 2, Canto 3, ll. 44.

223 *"Did not our great reformers use"*: Ibid., ll. 171–78.

223 *"The Knight, with various Doubts possest"*: Ibid., ll. 295–300.

223 *"business was to pump and wheedle"*: Ibid., Part 2, Canto 3, ll. 335–40, 371–72.

224 *"I resolve these"*: quoted in Thomas, *Religion and the Decline of Magic,* p. 383.

224 *"Thus Gallants, we like Lilly"*: quoted in McCaffery, *Astrology,* p. 323.

224 *"The utmost malice"*: *Annus Mirabilis,* Stanzas 291 and 292, ll. 1161–68, in *Poems and Fables,* p. 103.

225 *"observed with grief"*: Luke Broughton, quoted in McCaffery, *Astrology,* p. 322.

225 *"set him a double lesson"*: Ibid., p. 323.

226 *"a noonday star"*: Schechner Genuth, *Comets,* p. 84.

226 *"as soon as Born"*: Ibid.

226 *"was measured from a star"*: Preface to Morin, *Astrologia Gallica,* Book 21, p. ii.

227 *"to within ten hours"*: Brau et al., *Larousse Encyclopedia of Astrology,* p. 193.

227 *"The more the figure"*: Morin, *Astrologia Gallica,* Book 22, pp. 258–59.

228 *"had played an important role"*: Plant, "John Gadbury: Politics and the Decline of Astrology," p. 2.

228 *"indelibly tainted"*: Ibid.

228 *"democratize"*: Ibid.

228 *"to predict the weather"*: Thomas, *Religion and the Decline of Magic,* p. 334.

228 *"hang'd himself on expectation"*: quoted in Thomas, *Religion and the Decline of Magic,* p. 341.

228 *"Pamphlets which prognosticated"*: Ibid.

229 *"As for astrology"*: Hungad, *A Brief History of Astrology,* p. 7, in Lewis, *The Beginnings of Astrology in America.*

229 *"sober and regulated"*: quoted in Thomas, *Religion and the Decline of Magic,* p. 351.

229 *"believed in the astrological":* quoted in Thomas, *Religion and the Decline of Magic*, p. 355.
229 *"a true Astrology":* quoted in Curry, *Prophecy and Power*, p. 164.
230 *"we know planets only to know them":* quoted in Spencer, *True As the Stars Above*, p. 74.
230 *"celestial bodies":* quoted in Curry, *Prophecy and Power*, p. 63.
230 *"Dyed about 9":* Ibid., p. 140.
230 *"We have had of late":* Ibid., p. 51.
230 *"What it portends":* Ibid., p. 141.
230 *"we are governed":* quoted in Thomas, *Religion and the Decline of Magic*, p. 325.
230 *"had more than a passing":* North, *Horoscopes and History*, p. 181.
231 *"a crowd of ill directions":* Ibid., p. 329.
231 *"divine art":* Curry, *Prophecy and Power*, p. 35.
231 *"convinced that astrological":* Ibid.
231 *"planet-struck":* Thomas, *Religion and the Decline of Magic*, p. 633.
231 *"the Vulgar have esteemed":* quoted in Curry, *Prophecy and Power*, p. 141.
231 *"You know I put":* Ibid.
231 *"observ[ed] the punctual time":* Ibid.
231 *"if the Earth moved round":* Hunter and Gregory, eds., *An Astrological Diary*, p. 76.
232 *"July 16, Thursday":* Ibid., p. 104.
232 *"December 24 (1672)":* Ibid., p. 145.
232 *"May 14 (1677)":* Ibid., p. 150.
232 *"My stature short":* Ibid., p. 117.
233 *"March 1, Tuesday":* Ibid., p. 155.
233 *"a general treatise":* Holden and Hughes, *Astrological Pioneers of America*, p. 187.
233 *"the principal accidents":* Ibid., p. 188.
234 *"beacons, whose use":* quoted in Schechner Genuth, *Comets*, p. 151.
234 *"For if the Moles":* quoted in Plant, "John Gadbury," p. 2.
234 *"Having erected":* Ibid.
235 *"spurned horary work":* Curry, *Prophecy and Power*, p. 83.
235 *"divers Errors":* Partridge, *Opus Reformatum*, p. vi.
235 *"pretended to fetch":* Ibid., p. xi.
235 *"to promote or prevent":* Ibid.
235 *"with a Table of Directions":* Ibid., p. iii.
235 *"Saturn and Jupiter are both":* Ibid., p. 45.
236 *"for if all concur":* Ibid.
236 *"Under this Revolution":* Ibid.
237 *"an omniscient God":* Schechner Genuth, *Comets*, p. 111.
237 *"the quadratures of the Moon":* Thomas, *Religion and the Decline of Magic*, p. 355.
237 *"last but not least":* West and Toonder, *The Case for Astrology*, p. 81.
237 *"with the exception of the discovery":* Ibid., p. 82.
237 *"The astrologer himself careth not":* Heydon, *A Defence of Judiciall Astrologie*, p. 371.
237 *"assumed a relationship":* Thomas, *Religion and the Decline of Magic*, p. 326.
237 *astrology book:* Cowling, *Isaac Newton and Astrology*, p. 2.
238 *"used astrological concepts":* Curry, *Prophecy and Power*, p. 143.
238 *"he held that the best":* Ibid.
238 *"a spiritual force":* Cowling, *Isaac Newton*, p. 12.
238 *"The wild dance of shadows":* Koestler, *The Sleepwalkers*, p. 509.

Chapter Eleven

Page

241 *"But do you hear":* Quoted in Thomas, *Religion and the Decline of Magic*, pp. 353–54.
241 *"confessed on his death-bed":* Thomas, *Religion and the Decline of Magic*, p. 291.
241 *"Sowers of Discord":* quoted in Curry, *Prophecy and Power*, p. 105.
241 *"the practice of foretelling":* Ibid., p. 108.

242 *"a conjectural science"*: Ibid., p. 150.

242 *"Partridge the Almanack-maker"*: quoted in Russell, *History of Astrology & Prediction*, p. 72.

242 *"infallibly"*: Ibid.

242 *"Account of the death"*: Curry, *Prophecy and Power*, p. 90.

242 *"Here five feet deep"*: quoted in McCaffery, *Astrology*, p. 328.

242 *"Their observations and predictions"*: quoted in Thomas, *Religion and the Decline of Magic*, p. 336.

243 *"This month an eminent"*: Ibid.

243 *"It is not to be conceived"*: quoted in Curry, *Prophecy and Power*, p. 106.

243 *"A few days ago"*: Hood, *The World of Anecdote*, p. 314.

244 *"Immortal teacher"*: Worsdale, *Celestial Philosophy or Genethliacal Astronomy*, p. xix.

244 *"highly detailed mathematical"*: Curry, *Prophecy and Power*, p. 132.

244 *"This work contains"*: Worsdale, *Celestial Philosophy*, p. i.

244 *"the true method"*: Ibid.

244 *"thirty remarkable"*: Ibid.

245 *"When this unfortunate"*: Ibid., p. 127.

245 *"Venus applying"*: Ibid.

245 *"Jupiter in the tenth house"*: Hungad, "A Brief History of Astrology," p. 8, in Lewis, *The Beginnings of Astrology in America*.

246 *"A question of importance"*: Sibly, *An Illustration of the Celestial Science of Astrology*, p. 278.

247 *"So that from a question"*: Ibid., p. 277.

247 *"Jupiter in Scorpio"*: Ibid., p. 377.

247 *"Their disposition"*: Ibid., p. 456.

248 *"could not find a single"*: McIntosh, *Astrology*, pp. 115–16.

248 *"The newly-married pair"*: Ibid.

248 *"a total and eternal"*: quoted in Ovason, *The Secret Architecture of Our Nation's Capital*, p. 145.

248 *"Here is every prospect"*: quoted in Curry, *Prophecy and Power*, p. 135.

249 *"have an extensive"*: Sibly, *New and Complete Illustration*, p. 1053.

249 *"elevated and dignified"*: Curry, *Prophecy and Power*, p. 134.

249 *"That these United Colonies"*: Adams, *Familiar Letters of John Adams and His Wife*, p. 191.

250 *"at a little past meridian"*: Hazelrigg, "Metaphysical Astrology," p. 23, in Lewis, *The Beginnings of Astrology in America*.

250 *three wafers*: Stone, *The United States: Wheel of Destiny*, p. 53.

250 *"no person"*: Ibid.

250 *"This date [July 4, 1776]"*: quoted in Rudhyar, *The Astrology of America's Destiny*, p. 19.

251 *"the Declaration was graced"*: quoted in Stone, *The United States*, p. 57.

253 *"the minutes of divine"*: quoted in Schechner Genuth, *Comets*, p. 196.

253 *"funeral torches"*: quoted in Gleiser, *The Prophet and the Astronomer*, p. 63.

253 *"God helps them"*: quoted in Steel, *Marking Time*, p. 253.

253 *"I am particularly pleased"*: Ibid.

255 *"spent much time"*: quoted in McCaffery, *Astrology*, p. 335.

255 *"he regarded that"*: Ibid.

255 *"It was on the 28th"*: quoted in Sachs, *The Astrology File*, p. 17.

256 *"Ye stars! which are"*: Byron, *Childe Harold's Pilgrimage*, Canto 3, Stanza 88, in *Poetical Works*, p. 228.

256 *"Thou sun that sinkest"*: Sardanapalus, Act 2, Scene 1, ll. 5–7, in *Poetical Works*, p. 463.

256 *"The scheme projected"*: Hungad, "A Brief History," p. 8, in Lewis, *The Beginnings of Astrology in America*.

256 *"It occurred to me"*: quoted in McCaffery, *Astrology*, p. 338.

257 *"the founder of"*: Howe, *Astrology. A Recent History*, p. 29.

257 *"Mars and Saturn"*: Ibid., p. 334.

257 *"the peace would be associated"*: Whitfield, *Astrology*, p. 191.

258 *"a farrago of wretched":* quoted in Curry, "Astrology on Trial, and Its Historians: Reflections on the Historiography of 'Superstition,' " p. 49.
258 *"as a rogue":* Ibid.
258 *"various knights, lords":* Ibid., p. 50.
258 *"preposterous and mischievous":* Ibid., p. 51.
259 *"Of all the strange delusions":* Ibid.
259 *"What could be more absurd?":* Ibid.
260 *"before the summer of 1865":* Hazelrigg, "Metaphysical Astrology," p. 37, in Lewis, *The Beginnings of Astrology in America.*
260 *"to its own place":* Ibid.
260 *"Let the President be careful":* Ibid.
260 *"Should either of the luminaries":* Ibid.
260 *"some high official":* Craig, *Stars of Destiny,* p. 89.
260 *"a short, stocky man":* Holden, *Astrological Pioneers,* p. 29.
261 *"No one who does not understand":* quoted in McCaffery, *Astrology,* p. 348.
261 *"drilled with crashing boots":* O'Connor, *Jack London: A Biography,* p. 17.
261 *"I enjoyed the friendship":* quoted in Kingman, *A Pictorial Life of Jack London,* p. 23.
261 *"wheat city":* Ibid., p. 21.
262 *"astro-theology":* Ibid.
262 *"A very loose condition":* Ibid.
262 *"A Discarded Wife":* Ibid., p. 15.
262 *"Day before yesterday":* Perry, quoted in *Jack London: An American Myth,* p. 7.
263 *"Births: Chaney":* Ibid., p. 8.
263 *"College of Astrology":* Kingman, *A Pictorial Life,* p. 18.
263 *"I was impotent":* Ibid.
264 *"the estrangement":* quoted in O'Connor, *Jack London,* p. 20.
264 *"Had I followed my first":* quoted in Kingman, *A Pictorial Life,* p. 20.
265 *"a wound on the left side":* Ibid.
265 *"the pistol . . . had not":* Ibid., p. 21.
265 *"that Flora had taken up":* Ibid.
265 *"My own life has been":* Ibid.
265 *"a very learned man":* quoted in Perry, *Jack London,* p. 9.
265 *"Life's a skin game":* Ibid.

Chapter Twelve

Page
266 *"a traveling sweets salesman":* Spencer, *True as the Stars Above,* p. 79.
266 *"to accord with his Sun sign":* Ibid.
266 *without a trace:* Howe, *Astrology: A Recent History,* p. 57.
266 *penniless in Liverpool:* Ibid., p. 58.
267 *"held that the world's":* Spencer, *True as the Stars Above,* p. 81.
267 *"in touch with 'secret masters' ":* Ibid.
267 *"gave up his sales career":* Holden and Hughes, *Astrological Pioneers of America,* p. 201.
268 *"he made the Sun sign":* Spencer, *True as the Stars Above,* p. 83.
268 *"downgraded prediction":* Ibid.
268 *"Theosophy's teachings":* Ibid.
268 *"It is no exaggeration":* quoted in Howe, *Astrology,* p. 56.
268 *"the horoscope could be regarded":* Ibid.
269 *"to revise all his publications":* Holden and Hughes, *Astrological Pioneers of America,* p. 202.
269 *"He had the gift":* McCaffery, *Astrology,* p. 350.
269 *"The author, in offering":* Karr, "Dr. Karr's Guide," p. 17, in Lewis, *The Beginnings of Astrology in America.*

269 *"oriental electric"*: Ibid., p. 20.

270 *"ALLA RAGAH"*: Ibid., p. 14.

270 *"the sooner we bring"*: quoted in Howe, *Astrology,* p. 65.

270 *"Apex"*: Ibid.

270 *"Fixed signs show death"*: Sepharial, *The Manual of Astrology,* p. 128.

270 *"In this horoscope"*: Ibid., p. 134.

271 *"the perfect example"*: Ibid., p. 133.

271 *"Lord Byron was born"*: Ibid.

271 *"rogues and vagabonds"*: Omarr, *My World of Astrology,* p. 73.

271 *"disorderly persons"*: Ibid.

272 *"raised astrology"*: Coleman, *Astrology and the Law,* p. 46.

272 *"went through an absolutely mechanical"*: Ibid., p. 56.

273 *"the Montana copper king"*: Christino, *Foreseeing the Future: Evangeline Adams and Astrology in America,* p. 81.

274 *"Accidental Ascendant"*: Adams, *The Bowl of Heaven,* p. 9.

274 *"Astrology must be right"*: Ibid., p. 271.

276 *"my own name for the reconstruction"*: quoted in Zoller, "Marc Edmund Jones and the New Age Astrology in America," p. 51.

276 *"a psychological method for charting"*: Ibid.

276 *"Let us make hay"*: Jones, *Astrology: How and Why It Works,* p. 186.

276 *"Taurus is the fixed sign"*: Ibid.

277 *"a map of the psyche"*: Spencer, *True as the Stars Above,* p. 86.

277 *"The lunar symbol"*: Merlin, *Character and Fate,* p. 25.

278 *"[Yeats] believes entirely"*: quoted in Heine, "W. B. Yeats, Poet and Astrologer," p. 60.

278 *"The Moon reflects the ego"*: Kenton, *Astrology,* p. 31.

279 *"My evenings are taken up"*: Main, ed., *Jung on Synchronicity and the Paranormal,* p. 11.

279 *"In matters of occultism"*: quoted in Begg, "Jung, Astrology & the Millennium," p. 21.

279 *"thought transference"*: quoted in Campion, "Sigmund Freud's Investigation of Astrology," p. 52.

279 *"As a psychologist"*: Main, ed., *Jung on Synchronicity,* p. 11.

279 *"Whatever is born, or done"*: Ibid.

279 *"synchronicity"*: Ibid.

280 *"In the moments prior"*: Quoted in Stewart, *Astrology,* p. 122.

280 *"Jung was on the move"*: Hyde, *Jung and Astrology,* p. 89.

281 *"paid one premium"*: Holden and Hughes, *Astrological Pioneers of America,* p. 101.

281 *"The chief use of the horoscope"*: Lewi, *Heaven Knows What,* p. 13.

281 *"You belong to the positive"*: Ibid., p. 25.

282 *"While the Sun's position"*: Lewi, *Astrology for the Millions,* pp. 141–42.

282 *"Pisces is multifaceted"*: Aubin and Rifkin, *The Complete Book of Astrology,* pp. 67–70.

283 *"A statesman cannot create"*: quoted in Taylor, *Bismarck,* p. 115.

283 *"I am content when I see"*: Ibid.

283 *"with the great fixed star"*: Craig, *Stars of Destiny,* p. 100.

284 *"visible or total, or nearly so"*: Ibid.

284 *"Captivity, the ransacking"*: Ibid., p. 101.

284 *"Once before in history"*: Ibid., pp. 103–4.

285 *"We cannot allow"*: quoted in Avery, *The Rising Sign,* p. 8.

285 *"anything of importance"*: quoted in Howe, *Astrology,* p. 90.

285 *"A man of action"*: Ibid.

285 *"What on earth have women"*: Ibid., p. 93.

285 *"from explosive material"*: Ibid., p. 169.

286 *"working by the weak beam"*: "The Lone Assassin—George Elser (1900–1945)," http://www.joric.com/Conspiracy/Elser.htm.

286 *"wrapped it up"*: Ibid.

286 *"accurately predicted"*: Avery, *The Rising Sign,* p. 8.

287 *"unfavorable days"*: Howe, *The Black Game: British Subversive Operations Against the Germans During the Second World War,* p. 225.

287 *"consider the sad case"*: Ibid.

288 *"about the deterioration"*: Ibid., p. 215.

288 *"massive doses"*: Ibid.

288 *"My Fuhrer, I congratulate"*: from H. R. Trevor-Roper's *The Last Days of Hitler,* quoted in McIntosh, *Astrology,* pp. 94–95.

289 *"kept a weather eye"*: Stone, *The United States: Wheel of Destiny,* p. 33.

290 *"virtually every major move"*: Regan, *For the Record: From Wall Street to Washington,* p. 3.

290 *"break the jinx"*: quoted in Stearn, *A Time for Astrology,* p. 275.

291 *"is the only available"*: Brau, et al., *Larousse Encyclopedia of Astrology,* p. 187.

292 *"skewing of the meaning"*: Lehman, *The Book of Rulerships,* p. 8.

292 *"because of the modern practice"*: Ibid., p. 12.

293 *"These celestial bodies"*: quoted in Clydesdale, "Marsilio Ficino's Holistic Astrology," p. 26.

293 *"The saying that a wise man"*: quoted in McCaffery, *Astrology,* p. 251.

293 *"It was a clear day"*: Stearn, *A Time for Astrology,* pp. 13–14.

295 *"our native"*: Frawley, *The Real Astrology,* p. 14.

295 *"the cloth from which"*: Ibid., p. 12.

295 *"A strong, well-placed Mars"*: Ibid., p. 14.

296 *"in Hitler's chart"*: Ibid., p. 15.

296 *"piled-up corpses"*: Ibid., p. 14.

Chapter Thirteen

Page

297 *"an extremely powerful"*: Arens, "Attack on America," p. 7.

297 *"with Mars activating"*: Ibid., p. 7.

298 *"All the benefic aspects"*: Horton, "Attack on New York City and the United States," p. 14.

298 *"showed three very important"*: Ruiz, "Chart Comparisons of the U.S.A.," p. 12.

298 *"an emotional shock"*: Ibid.

298 *"this powerful transit"*: Ibid.

298 *"an astrologer can determine"*: Dobin, *Kabbalistic Astrology,* p. 173.

299 *"mundane astrological forecasts"*: Zoller, "Prediction and 11th September 2001," http://www.new-library.com/zoller/features.

299 *"If the U.S. does not stop"*: Ibid.

299 *"There is an increasing threat"*: Ibid.

299 *"I again draw attention"*: Ibid.

299 *"suffer major disruption"*: Ibid.

301 *"Of the nature of Mars"*: Robson, *The Fixed Stars and Constellations in Astrology,* p. 188.

301 *"piety and conservatism"*: Ibid., p. 150.

301 *"illegitimate preferment"*: Ibid., p. 207.

301 *"Time governs princes"*: quoted in Morley, *Jerome Cardan,* Vol. 2, p. 31.

301 *"Saturn obtains kingdoms"*: Author's papers.

302 *"They would say"*: quoted in Omarr, *My World of Astrology,* p. 127.

303 *"I find myself unable"*: letter to *The Humanist,* January/February 1976, reproduced in *Today's Astrologer,* June 21, 2001, p. 10.

303 *"The horoscope does not pronounce"*: Adams, *The Bowl of Heaven,* p. 271.

303 *"I, Sir, have studied"*: quoted in West and Toonder, *The Case for Astrology,* p. 96.

303 *"I have been able to cure you"*: Forman, *The Story of Prophecy,* p. 120.

304 *"Luna is with Antares"*: Lynch, *The Coffee Table Book of Astrology,* p. 32.

305 *"the child who is about to"*: quoted in Gauquelin, *Cosmic Influences on Human Behavior,* p. 186.

305 *"Of the career of the Prince":* Goodavage, *Astrology: The Space Age Science,* p. 9.

306 *"Mars was in the first house":* Stearn, *A Time for Astrology,* p. 100.

306 *"Almost 4,000 years ago":* Campion and Eddy, *The New Astrology,* p. 254.

307 *"In a business solar return chart":* Hay, "Business Astrology," p. 12.

307 *"when a group of traders":* Bates, "Cycles in the U.S. Stock Market," p. 51.

307 *"more reliable":* Ibid.

307 *"a win rate":* quoted in Thallon, "The Life and Work of W. D. Gann," p. 1.

309 *"naphthalene solidifies fastest":* Ibid., p. 248.

309 *"How do you know":* Hardy, "The Year's Awakening," 11. 1–10, in *The Complete Poems,* No. 275, p. 335.

309 *"So may we read":* quoted in McIntosh, *Astrology,* p. 116.

310 *"In the heavens you can see":* quoted in Woolley, *The Queen's Conjuror,* p. 256.

310 *"The desire to construct":* Aveni, *Conversing with the Planets,* p. 153.

310 *"We are just as much":* Burroughs, *The Heart of Burroughs's Journals,* December 2, 1860, p. 22.

BIBLIOGRAPHY

Abbott, John S. C. *Makers of American History: Christopher Columbus.* New York: Dodd & Mead, 1875.

Abū Maʿshar. *The Abbreviation of the Introduction to Astrology.* Trans. by Charles Burnett. New York: E. J. Brill, 1994. Reprint, Brewster, Mass.: Arhat Publications.

Aczel, Amir D. *The Riddle of the Compass.* New York: Harcourt, 2001.

Adams, Evangeline. *Astrology: Your Place Among the Stars.* New York: Dodd, Mead, & Co., 1930.

———. *Astrology: Your Place in the Sun.* New York: Dodd, Mead, & Co., 1928.

———. *The Bowl of Heaven.* New York: Dodd & Mead, 1926.

Adams, John, and Abigail Adams. *Familiar Letters of John Adams and His Wife.* Boston: Houghton Mifflin & Co., 1875.

A. E. *Voices of the Stones.* New York: Macmillan, 1925.

Al-Biruni, Muhammad Ibn Ahmad. *The Book of Instruction in the Elements of the Art of Astrology.* Trans. by R. Ramsay Wright. London: Luzac & Co., 1934.

Al-Khayyat, Abu Ali. *The Judgments of Nativities.* Trans. by J. H. Holden. Tempe, Ariz. American Federation of Astrologers, 1988.

Al-Masu'di, Abul Hasan Ali. *The Book of Golden Meadows.* Trans. by Paul Lunde and Caroline Stone. London: Kegan Paul, 1989.

Allen, Don Cameron. *The Star-Crossed Renaissance.* New York: Octagon, 1966.

Allison, Chantal. "The Ifriqiya Uprising Horoscope from *On Reception* by Masha'allah, Court Astrologer in the Early Abbasid Caliphate." *Culture and Cosmos,* Vol. 3, No. 1 (Spring/Summer 1999), pp. 35–56.

Amundsen, Karin. "The Duke's Devil and Doctor Lambe's Darling: A Case Study of the Male Witch in Early Modern England." Ph.D. diss.:, University of Nevada, Las Vegas, 2003.

Andrews, Luke. "Prediction and 11th September 2001." http://www.new-library.com/zoller/features.

Arens, Christine. "Attack on America." *Today's Astrologer,* October 16, 2001, pp. 7–13.

Arroyo, Stephen. *Astrology, Karma & Transformation.* Sebastopol, Calif.: CRCS, 1992.

———. *Astrology, Psychology, and the Four Elements.* Sebastopol, Calif.: CRCS, 1975.

———. *Chart Interpretation Handbook.* Sebastopol, Calif.: CRCS, 1989.

———. *Practicing the Cosmic Science.* Sebastopol, Calif.: CRCS, 1999.

———. *Relationships & Life Cycles.* Sebastopol, Calif.: CRCS, 1993.

Arroyo, Stephen, and Liz Greene. *New Insights in Modern Astrology.* Sebastopol, Calif.: CRCS, 1984.

"Astral Religion." In *New Catholic Encyclopedia.* Vol. 1. New York: McGraw-Hill, 1967.

The Astrolabe World Ephemeris 2001–2050 at Noon. Atglen, Penn.: Whitford Press, 1998.

Aubin, Ada, and June Rifkin. *The Complete Book of Astrology.* New York: St. Martin's Griffin, 1998.

Aubrey, John. *Aubrey's Brief Lives.* Ann Arbor: University of Michigan Press, 1962.

Augustine, Saint. *The City of God.* Trans. by Marcus Dods. New York: Modern Library, 1950.

Aveni, Anthony. *Conversing with the Planets.* New York: Kodansha, 1994.

Avery, Jeanne. *The Rising Sign.* New York: Doubleday, 1982.

Bach, Eleanor. *Astrology from A to Z.* New York: M. Evans & Co., 1990.

Balzac, Honoré de. *Catherine de Medici.* Trans. by K. P. Wormeley. Whitefish, Mont.: Kessinger Publishing Co., 2004.

Barrus, Clara, ed. *The Heart of Burroughs's Journals.* Boston: Houghton Mifflin, 1928.

Barton, Tamsyn. *Ancient Astrology.* London: Routledge, 1994.

Barz, Ellynor. *Gods and Planets.* Trans. by Boris Matthews. Willmette, Ill.: Chiron, 1991.

Bates, Graham. "Cycles in the U.S. Stock Market." *Astrological Journal,* Vol. 43, No. 3 (May/June 2001), pp. 51–54.

Begg, Ean. "Jung, Astrology & the Millennium." *Astrological Journal,* Vol. 10, No. 10 (May/June 2000), pp. 10–15.

Bender, Robert M., ed. *Five Courtier Poets of the English Renaissance.* New York: Washington Square Press, 1969.

Bentley, G. E., Jr. *The Stranger From Paradise: A Biography of William Blake.* New Haven: Yale University Press, 2001.

Berlinski, David. *The Secrets of the Vaulted Sky.* Orlando, Fla.: Harcourt, 2003.

Bezza, Giuseppe. "Astrological Considerations on the Length of Life in Hellenistic, Persian, and Arabic Astrology." *Culture and Cosmos,* Vol. 2, No. 2 (Autumn/Winter 1998), pp. 3–15.

Blaise, Clark. *Time Lord: Sir Sandford Fleming and the Creation of Standard Time.* New York: Pantheon, 2000.

Blish, James. *Doctor Mirabilis.* New York: Dodd, Mead, & Co., 1971.

Bonatti, Guido. *Liber Astronomiae, Part 1.* Trans. by Robert Zoller. Ed. by Robert Hand. Berkeley Springs, W.Va.: Golden Hind Press, 1994.

Bowden, Mary Ellen. "The Scientific Revolution in Astrology: The English Reformers, 1556–1686." Ph.D. diss., Yale University, 1974.

Brady, Bernadette. *The Eagle and the Lark: A Textbook of Predictive Astrology.* York Beach, Me.: Samuel Weiser, 1992.

Brau, Jean-Louis, Helen Weaver, and Allan Edmands. *Larousse Encyclopedia of Astrology.* New York: Plume, 1980.

A Briefe Description of the Notorious Life of John Lambe, otherwise called Doctor Lambe, together with his ignominious death. Amsterdam, 1628.

Brown, J. Wood. *Enquiry Into the Life and Legend of Michael Scot, 1897.* Whitefish, Mont.: Kessinger Publishing, 2001.

Burckhardt, Jacob. *The Civilization of the Renaissance in Italy.* New York: Penguin Classics, 1990.

Burke, Thomas. *The Streets of London Through the Centuries.* London: B. T. Batsford, 1943.

Burnet, Bishop Gilbert. *History of His Own Time.* New York: Charles E. Tuttle, 1992.

Burritt, Elijah H. *The Geography of the Heavens.* New York: Mason Bros., 1856.

Burroughs, John. *The Heart of Burroughs's Journals.* Ed. by Clara Barrus. Boston: Houghton Mifflin, 1928.

Burton, Robert. *The Anatomy of Melancholy.* New York: Farrar & Rinehart, 1927.

Butler, Samuel. *Hudibras.* Whitefish, Mont.: Kessinger Publishing Co., 2004.

Byron, George Gordon, Lord. *Poetical Works.* London: Oxford University Press, 1967.

Camilleri, Stephanie. *The House Book.* St. Paul, Minn.: Llewellyn, 1994.

Campion, Nicholas. *The Book of World Horoscopes.* Bristol, England: Cinnabar, 1996.

———. "Dorotheus of Sidon: His Life and Significance." Introduction to *Carmen Astrologicum,* by Dorotheus of Sidon, trans. by David Pingree. London: Ascella Books, 1993.

———. *The Great Year: Astrology, Millenarianism and History in the Western Tradition.* New York: Arkana, 1994.

———. *The Practical Astrologer.* Bristol, England: Cinnabar, 1993.

———. "Sigmund Freud's Investigation of Astrology." *Culture and Cosmos,* Vol. 2, No. 1 (Spring/Summer 1998), pp. 49–53.

Campion, Nicholas, Michael Baigent, and Charles Harvey. *Mundane Astrology.* London: Thorsons, 1995.

Campion, Nicholas, and Steve Eddy. *The New Astrology.* North Pomfret, Vt.: Trafalgar Square, 1999.

Cardan, Jerome [Girolamo Cardano]. *The Book of My Life.* Trans. by Jean Stoner. New York: Dover, 1929.

Carey, Hilary M. *Courting Disaster: Astrology at the English Court and University in the Later Middle Ages.* London: Macmillan, 1992.

Carter, C. E. O. *Astrological Aspects.* London: L. N. Fowler, 1967.

———. *The Astrology of Accidents.* Addington, England: Astrologers' Quarterly, n.d.

———. *An Encyclopaedia of Psychological Astrology.* London: Theosophical Publishing House, 1963.

———. *Essays on the Foundations of Astrology.* London: Theosophical Publishing House, 1947.

———. *The Principles of Astrology.* London: Theosophical Publishing House, 1939.

Casper, Max. *Kepler.* Trans. by C. Doris Hellman. New York: Dover, 1993.

Cassius Dio. *Roman History.* Books 67–70. Trans. by Earnest Cary. Boston: Loeb Classical Library, 1992.

Chapman, George. *The Plays and Poems of George Chapman.* Ed. by Thomas More Parrott. London: Macmillan, 1910.

Charpentier, Louis. *The Mysteries of Chartres Cathedral.* Wellington, England: Thorsons, 1966.

Chaucer, Geoffrey. *The Works of Geoffrey Chaucer.* Ed. by F. N. Robinson. New York: Houghton Mifflin, 1957.

Christianson, Gale E. *This Wild Abyss: The Story of the Men Who Made Modern Astronomy.* New York: Free Press, 1978.

Christino, Karen. *Foreseeing the Future: Evangeline Adams and Astrology in America.* Amherst, Mass.: One Reed Publications, 2002.

Cicero, Marcus Tullius. *Nine Orations and The Dream of Scipio.* Trans. by Palmer Bovie. New York: Mentor Books, 1967.

Clulee, Nicholas H. *John Dee's Natural Philosophy.* London: Routledge, 1988.

Clydesdale, Ruth. "The Guardians of the Months: Manilius and the Guardian Deities." *Astrological Journal,* Vol. 43, No. 6 (November/December 2001), pp. 53–57.

———. "Marsilio Ficino's Holistic Astrology." *Mountain Astrologer,* August/September 1996, pp. 22–28.

Coleman, Walter. *Astrology and the Law.* Greenlawn, N.Y.: Casa de Capricornio, n.d.

Collin, Rodney. *The Theory of Celestial Influence.* New York: Arkana, 1993.

Cornelius, Geoffrey, and Paul Devereux. *The Secret Language of the Stars and Planets.* San Francisco: Chronicle Books, 1996.

Cowling, T. G. *Isaac Newton and Astrology.* Leeds: Leeds University Press, 1977.

Craig, Anthony. "Gauquelin's Legacy: New Evidence for Planetary Types." C.U.R.A. 2001.

Craig, Katherine Taylor. *Stars of Destiny: The Ancient Science of Astrology and How to Make Use of It To-Day.* New York: E. P. Dutton & Co., 1916.

Cramer, Frederick H. *Astrology in Roman Law and Politics.* Philadelphia: American Philosophical Society, 1954.

Cumont, Franz. *Astrology and Religion Among the Greeks and Romans.* New York: Dover, 1960.

C.U.R.A. Centre Universitaire de Recherche en Astrologie *(Center for Astrological Research).* Editions 1–31, January 1999–January 2004. http://cura.free.fr/artic-en.html.

Curry, Patrick. "Astrology on Trial, and Its Historians: Reflections on the Historiography of 'Superstition.'" *Culture and Cosmos,* Vol. 4, No. 2 (Autumn/Winter 2000), pp. 47–56.

———. "Historical Approaches to Astrology." *Culture and Cosmos,* Vol. 4, No. 1 (Spring/ Summer 2000), pp. 2–9.

———. *Prophecy and Power: Astrology in Early Modern England.* Cambridge, England: Polity Press, 1989.

———, ed. *Astrology, Science and Society: Historical Essays.* Woodbridges, England: Boydell Press, 1987.

Dante Alighieri. *The Divine Comedy.* Trans. by John Ciardi. New York: W. W. Norton, 1970.

Davis, Geraldine. *Horary Astrology.* Los Angeles: Tate Printing, 1942.

Davison, Ronald C. *Astrology.* Sebastopol, Calif.: CRCS, 1987.

———. *Synastry.* Santa Fe, N.M.: Aurora, 1983.

———. *The Technique of Prediction.* Chadwell Heath, England: L. N. Fowler, 1955.

De Madariaga, Salvador. *Christopher Columbus.* New York: Macmillan, 1940.

Dill, Samuel. *Roman Society from Nero to Marcus Aurelius.* New York: Meridian Books, 1964.

Dobin, Joel C. *Kabbalistic Astrology: The Sacred Tradition of the Hebrew Sages.* Rochester, Vt.: Inner Traditions, 1999.

Dreyer, J. L. E. *Tycho Brahe.* New York: Dover, 1963.

Dryden, John. *Poems and Fables.* London: Oxford University Press, 1962.

Ebertin, Elsbeth. *Astrology and Romance.* New York: ASI, 1973.

Ebertin, Elsbeth, and Georg Hoffman. *Fixed Stars and Their Interpretation.* Trans. by Irmgard Banks. Tempe, Ariz.: American Federation of Astrologers, 1971.

Eggleston, Edward. *The Transit of Civilization from England to America in the Seventeenth Century.* New York, 1901.

Epstein, Meira B., trans. *The Correspondence Between the Rabbis of Southern France and Maimonides about Astrology.* Arhat, 1998.

Farnell, Kim. *The Astral Tramp: A Biography of Sepharial.* London: Ascella, 1998.

Ferris, Timothy. *Coming of Age in the Milky Way.* New York: Anchor Books, 1989.

Ficino, Marsilio. *The Book of Life.* Trans. by Charles Boer. Irving, Tex.: Spring Publications, 1980.

———. *Selected Letters (Meditations on the Soul).* Trans. by Members of the Language Dept., School of Economic Science, London. Rochester, Vt.: Inner Traditions, 1997.

Forman, Henry James. *The Story of Prophecy.* New York: Farrar & Rinehart, 1936.

Forrest, Steven. *The Inner Sky*. San Diego: ACS Publications, 1982.

Frank, Dennis. "Astrological Space." C.U.R.A., 2001.

———. "Astrology Considered as a Potential Science of Time," C.U.R.A., Edition 11, April 19, 2001.

———. "On the Metaphysical Basis of Astrology." C.U.R.A., Edition 5, May 17, 2000.

Frank, Tenney. *A History of Rome*. New York: Holt, Rhinehart, & Winston, 1961.

Frawley, David. *The Astrology of the Seers: A Guide to Vedic/Hindu Astrology*. Salt Lake City: Passage Press, 1990.

Frawley, John. *The Real Astrology*. London: Apprentice Books, 2000.

French, Peter. *John Dee: The World of an Elizabethan Magus*. New York: Dorset, 1972.

Gadbury, John. *The Nativity of the Most Valiant and Puissant Monarch Lewis the Fourteenth King of France and Navarre*. London, 1680.

Galileo Galilei. *Dialogue Concerning the Two Chief World Systems*. New York: Random House, 2001.

———. *Discoveries and Opinions: The Starry Messenger (1610); Letter to the Grand Duchess Christina (1615), Excerpts from Letters on Sunspots (1613), and The Assayer (1623)*. New York: Anchor Books, 1989.

Gardner, F. Leigh. *Bibliotheca Astrologica: A Catalog of Astrological Publications of the 15th through the 19th Centuries*. North Hollywood, Calif.: Symbols & Signs, 1977.

Garin, Eugenio. *Astrology in the Renaissance: The Zodiac of Life*. London: Routledge & Kegan Paul, 1982.

Garrett, Helen Adams. *Astrology and Metaphysics in the Bible*. Belleville, Ill.: A is A, 1997.

Gauquelin, Michel. *Cosmic Influences on Human Behavior*. Santa Fe, N.M.: Aurora, 1994.

———. *Planetary Heredity*. San Diego: ACS, 1966.

———. *The Scientific Basis of Astrology*. Trans. by James Hughes. New York: Stein & Day, 1969.

Gauquelin, Michel, and Francoise Gauquelin. "Birth and Planetary Data Gathered Since 1949." C.U.R.A., Edition 11, April 19, 2001.

Geddes, Sheila. *The Art of Astrology*. Wellingborough, England: Aquarian Press, 1980.

Geneva, Ann. *Astrology and the Seventeenth Century Mind: William Lilly and the Language of the Stars*. Manchester, England: Manchester University Press, 1995.

George, Demetra. "Manuel I Komnenos and Michael Glycas: A Twelfth-Century Defence and Refutation of Astrology." Part 1, "History and Background," *Culture and Cosmos,* Vol. 5, No. 1 (Spring/Summer 2001), pp. 3–47. Part 2, *Culture and Cosmos,* Vol. 5, No. 2 (Autumn/Winter 2001), pp. 23–51. Part 3, *Culture and Cosmos,* Vol. 6, No. 1 (Spring/Summer 2002), pp. 23–43.

George, Llewellyn. *The New A to Z Horoscope Maker and Delineator*. St. Paul, Minn.: Llewellyn Publications, 1997.

Gettings, Fred. *Fate & Prediction*. New York: Exeter Books, 1980.

Gibbon, Edward. *The Decline and Fall of the Roman Empire*. Abridged edition. New York: Penguin, 2001.

Gillman, Ken. "Twelve Gods and Seven Planets," C.U.R.A., Edition 10, November 2, 2001.

Glazerson, M. *Astrology and Kabbala*. Moscow: Gesharim, 1998.

Gleadow, Rupert. *The Origin of the Zodiac*. London: Jonathan Cape, 1968.

Gleiser, Marcelo. *The Prophet and the Astronomer*. New York: W. W. Norton & Co., 2002.

Goodavage, Joseph F. *Astrology: The Space Age Science*. West Nyack, N.Y.: Parker, 1966.

Grafton, Anthony. *Cardano's Cosmos: The Worlds and Works of a Renaissance Astrologer*. Cambridge, Mass.: Harvard University Press, 1999.

Grafton, Anthony, and William R. Newman, eds. *Secrets of Nature: Astrology and Alchemy in Early Modern Europe.* Cambridge, Mass.: MIT Press, 2001.

"The Great Plague 1665." http://www.historic-uk.com/HistoryUK/England-History/GreatPlague.htm.

Greene, Liz. *Astrology for Lovers.* York Beach, Me.: Samuel Weiser, 1989.

———. *The Astrological Neptune.* York Beach, Me.: Samuel Weiser, 2000.

———. *The Astrology of Fate.* York Beach, Me.: Samuel Weiser, 1984.

———. *Saturn: A New Look at an Old Devil.* York Beach, Me.: Samuel Weiser, 1976.

Greene, Liz, and Howard Sasportas. *The Inner Planets.* York Beach, Me: Samuel Weiser, 1993.

———, and Howard Sasportas. *The Luminaries.* York Beach, Me.: Samuel Weiser, 1992.

Guinard, Patrice. "L'astrologie francaise au XXeme siècle," C.U.R.A., Edition 10, November 2, 2001.

Hale, Edward Everett. *The Life of Christopher Columbus from his own Letters and Journals and Other Documents of His Time.* McLean, Va., 1891.

Halevi, Z'ev ben Shimon. *Astrology & Kabbalah.* London: Urania Trust, 1999.

Hall, Manly P. *Astrological Keywords.* Los Angeles: Philosophical Research Society, 1995.

———. *The Philosophy of Astrology.* Los Angeles: Philosophical Research Society, 1976.

———. *Planetary Influence and the Human Soul.* Los Angeles: Philosophical Research Society, 1957.

———. *Psychoanalyzing the Twelve Zodiacal Types.* Los Angeles: Philosophical Research Society, 1982.

———. *The Secret Teachings of All Ages.* Los Angeles: Philosophical Research Society, 1997.

———. *The Story of Astrology.* Los Angeles: Philosophical Research Society, 1975.

Hamaker-Zontag, Karen. *Aspects and Personality.* York Beach, Me.: Samuel Weiser, 1990.

———. *The House Connection.* York Beach, Me.: Samuel Weiser, 1994.

———. *Planetary Symbolism in the Horoscope.* York Beach, Me.: Samuel Weiser, 1996.

———. *Psychological Astrology.* York Beach, Me.: Samuel Weiser, 1990.

———. *The Twelfth House.* York Beach, Me.: Samuel Weiser, 1992.

Hand, Robert. *Planets in Transit.* Atglen, Pa: Whitford Press, 1976.

Harding, Mike. "Prejudice in Astrological Research," C.U.R.A., Edition 11, April 19, 2001.

Hardy, Thomas. *The Complete Poems.* New York: Macmillan, 1979.

Hare, Christopher. *The Romance of a Medici Warrior.* New York: Charles Scribner's Sons, 1910.

Harvey, Charles, and Suzi Harvey. *Principles of Astrology.* London: Thorsons, 1999.

———. *Sun Sign, Moon Sign.* Northampton, England: Thorsons, 1994.

Haskins, Charles Homer. *The Renaissance of the 12th Century.* New York: Meridian, 1960.

———. *Studies in the History of Medieval Science.* Harvard Historical Studies, No. 27. Cambridge, Mass.: Harvard University Press, 1924.

Hay, Clement. "Business Astrology." *Today's Astrologer,* October 20, 1941. Reprinted in *Today's Astrologer,* August 18, 2001, pp. 12–16.

Heilbron, J. L. *The Sun in the Church: Cathedrals as Solar Observatories.* Cambridge, Mass.: Harvard University Press, 1999.

Heindel, Max. *Simplified Scientific Astrology.* London: L. N. Fowler & Co., 1928.

Heine, Elizabeth. "W. B. Yeats: Poet and Astrologer." *Culture and Cosmos,* Vol. 1, No. 2 (Winter/Autumn 1997), pp. 60–75.

Herodotus. *The Histories.* Trans. by Aubrey de Selincourt. New York: Penguin, 1960.

Heydon, Sir Christopher. *A Defence of Judiciall Astrologie.* Cambridge, England, 1603.

Hickey, Isabel. *Astrology: A Cosmic Science.* Sebastopol, Calif.: CRCS, 1992.

———. *Minerva/Pluto: The Choice Is Yours.* Watertown, Mass.: Fellowship House Bookshop, 1977.

Hodges, Richard, and David Whitehouse. *Mohammed, Charlemagne & the Origins of Europe: Archeology and the Pirenne Thesis.* Ithaca, N.Y.: Cornell University Press, 1983.

Holden, James Herschel. "Arabian Astrology." C.U.R.A., Edition 23, March 12, 2002.

———. "Early Horoscopes of Jesus," C.U.R.A., Edition 24, May 1, 2003.

———. "The Foundation Chart of Baghdad," C.U.R.A., Edition 25, March 12, 2003.

———. *A History of Horoscopic Astrology.* Tempe, Ariz.: American Federation of Astrologers, 1996.

Holden, James Herschel, and Robert A. Hughes. *Astrological Pioneers of America.* Tempe, Ariz.: American Federation of Astrologers, 1988.

Holt, Peter. *Stars of India: Travels in Search of Astrologers and Fortune-Tellers.* London: Mainstream Publishing, 1998.

Hone, Margaret. *Applied Astrology.* Chadwell Heath, England: L. N. Fowler, 1953.

Hood, Paxton. *The World of Anecdote.* London: Hodder and Stoughton, 1897.

———. *The Modern Textbook of Astrology.* London: L. N. Fowler, 1951.

Hoppmann, Jurgen G. H. "The Lichtenberger Prophecy and Melanchthon's Horoscope for Luther." *Culture and Cosmos,* Vol. 1, No. 2 (Autumn/Winter 1997), pp. 49–53.

Hort, G. M. *Dr. John Dee: Elizabethan Mystic and Astrologer.* Whitefish, Mont.: Kessinger Publishing, 2000.

Horton, Susan. "Attack on New York City and the United States," *Today's Astrologer,* October 16, 2001, pp. 14–16.

Houlding, Deborah. "John Gadbury: Of Thefts, Fugitives, and Strays, &c. Doctrine of Horary Questions, Chapter 9, Part 3." Transcribed and annotated 2004. http://www.skyscript.co.uk.

Howe, Ellic. *Astrology: A Recent History Including the Untold Story of Its Role in World War II.* New York: Walker & Co., 1967.

———. *The Black Game: British Subversive Operations Against the Germans During the Second World War.* London: Michael Joseph, 1982.

Huber, Bruno, and Louise Bruno. *Moon-Node Astrology.* York Beach, Me.: Samuel Weiser, 1995.

Hunter, Michael, and Annabel Gregory, eds. *An Astrological Diary of the Seventeenth Century: Samuel Jeake of Rye, 1652–1699.* Oxford: Clarendon Press, 1988.

Huxley, Julian. *From an Antique Land.* London: Max Parrish, 1954.

Hyde, Maggie. *Jung and Astrology.* Northampton, England: Thorsons, 1992.

Idemon, Richard. *Through the Looking Glass.* York Beach, Me.: Samuel Weiser, 1992.

Innes, Brian. *Horoscopes.* New York: Crescent Books, 1976.

Irving, Washington. *The Life and Voyages of Christopher Columbus.* New York: T. Nelson & Sons, 1892.

J. Gaffarel's Unheard-of Curiosities concerning the Talismanical Sculpture of the Persians. London, 1650.

The Jewish Encyclopedia. Vol. 2. KTAV Publishing House. n.d.

Jocelyn, John. *Meditations on the Signs of the Zodiac.* Blauvelt, N.Y.: Multimedia, 1975.

Jones, Marc Edmund. *Astrology: How and Why It Works.* Baltimore: Penguin, 1972.

———. *The Essentials of Astrological Analysis.* New York: Sabian Publishing, 1960.

———. *The Guide to Horoscope Interpretation.* Philadelphia: David McKay, 1946.

———. *How to Learn Astrology.* Boulder, Colo.: Shambhala, 1969.

Jonson, Ben. *The Alchemist.* Cambridge, England: Cambridge University Press, 1996.

————. *The Devil Is An Ass, and Other Plays.* Oxford, England: Oxford University Press, 2000.

Jordan, Shelley. "An Astrological Saga of Wolfgang Amadeus Mozart: Genius before His Time." C.U.R.A., Edition 14, October 14, 2001.

Jung, Carl Gustav. *Synchronicity.* Trans. by R. F. C. Hull. Princeton, N.J.: Princeton University Press, 1973.

Kay, Richard. *Dante's Christian Astrology.* Philadelphia: University of Pennsylvania Press, 1994.

Kennedy, E. S., and David Pingree. *Astrological History of Masha'allah.* Cambridge, Mass.: Harvard University Press, 1971.

Kenton, Warren. *Astrology: The Celestial Mirror.* New York: Thames & Hudson, 1997.

Kepler, Johannes. *Concerning the More Certain Fundamentals of Astrology.* New York: Holmes Publishing Group, 1987.

————. *Epitome of Copernican Astronomy and Harmonies of the World.* Trans. by Charles Glenn Wallis. New York: Prometheus Books, 1995.

————. *Mysterium Cosmographicum.* Trans. by E. J. Aiton and A. M. Duncan. New York: Opal, 1979.

————. *Somnium: The Dream.* Trans. by Edward Rosen. Madison: University of Wisconsin Press, 1967.

Kidger, Mark. *The Star of Bethlehem: An Astronomer's View.* Princeton: Princeton University Press, 1999.

Kingman, Russ. *A Pictorial Life of Jack London.* New York: Crown, 1979.

Kitney, Elizabeth. "Culpeper's Herbal Medicine and Mesmerism as Healing Modalities." *Astrological Journal,* Vol. 44, No. 33 (May/June 2002), pp. 11–14.

Kitson, Annabella, ed. *History and Astrology: Clio and Urania Confer.* London: Mandala, 1989.

Koestler, Arthur. *The Sleepwalkers.* New York: Grosset & Dunlap, 1963.

Kohout, Edward. "The Riddle of the Sibly Chart for American Independence." C.U.R.A., Edition 22, September 28, 2002.

Kolev, Rumen. "Some Reflections about Babylonian Astrology." C.U.R.A., Edition 9, November 22, 2000.

Kollerstrom, Nick. "The Star Zodiac of Antiquity." *Culture and Cosmos,* Vol. 1, No. 2 (Autumn/Winter 1997), pp. 5–22.

————. *Beyond the Blue Horizon: Myths and Legends of the Sun, Moon, Stars, and Planets,* London: Oxford University Press, 1992.

Krupp, E. C. *Skywatchers, Shamans & Kings: Astronomy and the Archaeology of Power.* New York: John Wiley & Sons, 1997.

Kugel, Herb. "The Roman Empire and Astrology: A Contemporary View." *Astrological Journal,* Vol. 43, No. 2 (March/April 2001), pp. 6–11.

Laird, Edgar. "Christine de Pizan and Controversy Concerning Star-Study in the Court of Charles V." *Culture and Cosmos,* Vol. 1, No. 2 (Winter/Autumn 1997), pp. 34–48.

Larkey, Sanford V. "Astrology and Politics in the First Years of Elizabeth's Reign." *Bulletin of the Institute of the History of Medicine,* Vol. 3, No. 3 (March 1935), pp. 171–86.

Latitudes 0 to 60. New York: ASI, 1975.

Leek, Sybil. *My Life in Astrology.* Englewood Cliffs, N.J.: Prentice Hall, 1972.

Lehman, J. Lee. *The Book of Rulerships.* West Chester, Pa.: Whitford Press, 1992.

————. *Classical Astrology for Modern Living.* Atglen, Pa.: Whitford Press, 1996.

Leo, Alan. *Casting the Horoscope.* London: L. N. Fowler, 1970.

————. *Esoteric Astrology.* Rochester, Vt.: Destiny Books, 1989.

Lewi, Grant. *Astrology for the Millions.* Garden City, N.Y.: Garden City Publishing Co., 1942.

———. *Heaven Knows What.* St. Paul, Minn.: Llewellyn Publications, 1967.

Lewis, Bernard. *The Arabs in History.* Oxford: Oxford University Press, 1993.

Lewis, James R. *The Beginnings of Astrology in America.* New York: Garland Publishing, 1990.

———. *Encyclopedia of Astrology.* London: Visible Ink Press, 1994.

Lilly, William. *Christian Astrology.* Issaquah, Wash.: Justus & Associates, 1997.

———, ed. *Anima Astrologiae: Or, A Guide for Astrologers, Being the Considerations of Guido Bonatus . . . and the Choicest Aphorisms of Cardan.* London: B. Harris, 1676.

———. *The Astrological Aphorisms of Cardan, 1675.* Edmonds, Wash.: Sure Fire Press, 1989.

Lilly, William, and Elias Ashmole. *Lives of Those Eminent Antiquaries Elias Ashmole and William Lilly, Written by Themselves.* London: T. Davies, 1774.

Lineman, Rose, and Jan Polelka. *Compendium of Astrology.* Atglen, Pa.: Whitford Press, 1984.

"The Lone Assassin—George Elser (1900–1945)." http://www.joric.com/Conspiracy/Elser.htm.

Longfellow, Henry Wadsworth. *The Complete Poems.* Boston: Houghton Mifflin, 1908.

Longitudes and Latitudes in the U.S. Washington, D.C.: American Federation of Astrologers, 1945.

Louis, Anthony. *Horary Astrology Plain & Simple.* St. Paul, Minn.: Llewellyn Publications, 1998.

Lynch, John, ed. *The Coffee Table Book of Astrology.* New York: Viking, 1967.

MacNeice, Louis. *Astrology.* London: Bloomsbury Books, 1989.

Main, Roderick, ed. *Jung on Synchronicity and the Paranormal.* Princeton, N.J.: Princeton University Press, 1997.

Manilius, Marcus. *Astronomica.* Trans. by G. P. Goold. Loeb Classical Library. Cambridge, Mass.: Harvard University Press, 1977.

Mann, Thomas. *Joseph and His Brothers.* New York: Everyman's Library, 2005.

Marks, Tracy. *The Art of Chart Interpretation.* Sebastopol, Calif.: CRCS, 1986.

Mason, Herbert M. Jr. *Hitler Must Die.* New York: Penguin, 1985.

Mason, Sophia. *Delineation of Progressions.* Tempe, Ariz.: American Federation of Astrologers, 1985.

Masonic Libraries. Silver Spring, Md.: Masonic Service Association, 1998.

Maternus, Firmicus. *Ancient Astrology: Theory and Practice.* Trans. by Jean Rhys Bram. Park Ridge, N.J.: Noyes Press, 1975.

Maunder, A., and E. W. Maunder. *The Heavens and Their Story.* London: Estes, 1908.

Maunder, E. W. *The Astronomy of the Bible.* London: Epworth Press, 1922.

Mayo, Jeff. *Astrology: A Key to Personality.* Saffron Walden, England: C. W. Daniel, 2000.

———. *Teach Yourself Astrology.* Chicago: NTC Publishing Group, 1992.

McCaffery, Ellen. *Astrology: Its History and Influence in the Western World.* New York: Charles Scribner's Sons, 1942.

———. *Graphic Astrology.* Richmond, Va.: Macoy, 1952.

McCaffery, Ellen Conroy. *An Astrological Key to Biblical Symbolism.* New York: Samuel Weiser, 1975.

McIntosh, Christopher. *Astrologers and Their Creed: An Historical Outline.* New York: Frederick A. Praeger, 1969.

———. *Astrology: The Stars and Human Life.* New York: Castle Books, 1973.

McMinn, David. "Astro Economics & The 56 Year Cycle." C.U.R.A., Edition 14, October 14, 2001.

Merlin, Katharine. *Character and Fate: The Psychology of the Birthchart.* New York: Arkana, 1989.

Messina, Pauline. *Pick Your Mate by Astrology.* New York: Curtis Books, 1970.

Miller, Anistatia R., and Jared M. Brown. *The Complete Astrological Handbook for the Twenty-First Century.* New York: Schocken Books, 1999.

Mitchell, C. E. *Foretold by the Stars.* Halifax, England: Fawcett, Greenwood & Co., 1936.

Molnar, Michael. "Blood on the Moon in Aquarius: The Assassination of Domitian." *Celator,* May 1995, pp. 6–12.

———. *The Star of Bethlehem: The Legacy of the Magi.* New Brunswick, N.J.: Rutgers University Press, 1999.

Moore, Thomas. *The Planets Within: The Astrological Psychology of Marsilio Ficino.* Hudson, N.Y.: Lindisfarne Books, 1982.

Mori, Takeo, and Dragan Milenkovic. *Secrets of Japanese Astrology.* New York: Tengu Books, 1993.

Morin, J. B. *Astrologica Gallica.* Book 21, trans. by R. S. Baldwin. Washington, D.C.: American Federation of Astrologers, 1974. Book 22, trans. by J. H. Holden. Tempe, Ariz.: American Federation of Astrologers, 1994.

Morley, Henry. *Jerome Cardan: The Life of Girolamo Cardano, of Milan, Physician.* 2 vols. London: Chapman & Hall, 1854.

Negus, Kenneth G. "Kepler's Astrology." C.U.R.A., Edition 15, November 23, 2001.

———. "Kepler's *Tertius Interveniens.*" *Culture and Cosmos,* Vol. 1, No. 1 (Spring/Summer 1997), pp. 51–54.

Neugebauer, Otto. "The Study of Wretched Subjects." *Isis,* Vol. 42 (1951), pp. 111–21.

Neville, E. W. *Planets in Synastry.* Atglen, Pa.: The Whitford Press, 1990.

New Catholic Encyclopedia. 16 vols. New York: McGraw-Hill, 1967.

North, J. D. *The Ambassadors' Secret.* London: Hambledon and London, 2002.

———. *Horoscopes and History.* London: Warburg Institute, 1986.

Notker the Stammerer and Einhard. *Two Lives of Charlemagne.* Trans. by Lewis Thorpe. New York: Penguin Classics, 1984.

Oakes-Smith, E. *The Shadow Land, or the Seer.* London, 1852.

O'Connor, Richard. *Jack London: A Biography.* Boston: Little Brown & Co., 1964.

Ogg, F. A., ed. *A Source Book of Medieval History.* New York: Cooper Square, 1972.

Oken, Alan. *Houses of the Horoscope.* Freedom, Calif.: Crossing Press, 1999.

———. *Pocket Guide to Horoscope Interpretation.* Freedom, Calif.: Crossing Press, 1996.

———. *Rulers of the Horoscope.* Freedom, Calif.: Crossing Press, 2000.

Omarr, Sydney. *Answer in the Sky . . . Almost.* Charlottesville, Va.: Hampton Roads, 1995.

———. *My World of Astrology.* New York: Fleet Publishing, 1965.

Ovason, David. *The Secret Architecture of Our Nation's Capital.* New York: Perennial, 2000.

Pagan, Isabelle M. *From Pioneer to Poet, or The Twelve Great Gates.* London: Theosophical Publishing House, 1911.

Papus. *Astrology for Initiates.* Trans. by J. Lee Lehman. York Beach, Me.: Samuel Weiser, 1996.

Parker, Derek. *Familiar to All: William Lilly and Astrology in the Seventeenth Century.* London: Jonathan Cape, 1975.

Parker, Else. *Astrology and Its Practical Application.* Trans. by Coba Goedhart. North Hollywood, Calif.: Newcastle, 1977.

Parr, Johnstone. *Tamburlaine's Malady.* Kingsport: University of Alabama Press, 1953.

Partridge, John. *Opus Reformatum, or Treatise of Astrology in which The Common Errors of that Art are Modestly Exposed and Rejected.* London: John Churchill, 1693.

Paul, Haydn. *The Astrological Moon.* York Beach, Me.: Samuel Weiser, 1998.

Perry, John. *Jack London: An American Myth.* Chicago: Nelson-Hall, 1981.

Peters, Fritz. *Gurdjieff Remembered.* London: Victor Gollancz, 1965.

Phillips, H. L. *Translators and Translations.* Bloomington: Indiana University Press, 1958.

Plant, David. "John Gadbury: Politics and the Decline of Astrology." *Traditional Astrologer,* No. 11, 1996.

———. "Vettius Valens: An Ancient Judgement of Wealth." Cumberland, Md: Golden Hind Press, 2004. http://www.skyscript.co.uk.

Poss, Richard L. "Stars and Spirituality in the Cosmology of Dante's *Commedia.*" *Culture and Cosmos,* Vol. 5, No. 1 (Spring/Summer 2001), pp. 49–70.

Potter, David. *Prophets and Emperors.* Cambridge, Mass.: Harvard University Press, 1994.

Psellus, Michael. *Fourteen Byzantine Rulers.* Trans. by E. R. A. Sewter. New York: Penguin, 1979.

Ptolemy. *Tetrabiblos.* Trans. by J. M. Ashmand. North Hollywood, Calif.: Symbols & Signs, 1976.

Quigley, Joan. *What Does Joan Say? My Seven Years as White House Astrologer to Nancy and Ronald Reagan.* New York: Birch Lane Press, 1990.

Reagan, Nancy, with William Novak. *My Turn: The Memoirs of Nancy Reagan.* New York: Random House, 1989.

Regan, Donald T. *For the Record: From Wall Street to Washington.* New York: Harcourt Brace Jovanovich, 1988.

Reinhart, Melanie. *Chiron and the Healing Journey.* New York: Penguin, 1998.

Reston, James Jr. *Galileo: A Life.* New York: HarperCollins, 1994.

Rice, Hugh S. *American Astrology Table of Houses.* Philadelphia: David McKay, 1944.

Roberts, Peter, and Helen Greengrass. *The Astrology of Time Twins.* Durham, England: Pentland Press, 1994.

Robson, Vivian. *The Fixed Stars and Constellations in Astrology.* New York: Astrology Classics Reprint, 2001.

Rochberg, Francesca. *Babylonian Horoscopes.* Philadelphia: American Philosophical Society, 1998.

———. *Divination, Horoscopy, and Astronomy in Mesopotamian Culture.* Cambridge, England: Cambridge University Press, 2004.

Roosevelt, Theodore. *History as Literature.* New York: Charles Scribner's Sons, 1913.

Rosen, Edward, trans. *Three Copernican Treatises: The Commentariolus of Copernicus, The Letter Against Werner, The Narratio Prima of Rheticus.* New York: Dover, 1959.

Ross, David J. "The Bird, the Cross, and the Emperor: Investigations into the Antiquity of the Cross in Cygnus." *Culture and Cosmos,* Vol. 4, No. 2 (Autumn/Winter 2000), pp. 10–15.

Rousseau, Claudia. "An Astrological Prognostication to Duke Cosimo I de' Medici of Florence." *Culture and Cosmos,* Vol. 3, No. 2 (Autumn/Winter 1999), pp. 31–53.

Rowse, A. L. *Sex and Society in Shakespeare's Age: Simon Forman the Astrologer.* New York: Charles Scribner's Sons, 1974.

Rudhyar, Dane. *The Astrology of America's Destiny.* New York: Vintage, 1974.

———. *The Astrology of Personality.* Santa Fe, N.M.: Aurora, 1991.

Ruiz, Ana. "Chart Comparisons of the U.S.A." *Today's Astrologer,* January 13, 2002, pp. 11–15.

Russell, Eric. *History of Astrology & Prediction.* London: New English Library, 1972.

Ryan, W. F. *The Bathhouse at Midnight.* Phoenix Mill, England: Sutton, 1999.

Sachs, Gunter. *The Astrology File.* London: Orion, 1997.

Sakoian, Frances, and Louis Acker. *The Astrologer's Handbook.* New York: Harper Perennial, 1989.

———. *The Astrology of Human Relationships.* New York: Harper Perennial, 1989.

———. *The Inconjunct-Quincunx: The Not So Minor Aspect.* Boston: New England School of Astrology, 1972.

———. *Those Inconjunct Quincunx.* Boston: New England School of Astrology, 1978.

———. *Major and Minor Approaching and Departing Aspects.* Boston: New England School of Astrology, 1974.

———. *The Minor Aspects.* Boston: New England School of Astrology, 1978.

———. *Predictive Astrology.* New York: Harper Perennial, 1989.

———. *Transits Simplified.* Boston: New England School of Astrology, 1976.

———. *The Zodiac Within Each Sign.* Boston: New England School of Astrology, 1975.

Sakoian, Frances, and Betty Caulfield. *Astrological Patterns: The Key to Self-Discovery.* New York: Harper & Row, 1980.

Sasportas, Howard. *The Gods of Change.* New York: Arkana, 1989.

———. *The Twelve Houses.* London: Thorsons, 1998.

Schechner Genuth, Sara J. *Comets, Popular Culture, and the Birth of Modern Cosmology.* Princeton, N.J.: Princeton University Press, 1997.

Schulman, Martin. *The Astrology of Sexuality.* York Beach, Me.: Samuel Weiser, 1982.

Scofield, Paul. *User's Guide to Astrology.* Amherst, Mass.: One Reed Publications, 1997.

Scott, Sir Walter. *The Lay of the Last Minstrel,* www.theotherpages.org/poems/minstrel.html.

cramuzza, V. M. *The Emperor Claudius.* London: Macmillan, 1940.

Sedgwick, Henry Dwight. *A Short History of Spain.* Boston: Little, Brown, & Co., 1925.

Sepharial [Walter Gorn Old]. *The Manual of Astrology.* London: W. Foulsham & Co., 1962.

Seward, A. F. *The Zodiac and Its Mysteries.* Chicago: A. F. Seward & Co., 1923.

Seymour-Smith, Martin. *The New Astrologer.* New York: Collier, 1981.

Shakespeare, William. *The Complete Works.* London: Oxford University Press, 1965.

Sibly, Ebenezer. *An Illustration of the Celestial Science of Astrology, or The Art of Foretelling Future Events and Contingencies by the Aspects, Positions, and Influences of the Heavenly Bodies.* 3 vols. London, 1798.

Simmonite, W. J. *The Celestial Philosopher, or The Complete Arcana of Astral Philosophy.* New York: John Story, 1890.

Smoller, Laura Ackerman. *History, Prophecy, and the Stars: The Christian Astrology of Pierre d'Ailly, 1350–1420.* Princeton, N.J.: Princeton University Press, 1994.

Sobel, Dava. *Galileo's Daughter.* New York: Penguin, 2000.

———. *Longitude.* New York: Penguin, 1995.

Spencer, Katharine Q. *The Zodiac Looks Westward.* Philadelphia: David McKay, 1943.

Spencer, Neil. *True as the Stars Above.* London: Victor Gollancz, 2000.

Stahlman, William D. "Astrology in Colonial America." *William and Mary Quarterly,* 3rd ser. 13 (1956).

Stearn, Jess. *A Time for Astrology.* New York: Signet, 1972.

Steel, Duncan. *Marking Time.* London: John Wiley & Sons, 2000.

Stewart, J. V. *Astrology: What's Really in the Stars.* Amherst, N.Y.: Prometheus Books, 1996.

Stone, Diana Bills. *The United States: Wheel of Destiny.* Tempe, Ariz.: American Federation of Astrologers, 1976.

Suetonius. *The Twelve Caesars.* Trans. by Robert Graves. New York: Penguin, 1967.

Sullivan, Erin. *Retrograde Planets.* York Beach, Me.: Samuel Weiser, 2000.

Sutherland, James. *English Literature of the Late Seventeenth Century.* Oxford: Oxford University Press, 1969.

Tables of Houses: Latitude, 0 to 66 Degrees. Oceanside, Calif.: Rosicrucian Fellowship, 1949.

Tacitus. *The Annals of Imperial Rome.* Trans. by Michael Grant. New York: Penguin, 1956.

Tamsyn, Barton. *Ancient Astrology.* London: Routledge, 1994.

Taylor, A. J. P. *Bismarck: The Man and the Statesman.* New York: Alfred A. Knopf, 1955.

Teissier, Elizabeth. *Les Etoiles de L'Elysée.* Paris: Edition 1, 1995.

———. *Sous Le Signe De Mitterand.* Paris: Edition 1, 1997.

Tester, S. J. *A History of Western Astrology.* Suffolk, England: Boydell Press, 1987.

Thallon, Solomon. "The Life and Work of W. D. Gann." http://www.afsd.com.au/article/hottrader/soloman6a.htm.

Thomas, Keith. *Religion and the Decline of Magic.* Oxford: Oxford University Press, 1971.

Thompson, C. J. S. *The Mystery and Romance of Astrology.* New York: Causeway Books, 1973.

Thompson, R. Campbell. *The Reports of the Magicians and Astrologers of Nineveh and Babylon.* 1900.

Thorndike, Lynn. *A History of Magic and Experimental Science.* Vols. 5 and 6. New York: Columbia University Press, 1941.

Tourneur, Cyril. *The Revenger's Tragedy.* New York: Hill & Wang, 1967.

Traister, Barbara Howard. *The Notorious Astrological Physician of London.* Chicago: University of Chicago Press, 2000.

Travers, P. L. "Letter to a Learned Astrologer." *Parabola,* Vol. 3 No. 4 (November 1978), pp. 58–65.

Trever-Roper, H. R., ed. *Hitler's Table Talk, 1941–1944.* New York: Enigma Books, 2002.

Turnbull, Coulson. *The Divine Language of Celestial Correspondences.* Orange, Calif.: Gnostic Press, 1926.

Tyl, Noel, ed. *Astrology Looks at History.* St. Paul, Minn.: Llewellyn Publications, 1995.

Vasiliev, A. A. *History of the Byzantine Empire.* 2 vols. Madison: University of Wisconsin Press, 1952.

Vaughan, Richard. *Astrology in Modern Language.* Sebastopol, Calif.: CRCS, 1992.

Vescovini, Graziella Federici. "Biagio Pelacani's Astrological History for the Year 1405." *Culture and Cosmos,* Vol. 2, No. 1 (Spring/Summer 1998), pp. 24–32.

Vitruvius. *The Ten Books On Architecture.* Trans. by Morris Hicky Morgan. New York: Dover, 1960.

Vorel, Iris. *Be Your Own Astrologer.* New York: Ives Washburn, 1935.

Voss, Angela. "The Astrology of Marsilio Ficino: Divination or Science?" *Culture and Cosmos,* Vol. 4, No. 2 (Autumn/Winter 2000), pp. 29–45.

———. "The Music of the Spheres: Marsilio Ficino and Renaissance Harmonia." *Culture and Cosmos,* Vol. 2, No. 2 (Autumn/Winter 1998), pp. 16–38.

Walker, D. P. *Spiritual and Demonic Magic from Ficino to Campanella.* University Park: Pennsylvania State University Press, 2000.

Warren-Davis, Dylan. *Astrology and Health: A Beginner's Guide.* London: Hodder & Stoughton, 1998.

———. "Nicholas Culpeper: Herbalist of the People." http://www.skyscript.co.uk/culpeper.html.

Waterfield, Robin. "The Evidence for Astrology in Classical Greece." *Culture and Cosmos,* Vol. 3, No. 2 (Autumn/Winter 1999), pp. 2–15.

Watts, Pauline Moffitt. "Prophecy and Discovery: On the Spiritual Origins of Christopher Columbus's 'Enterprise of the Indies.' " *American Historical Review,* Vol. 90, No. 1 (February 1985), pp. 73–102.

Webster, John. *The Duchess of Malfi, and Other Plays.* Oxford, England: Oxford University Press, 1998.

Wedel, Theodore Otto. *The Medieval Attitude Toward Astrology.* New Haven, Conn.: Yale University Press, 1920.

West, John Anthony, and Jan Gerhard Toonder. *The Case for Astrology.* Baltimore: Penguin, 1973.

Whitfield, Peter. *Astrology: A History.* New York: Harry N. Abrams, 2001.

Whitman, Edward W. *Aspects and their Meanings: Astrol-Kinetics, Vol. 3.* London: L. N. Fowler, 1970.

———. *The Influence of the Houses: Astro-Kinetics, Vol. 1.* London: L. N. Fowler, 1970.

———. *The Influence of the Planets: Astro-Kinetics, Vol. 2.* London: L. N. Fowler, 1970.

Wiet, Gaston. *Baghdad: Metropolis of the Abbasid Caliphate.* Tulsa: University of Oklahoma Press, 1971.

Wigzell, Faith. *Reading Russian Fortunes.* Cambridge, England: Cambridge University Press, 1998.

Woolley, Benjamin. *The Queen's Conjurer: The Science and Magic of Dr. John Dee, Adviser to Queen Elizabeth I.* New York: Henry Holt, 2001.

World Ephemeris for the 20th Century 1900 to 2000 at Noon. Atglen, Pa.: Whitford Press, 1983.

Worsdale, John. *Celestial Philosophy or Genethliacal Astronomy.* London: Longman & Co., n.d.

Wright, Louis B. *Middle-Class Culture in Elizabethan England.* Ithaca, N.Y.: Cornell University Press, 1935.

Yates, Frances A. "The Hermetic Tradition in Renaissance Science." In *Art, Science, and History in the Renaissance,* ed. C. S. Singleton, Baltimore: Johns Hopkins University Press, 1967.

———. *Theatre of the World.* Chicago: University of Chicago Press, 1969.

Zambelli, Paolo, ed. *Astrologi Hallucinati: Stars and the End of the World in Luther's Time.* New York: Walter de Gruyter, 1986.

Zoller, Robert. *The Arabic Parts in Astrology.* Rochester, Vt.: Inner Traditions International, 1989.

———. *Fate, Freewill & Astrology.* Mansfield, England: Ascella, n.d.

———. *The Fifth House.* Mansfield, England: Ascella, n.d.

———. "Marc Edmund Jones and the New Age Astrology in America." *Culture and Cosmos,* Vol. 2, No. 2 (Autumn/Winter 1998), pp. 39–57.

———. "A Study of the Development of Astrology in the Nineteenth Century." http://www.new-library.com/zoller.

———. *Tools & Techniques of the Medieval Astrologers.* Mansfield, England: Ascella, n.d.

INDEX